Annals of Mathematics Studies

Number 116

Gauss Sums, Kloosterman Sums, and Monodromy Groups

by

Nicholas M. Katz

PRINCETON UNIVERSITY PRESS

PRINCETON, NEW JERSEY

1988

Clothbound editions of Princeton University Press books are printed on acid-free paper, and binding materials are chosen for strength and durability. Paperbacks, while satisfactory for personal collections, are not usually suitable for library rebinding

Printed in the United States of America
by Princeton University Press, 41 William Street
Princeton, New Jersey

Library of Congress Cataloging-in-Publication Data

Katz, Nicholas M., 1943-
Gauss sums, Kloosterman sums, and monodromy groups

(The Annals of mathematics studies ; 116)
Bibliography: p.
1. Gaussian sums. 2. Kloosterman sums. 3. Homology theory. 4. Monodromy groups. I. Title. II. Series: Annals of mathematics studies ; no. 116.
QA246.8.G38K37 1987 512'.7 87-45525

ISBN 0-691-08432-7 (alk. paper)
ISBN 0-691-08433-5 (pbk.)

Contents

Leitfaden

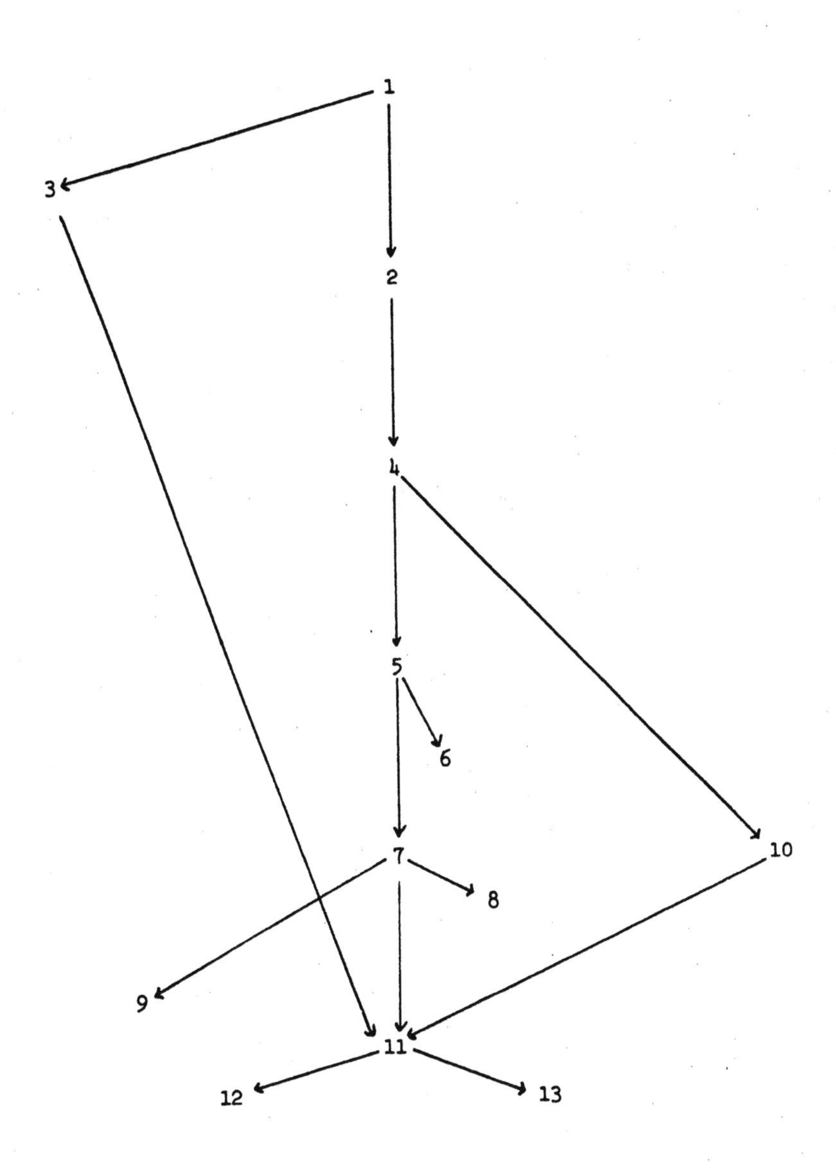

Introduction

This book is simultaneously an account of some of the main cohomological and representation-theoretic ideas, techniques, and results which can fruitfully be brought to bear upon the study of general exponential sums over finite fields, and an account of their concrete application to particular questions concerning the equidistribution properties of Kloosterman sums and Gauss sums. The idea of combining, in a single book, both general theory and concrete application, neither of which could stand alone except in the utmost aridity, is certainly not a new one. Ideally, the two parts reinforce each other, the applications both illustrating and motivating the general techniques, and bring the reader to a deeper understanding of the subject: this is, in any case, the author's intention.

Let us now turn to a more technical discussion of the ideas, techniques, and results to be applied.

That there is an intimate connection between exponential sums on varieties over finite fields and the theory of zeta and L-functions of such varieties was first realized by Hasse [Ha] and Hasse–Davenport [H-D] some half-century ago. This connection was exploited by Weil, who gave [We] as application of his proof of the "Riemann Hypothesis" for curves over finite fields the best possible estimates for one-variable sums.

The 1973 proof of the Weil Conjectures for projective smooth varieties of arbitrary dimension over finite fields by Deligne in [De-6] had proportionally less impact on the theory of exponential sums in several variables, for some seemingly technical reasons. An exponential sum naturally occurs on an affine variety V, and it is cohomologically accounted for by a suitable direct factor of the compact cohomology of a suitable Artin–Schreier covering W of V. The variety W will be smooth if V is, but it is affine. If one knew explicitly how to compactify W to a projective smooth variety $\overline{W}$, one could apply the Weil Conjectures directly to $\overline{W}$. The problem is that one does not know in general how to construct explicitly a projective smooth compactification of W, (nor even if one exists!). Of course if V, and hence W, is a smooth curve, then W has a canonical smooth compactification, and so this problem does not arise in studying exponential sums in one variable.

The real advance came with Deligne's second proof, Weil II, [De-5], which gives estimates for the weights of the compact cohomology groups of an arbitrary variety with coefficients in an arbitrary sheaf whose weights are already known. (From this point of view, Weil I treats the case of a projective smooth variety and the constant sheaf.) The full power of this result, especially when it is combined with various cohomology techniques (e.g., fibering by curves, fibering over cruves, Leray spectral sequence, analysis of vanishing cycles), is far from having been fully exploited.

In many case, problems concerning exponential sums are completely transformed by this sheaf-theoretic perspective: rather than struggling with a single exponential sum, one is led naturally to construct an entire family of sums which includes the original sum, and then to construct a complex of sheaves on the parameter variety whose local traces are the sums in question. In favorable cases, a cohomological analysis will allow one to prove that the complex of sheaves is but a single sheaf, and that this single sheaf is both lisse and pure.

When the parameter variety is itself a group, one can perform upon its sheaves an l-adic analogue of convolution, and this operation, applied to sheaves whose local traces are exponential sums, will yield new sheaves whose local traces are again exponential sums. This idea is already implicit in Deligne's 1974 discussion [De-3] of Kloosterman sums. Another idea of the same sort, also due originally to Deligne, was developed and exploited by Brylinski and Laumon, is that of the l-adic Fourier transform of sheaves on the additive group $\mathbf{A}^1$, or more generally on $\mathbf{A}^n$. This is another important tool in the construction and analysis of sheaves of exponential sums.

The overall moral is that difficulties about compactifying Artin–Schreier coverings often become irrelevant if one systematically studies sheaves of exponential sums on parameter varieties, rather than individual sums one at a time.

Another advantage of the sheaf-theoretic approach to exponential sums is the incredible simplification it leads to in the analysis of equidistribution questions. The traditional approach to equidistribution involves computing all the moments. Even if one starts with a one-parameter family of one-variable sums, the n-th moment will be an $n+1$-variable sum, which often becomes intractable as n grows, due to the above-mentioned difficulties in compactifying Artin–Schreier varieties of higher and higher dimension.

However, if one is given any lisse pure l-adic sheaf on a curve over a finite field, it results from Weil II that the equidistribution theory of its local traces is governed in a very simple way by a certain semi-simple algebraic group over $\overline{\mathbf{Q}}_l$, namely the Zariski closure G_{geom} of the image of the geometric fundamental group of the curve in the l-adic representation which "is" the sheaf. Under a mild hypothesis, one shows that after embedding $\overline{\mathbf{Q}}_l$ into $\mathbf{C}$ and picking a maximal compact subgroup K of the complex Lie group $G_{\text{geom}}(\mathbf{C})$, the "angles of Frobenius" of the local traces make sense as points in the space $K^\natural$ of conjugacy classes of K, and that they are equidistributed in $K^\natural$ with respect to the measure $\mu^\natural$ which is the direct image of normalized Haar measure on K. This is discussed in detail in Chapter III.

From this point of view, then, there is always an equidistribution theorem for nice equicharacteristic families of pure exponential sums; the "only" problem is to decipher it in whatever particular case interests us by actually computing G_{geom}. (The relation to the method of moments is that the n-th moment tells us the multiplicity of the trivial representation in the n-th tensor power of the standard representation of G_{geom} resulting from its definition as a Zariski closure.)

However, in practice it is not always so easy to compute G_{geom}, even when the parameter space is a curve. We often have only meager global information about the sheaf in question, and so we try first to extract and then to exploit information about its local monodromy around each of the points at infinity of the parameter curve. One striking way in which pure lisse sheaves arising from exponential sums differ from the more traditional pure lisse sheaves arising as "cohomology along the fibres, with constant coefficients, of a proper smooth morphism" is that the local monodromy of the former can be quite wildly ramified, and can be so in quite interesting ways. This possibility can often be exploited to impose some very severe restrictions on G_{geom}. The underlying mechanisms of wild ramification and the restrictions it can impose are discussed in Chapter I.

One way in which the invariants and covariants of local monodromy can be detected and analyzed is through their interpretation as the difference between the compactly supported and the ordinary cohomology groups of the parameter curve with coefficients in the sheaf under discussion. This relation, together with a thorough discussion of the basic general facts about curves and their cohomology, is given in Chapter II, and systematically exploited in Chapter VII.

Once we have marshalled together as much local and global information as we can about our sheaf, we can often reduce the problem of determining

G_{geom} to that of determining its Lie algebra. This is a problem in the classification theory of pairs $(\mathcal{G}, \rho)$ where $\mathcal{G}$ is a semi-simple Lie algebra and where ρ is a faithful representation of $\mathcal{G}$ about which we have limited information. For example, in the case of Kloosterman sheaves, the basic information is that $\mathcal{G}$ contains a nilpotent element n such that $\rho(n)$ has a one-dimensional kernel. The classification of these $(\mathcal{G}, \rho)$ is presented in Chapter XI, in a manner intended to be accessible to the non-specialist in Lie theory.

Another question which naturally arises from the sheaf-theoretic point of view is whether the actual image of π_1^{geom} in the l-adic representation is as big as possible, given that its Zariski closure is G_{geom}. That it is so for all $l \gg 0$ satisfying a suitable congruence condition can sometimes be proven by an ingenious argument of Ofer Gabber (cf. Chapter XII) which reduces the question to a quite general result of his concerning subgroups of simply-connected groups which are generated by enough unipotent elements. (A variant of this last result is due independently to M. Nori [No].)

One of the main themes of this book is the study of the convolution of suitable lisse sheaves on the multiplicative group $\mathbf{G}_m = \mathbf{P}^1 - \{0, \infty\}$ over a finite, or algebraically closed, field of characteristic $p > 0$ (cf. Chapters 5, 6, 7, 8). It may seem strange at first sight to devote so much attention to lisse sheaves on so particular a curve as $\mathbf{G}_m$; it would certainly be so if we were in characteristic zero.

However, if X is any smooth geometrically curve over a perfect field k of characteristic $p > 0$, then there exists a non-empty Zariski open set U in X and a finite étale k-morphism $U \to \mathbf{G}_m$. This analogue of Belyi's striking theorem [Be] about open sets of curves over $\overline{\mathbf{Q}}$ being finite étale over $\mathbf{P}^1 - \{0, 1, \infty\}$ is easily proven (take a general function $f : X \to \mathbf{P}^1$ to reduce to the case when X is open in $\mathbf{P}^1$; shrinking X, we may assume $X = \mathbf{A}^1 - \Gamma$, where Γ is *finite* étale additive subgroup of $\mathbf{A}^1$—this is where we use characteristic p—the quotient group $\mathbf{A}^1/\Gamma$ is again $\mathbf{A}^1$, via the explicit projection $\pi : t \to \Pi_\gamma(t - \gamma)$, and this same π is the required finite étale map of $\mathbf{A}^1 - \Gamma$ to $\mathbf{G}_m$). Its moral is that in characteristic p, whatever can happen for lisse sheaves on general curves already happens for lisse sheaves on $\mathbf{G}_m$.

In the same chain of ideas, it is natural to try to classify all lisse sheaves on $\mathbf{G}_m$ with specified local monodromy at both zero and infinity. It turns out (cf. 8.7) that the Kloosterman sheaves can be specified among all lisse sheaves on $\mathbf{G}_m$ exactly by specifying their local monodromy at both zero and infinity. This intrinsic characterization suggests that Kloosterman sheaves are more intrinsic than their construction in terms of exponential sums might suggest. The proof of this characterization relies heavily upon the l-adic Fourier transform on $\mathbf{A}^1$ as developed by Brylinski and Laumon, which we discuss at some length (cf. 8.2–8.5).

After this brief survey of the ideas, techniques, and results to be applied, we now turn to the actual concrete problems to which they will be applied.

We will begin by explaining the original motivating problem. Given a prime number p and an integer a which is prime to p, the Kloosterman sum is defined as the complex number

$$\mathrm{Kl}(p,a) = \sum_{\substack{xy \equiv a \mod p \\ x,y \in \mathbf{Z}/p\mathbf{Z}}} \exp\left(\frac{2\pi i}{p}(x+y)\right).$$

This sum is real (replace (x, y) by $(-x, -y)$), and its absolute value is at most $2\sqrt{p}$, this last estimate due to Weil as a consequence of the "Riemann Hypothesis" for curves over finite fields. Therefore, there exists a unique angle $\theta(p, a) \in [0, \pi]$ for which

$$-\mathrm{Kl}(p,a) = 2\sqrt{p}\,\cos\theta(p,a).$$

The problem is to understand how the angles $\theta(p, a)$ are distributed in $[0, \pi]$ for a fixed non-zero integer a, as p varies over all the primes not dividing a. Although there is no compelling conceptual justification, and not a great deal of "computer evidence," it is nonetheless tempting to believe that for fixed $a \neq 0$ in $\mathbf{Z}$, the angles $\{\theta(p, a)\}_p$ are equidistributed in $[0, \pi]$ with respect to the "Sato–Tate measure" $(2/\pi)\sin^2\theta\, d\theta$. Unfortunately, we have absolutely no contribution to make to this question.

Our results are concerned rather with the distribution of the angles $\theta(p, a)$ for *fixed* p as a varies over $\mathbf{F}_p^\times$. We prove (cf. Chapter 13) that as $p \uparrow \infty$, the $p-1$ angles $\{\theta(p,a)\}_{a\in\mathbf{F}_p^\times}$ "become" equidistributed in $[0, \pi]$ for the Sato–Tate measure, and we give bounds for the error term. In a given characteristic $p > 0$, once we pick a prime number $l \neq p$ and an l-adic place λ of the field $\mathbf{Q}(\zeta_p) = E$, there is a natural lisse, rank two E_λ-sheaf, "$\mathrm{Kl}_\psi(2)$", on the multiplicative group $\mathbf{G}_m \otimes \mathbf{F}_p$ in characteristic p, whose local "traces of Frobenius" are (minus) the Kloosterman sums in question. The equidistribution property of the angles $\{\theta(p,a)\}_{a\in\mathbf{F}_p^\times}$ then results (via Weil II) from the fact the geometric monodromy group of the rank-two sheaf $\mathrm{Kl}_\psi(2)$ is a Zariski-dense subgroup of $\mathrm{Sl}(2)$, cf. Chapter 11.

For any integer $n \geq 2$, one may consider the more general Kloosterman sums

$$\mathrm{Kl}(p,n,a) = \sum_{\substack{x_1\dots x_n \equiv a \mod p \\ x_1,\dots,x_n \mod p}} \exp\left(\frac{2\pi i}{p}(x_1+\dots+x_n)\right)$$

and their analogues over finite extension fields $\mathbf{F}_q$ of $\mathbf{F}_p$; one again constructs a natural lisse rank n E_λ-sheaf "$\mathrm{Kl}_\psi(n)$" on $\mathbf{G}_m \otimes \mathbf{F}_p$ whose "local traces of Frobenius" are ($(-1)^{n-1}$ times) the Kloosterman sums in question. The angle $\theta(p,a) \in [0,\pi]$ which we encountered for $n=2$ is now interpreted as a conjugacy class $\theta(p,n,a)$ in a compact form K of the Zariski closure G_{geom} of the geometric monodromy group of the sheaf $\mathrm{Kl}_\psi(n)$ on $\mathbf{G}_m \otimes \mathbf{F}_p$ (for $n=2$ this is the parametrization

$$\theta \mapsto \begin{pmatrix} e^{i\theta} & 0 \\ 0 & e^{-i\theta} \end{pmatrix}$$

of conjugacy classes in $K = \mathrm{SU}(2)$ by $[0,\pi]$). It follows from Weil II that the conjugacy classes $\{\theta(p,n,a)\}_{a\in\mathbf{F}_p^\times}$ have an approximate equidistribution, with explicit error estimate, for the measure on $K^\natural = \{$conjugacy classes in $K\}$ which is the direct image of Haar measure on K (for $K = \mathrm{SU}(2)$, this measure on $K^\natural \approx [0,\pi]$ is Sato–Tate measure $(2/\pi)\sin^2\theta\, d\theta$). The problem is to compute the Zariski closure G_{geom} of the geometric monodromy group of the lisse E_λ-sheaf $\mathrm{Kl}_n(\psi)$ on $\mathbf{G}_m \otimes \mathbf{F}_p$.

A priori, the group G_{geom} "depends" on three parameters (p,n,λ). In fact, it is independent of λ, and very nearly independent of p as well.

For $n \geq 2$ even, the sheaf $\mathrm{Kl}_\psi(n)$ carries an alternating autoduality respected by π_1^{geom}, and we prove that, in fact

$$G_{\mathrm{geom}} = \mathrm{Sp}(n) \quad \text{for } n \text{ even, } p \text{ arbitrary.}$$

For $n \geq 3$ odd, and p odd, we prove that

$$G_{\mathrm{geom}} = \mathrm{SL}(n) \quad \text{if } pn \text{ is odd.}$$

For $n \geq 3$ odd, and $p = 2$, the sheaf $\mathrm{Kl}_\psi(n)$ carries a symmetric autoduality respected by π_1^{geom}, and we prove that

$$G_{\mathrm{geom}} = \mathrm{SO}(n) \quad \text{if } p=2,\ n\geq 3 \text{ odd},\ n \neq 7.$$

For $n = 7$ and $p = 2$, a general argument based on classification shows that *either* $G_{\mathrm{geom}} = \mathrm{SO}(7)$, *or* $G_{\mathrm{geom}} =$ the subgroup G_2 of $\mathrm{SO}(7)$, where $G_2 \hookrightarrow \mathrm{SO}(7)$ is the seven-dimensional irreducible representation of G_2.

We initially expected to find $SO(7)$ as the answer. To our surprise, we found instead that

$$G_{\text{geom}} = G_2 \quad \text{for } p = 2,\ n = 7.$$

Perhaps exceptional Lie groups aren't so exceptional after all.

All of these results on G_{geom} are proven in Chapter 11, as consequences of some general classification theorems (11.6, 11.7) which are of independent interest.

In Chapter 12, we show that for given p and n, the actual λ-adic image of π_1^{geom} (i.e., rather than the Zariski closure) is "as big as possible," cf. 12.1, 12.2, 12.5.2, and 12.6.2. The ideas and arguments of this chapter are due entirely to Ofer Gabber.

In Chapter 13, we apply the results of Chapter 11 to the equidistribution of Kloosterman angles. The results of this chapter make it reasonable to ask whether for $n \geq 2$ a fixed integer, and $a \neq 0$ a fixed non-zero integer, the conjugacy classes $\{\theta(p,n,a)\}_p$, as p runs over odd primes which are prime to a, are equidistributed for Haar measure in $K^\natural$, for K a compact form of G_{geom} ($=$ Sp(n) if n is even, SL(n) if odd). For $n = 2$, this is the motivating problem with which we began.

A second problem (cf. Chapter 9) which we consider is the following. Let $\mathbf{F}_q$ be a finite field, ψ a non-trivial $\mathbf{C}$-valued additive character of $\mathbf{F}_q$, and χ a generator of the group of all $\mathbf{C}$-valued multiplicative characters of $\mathbf{F}_q^\times$. For each integer $1 \leq a \leq q-2$, the gauss sum

$$g(\psi, \chi^a) = \sum_{x \in \mathbf{F}_q^\times} \psi(x)\chi^a(x)$$

has absolute value $\sqrt{q}$. Let us temporarily denote by $\theta(a)$

$$\theta(a) \stackrel{\text{def}}{=} \frac{g(\psi,\chi^a)}{\sqrt{q}} \in S^1$$

the corresponding angle, viewed as a point on the unit circle. It is well known (cf. [Ka-1]) that as $q \uparrow \infty$, the $q-2$ points $\theta(a) \in S^1$ "become" equidistributed in S^1 with respect to usual Haar measure; this equidistribution results from the fact that the Kloosterman sheaves $\mathrm{Kl}_\psi(n)$ on $\mathbf{G}_m \otimes \mathbf{F}_p$ are lisse sheaves of rank n which are pure of weight $n-1$. Motivated by an early paper of Davenport [Dav], we ask about the distribution, for a given integer $r \geq 1$, of the r-tuples of "successive" angles

$$(\theta(a+1), \theta(a+2), \ldots, \theta(a+r)) \in (S^1)^r,$$

as a runs over the interval $0 \leq a \leq q-2-r$.

We prove that as $q \uparrow \infty$, these $q-1-r$ points in $(S^1)^r$ "become" equidistributed with respect to Haar measure. The proof is based on the fact that monomials in gauss sums (which arise in applying the classical Weyl criterion) are the multiplicative Fourier transform of certain "Kloosterman sums with multiplicative characters" of the form

$$\sum_{\substack{x_1 \ldots x_n = a \\ x_i \in \mathbf{F}_q^\times}} \psi(x_1 + \cdots + x_n)\chi_1(x_1)\ldots\chi_n(x_n),$$

where $\chi_1, \ldots, \chi_n$ are multiplicative characters of $\mathbf{F}_q^\times$.

For fixed ψ and $\chi_1, \ldots, \chi_n$, there is a natural lisse, rank n E_λ-sheaf $\mathrm{Kl}(\psi; \chi_1, \ldots, \chi_n)$ on $\mathbf{G}_m \otimes \mathbf{F}_q$ (here $E = \mathbf{Q}(\zeta_p, \zeta_{q-1})$, λ is a finite place of E of residue characteristic $l \neq p = \mathrm{char}(\mathbf{F}_q)$) whose local traces are $(-1)^{n-1}$ times the above Kloosterman sums with multiplicative characters. The equidistribution result on tuples of gauss sum angles results (by Weil II, the Lefschetz trace formula, and Parseval's identity) from the fact that if $\mathrm{Kl}(\psi; \chi_1, \ldots, \chi_n)$ is geometrically isomorphic to $\mathrm{Kl}(\psi; \chi_1', \ldots, \chi_n')$, then, after renumbering the χ_i', we have $\chi_i' = \chi_i$ for $i = 1, \ldots, n$.

In fact, we prove that given the sheaf $\mathrm{Kl}(\psi; \chi_1, \ldots, \chi_n)$, we may recover $(\chi_1, \ldots, \chi_n)$ entirely in terms of its local monodromy at zero. In trying to prove this result, it became natural to view the sheaf $\mathrm{Kl}(\psi; \chi_1, \ldots, X_n)$ as the n-fold "convolution" of the simpler rank one sheaves $\mathrm{Kl}(\psi; \chi_i)$.

The formalism of convolution of $\mathbf{G}_m$ took on a interest in its own right with our realization that over a finite field, the category of lisse λ-adic sheaves of $\mathbf{G}_m \otimes \mathbf{F}_q$ which are pure, tame at zero, and completely wild at infinity, is stable under convolution. Moreover, given two such sheaves, the weight, rank, local monodromy at zero and swan conductor at infinity of their convolution is easily expressed in terms of the same data for the convolvees. Ofer Gabber showed us how to drop the hypotheses "over a finite field, and pure" from our original weight-theoretic arguments concerning local monodromy at zero. His method is explained in the appendix to Chapter 7, and the "weight-free" proof we give of 5.2.1(3) is due to him as well: our original proof of 5.2.1(3) was based on the dimension count given in 7.1.6, "read backwards."

By combining Gabber's "topological" approach to local monodromy with the theory of Laumon et al. of Fourier transform on the additive group, we are able to give a complete description of all lisse λ-adic sheaves on $\mathbf{G}_m \otimes \mathbf{F}_q$ which are tame at zero and totally wild at infinity with swan conductor one.

We prove that any such sheaf is, up to a twist, a sheaf of Kloosterman type. In particular, up to a twist, any such λ-adic sheaf is pure of some weight, and part of a "compatible system." It would be interesting to compare this result with the conjectural description of such sheaves, provided by the Langlands philosophy, in terms of automorphic forms.

We now turn to a discussion of some open problems related to the above results.

The first is the most vague, but also the most interesting. Is there a reasonable theory "over $\mathbf{Z}$" of exponential sums? For example, we found that for $p > 2$, the group G_{geom} attached to n-variable Kloosterman sums was a function of n alone (i.e., independent of $p \neq 2$ and of λ). This phenomenon is presumably typical of quite general families of exponential sums. Can one prove it? Can one "explain" it, even conjecturally?

There are quite a few general problems concerning convolution. What operation on automorphic representations should correspond it? Given two lisse sheaves in our convolution category $\mathcal{C}$ (tame at zero, totally wild at infinity) what information about them does one need in order to compute G_{geom} for their convolution? to insure the irreducibility of their convolution? to compute the breaks at infinity of their convolution (cf. 7.6)?

Another question which bears looking into is the following. Are there Kloosterman sheaves whose geometric monodromy group is an "interesting" finite group, i.d., are there finite monodromy "surprises"analogous to the $p = 2$, $n = 7$ surprise of finding G_2 instead of $\mathrm{SO}(7)$? It might be interesting to examine Kloosterman sheaves $\mathrm{Kl}(\psi; \chi_1, \ldots, \chi_n)$ with "given" $\chi_1, \ldots, \chi_n$ which only have finite geometric monodromy in exceptional characteristics.

There exists a close analogy between the Kloosterman sheaves $\mathrm{Kl}(\psi; \chi_1, \ldots, \chi_n)$ on $\mathbf{G}_m \otimes \mathbf{F}_q$ and the differential equations on $\mathbf{G}_m \otimes \mathbf{C}$ given by

$$(*) \qquad \left(\prod_{i=1}^{n} \left(x\frac{d}{dx} - \alpha_i\right) - \pi x\right) f = 0,$$

where the α_i are rational numbers with denominator dividing $q - 1$, and π is an arbitrary non-zero constant. In this analogy, the differential equation corresponding to $\mathrm{Kl}_\psi)n)$ (i.e., all $\chi_i = 1$) is the one with all $\alpha_i = 0$:

$$(**) \qquad \left(\left(x\frac{d}{dx}\right)^n - \pi x\right) f = 0.$$

The arguments developed here to compute G_{geom} for $\mathrm{Kl}_\psi(n)$ can also be used to show that for $n \geq 2$, the differential galois group of the above differential equation (**) is Sp(n) for n even, and SL(n) for n odd (cf. [Ka-7]).

The p-adic analytic version of the above analogy, begun by Dwork [Dw], has been developed extensively by Sperber [Sp-1], [Sp-2]. One of Sperber's most striking results (cf. [Sp-1]) is the determination of the p-adic Newton polygon of the local characteristic polynomial at each closed point x of Frobenius F_x on the sheaf $\mathrm{Kl}_\psi(n)$ on $\mathbf{G}_m \otimes \mathbf{F}_q$. He shows that the suitably normalized p-adic valuations of the n eigenvalues $\alpha_i(x)$ of each Frobenius F_x are $0, 1, 2, \ldots, n-1$;

$$\alpha_i(x) = q^{i \deg(x)} u_i(x), \qquad u_i(x) \in (\mathbf{Z}_p[\zeta_p])^\times,$$

for $i = 0, \ldots, n-1$. From the theory of the Newton–Hodge filtration [Ka-3] of Sperber's F-crystal analogue of $\mathrm{Kl}_\psi(n)$, it follows that for each $0 \le i \le n-1$, there is a continuous character χ_i

$$\chi_i : \pi_1(\mathbf{G}_m \otimes \mathbf{F}_q, \bar{\eta}) \to (\mathbf{Z}_p[\zeta_p])^\times$$

such that for every closed point x of $\mathbf{G}_m \otimes \mathbf{F}_q$ we have

$$\chi_i(F_x) = u_i(x).$$

Nearly nothing is known about these p-adic characters χ_i and their interrelations, except for the "obvious" relation $\chi_0\chi_1 \cdots \chi_{n-1} = 1$, and, for n even, the relations $\chi_i\chi_{n-1-i} = 1$. What is the image of χ_i? What is the image of $(\chi_0, \ldots, \chi_{n-1}) : \pi_1 \to ((\mathbf{Z}_p[\zeta_p]^\times)^n)$? What is the Zariski closure in $(\mathbf{G}_m)^n$ of the image of $(\chi_0, \ldots, \chi_{n-1})$?

It is a pleasure to acknowledge the fundamental influence on this book of Deligne's work, especially [De-3] and [De-5]. Much of this material was presented in a course at Orsay in the fall of 1983, and parts were presented earlier in 1983 in lectures at the Universities of Minnesota and Tokyo, and at Princeton. My thanks to all of these institutions for their hospitality, and to the IHES where this manuscript was written. As will be obvious to the reader, I benefitted greatly from discussions with T. Ekedahl, O. Gabber, and G. Laumon, whose comments, suggestions and questions considerably clarified a number of the topics treated here. I would also like to thank J.-P. Serre for his helpful comments concerning G_2, and Benji Fisher for his help in proofreading.

The overall structure of the book is:
Chapters 1–3: Fundamental Background
Chapter 4: The Basic Example
Chapters 5–8: Detailed Study of Sheaves on $\mathbf{G}_m$

Chapters 9–13: Analysis of Monodromy, and Applications

For the readers' convenience, there is a Leitfaden and a fairly detailed Table of Contents.

CHAPTER 1

Breaks and Swan Conductors

1.0. The Basic Setting (cf. [Se-1], pp. 80–82). Let K be the fraction field of a henselian discrete valuation ring R whose residue field k is perfect of characteristic $p > 0$, K^{sep} a separable closure of K, $D = \text{Gal}(K^{\text{sep}}/K)$ the galois group (D for "decomposition group," which is how this group occurs in the global theory), $I \subset D$ the inertia subgroup, and $P \subset I$ the p-sylow subgroup of I. The groups P and I are both normal in D, and they sit in standard short exact sequences

$$1 \to I \to D \to \text{Gal}(k^{\text{sep}}/k) \to 1$$
$$1 \to P \to I \to \prod_{l \neq p} \mathbf{Z}_l(1) \to 1.$$

The quotient I/P is sometimes also denoted I^{tame}.

The "upper numbering filtration" on I is a decreasing filtration of I by closed subgroups $I^{(r)}$, indexed by real numbers $r \geq 0$. Each $I^{(r)}$ is normal in D, $I^{(0)}$ is I itself, and $I \supset P \supset I^{(r_1)} \supset I^{(r_2)}$ for $0 < r_1 < r_2$. Furthermore,

$$P = \text{the closure of } \bigcup_{r>0} I^{(r)},$$
$$\{1\} = \bigcap_{r>0} I^{(r)}$$
$$I^{(r)} = \bigcap_{0<x<r} I^{(x)} \quad \text{for } r > 0.$$

1.1. Proposition. *Let M be a $\mathbf{Z}[1/p]$ module on which P operates through a finite discrete quotient, say by $\rho : P \to \text{Aut}_{\mathbf{Z}}(M)$. Then:*

(1) *M has a unique direct-sum decomposition $M = \bigoplus M(x)$ into P-stable submodules $M(x)$, indexed by real numbers $x \geq 0$, such that $M(0) = M^P$ and*

$$(M(x))^{I^{(x)}} = 0 \quad \textit{for } x > 0,$$
$$(M(x))^{I^{(y)}} = M(x) \quad \textit{for all } y > x.$$

(2) *If* $x > 0$, *then* $M(x) = 0$ *for all but the finitely many values of* x *for which*

$$\rho(I^{(x)}) \supsetneqq \bigcup_{y>x} \rho(I^{(y)}).$$

(3) *For variable* M *but fixed* x, *the construction* $M \mapsto M(x)$ *is an exact functor.*

(4) *For* M, N *as above, we have* $\mathrm{Hom}_{P-\mathrm{mod}}(M(x), N(y)) = 0$ *if* $x \neq y$.

Proof. If $P \twoheadrightarrow G$ is any finite discrete quotient of P through which ρ factors, define subgroups $G(x) \subset G$ for $x > 0$ by

$$G(x) = \mathrm{image}(I^{(x)})$$

and subgroups $G(x+) \subset G$ for $x \geq 0$ by

$$G(x+) = \bigcup_{y>x} \mathrm{image}(I^{(y)}).$$

For each x, define the corresponding "projection onto the invariants,"

$$\pi(x) = \frac{1}{\#G(x)} \sum_{g \in G(x)} g$$

$$\pi(x+) = \frac{1}{\#G(x+)} \sum_{g \in G(x+)} g.$$

Because the groups $G(x)$ and $G(x+)$ are normal in G, these operators are all central idempotents in the $\mathbf{Z}[1/p]$-group-ring of G.

For $0 \leq x < y$ we have $G \supset G(x+) \supset G(y) \supset G(y+)$

$$G(y) = G(y - \varepsilon) \quad \text{for all sufficiently small } \varepsilon > 0,$$

$$G(x+) = G(x + \varepsilon) \quad \text{for all sufficiently small } \varepsilon > 0$$

and

$$G(x) = \{1\} \quad \text{for } x \text{ sufficiently large.}$$

One verifies easily that the central idempotents $\pi(0+)$ and

$$\pi(x+)(1 - \pi(x)), \quad \text{for each } x > 0,$$

are orthogonal, that all but finitely many vanish, that they sum to 1, and that the corresponding decomposition of M is the unique one satisfying condition (1) of the lemma. Taking $G = \rho(P)$ shows that (2) holds. Assertions (3) and (4) are clear from the construction by idempotents. ∎

1.2. Definition. The decomposition $M = \bigoplus M(x)$ of the lemma is called the break-decomposition of M. The values of $x \geq 0$ for which $M(x) \neq 0$ are called the breaks of M.

1.3. Lemma. *For M and N two $\mathbf{Z}[1/p]$ modules on which P acts through a finite discrete quotient, $M \otimes N$ and $\mathrm{Hom}_{\mathbf{Z}}(M, N)$ are again such modules (with action $g \otimes g$ on $M \otimes N$, $\psi \mapsto g\psi g^{-1}$ on $\mathrm{Hom}_{\mathbf{Z}}(M, N)$), and their break-decompositions are related by*

$$\begin{aligned} M(x) \otimes N(y) &\subset (M \otimes N)(\sup(x, y)) \quad \text{if } x \neq y, \\ M(x) \otimes N(x) &\subset \sum_{y \leq x} (M \otimes N)(y), \\ \mathrm{Hom}_{\mathbf{Z}}(M(x), N(y)) &\subset \mathrm{Hom}_{\mathbf{Z}}(M, N)(\sup(x, y)) \quad \text{if } x \neq y, \\ \mathrm{Hom}_{\mathbf{Z}}(M(x), N(x)) &\subset \sum_{y \leq x} \mathrm{Hom}_{\mathbf{Z}}(M, N)(y). \end{aligned}$$

Proof. Let G be a finite p-group quotient of P through which the action of P on both M and N factors. Then the action of P on both $M \otimes N$ and $\mathrm{Hom}_{\mathbf{Z}}(M, N)$ also factors through G.

If $x < y$, then $G(y)$ acts trivially on $M(x)$, but $(N(y))^{G(y)} = 0$. By the exactness of the functor "$G(y)$-invariants" on $\mathbf{Z}[1/p][G(y)]$-modules, we see that

$$(M(x) \otimes N(y))^{G(y)} = M(x) \otimes (N(y))^{G(y)} = 0.$$

The group $G(y+)$ operates trivially on both $M(x)$ and on $N(y)$, so

$$(M(x) \otimes N(y))^{G(y+)} = M(x) \otimes (y),$$

whence the unique break of $M(x) \otimes N(y)$ is $\sup(x, y)$ for $x \neq y$.

If $x < y$, then $G(y+)$ acts trivially on $\mathrm{Hom}_{\mathbf{Z}}(M(x), N(y))$, whereas $G(y)$ acts trivially on $M(x)$ but with no non-zero invariants on $N(y)$. We must show $G(y)$ has no non-zero invariant in $\mathrm{Hom}_{\mathbf{Z}}(M(x), N(y))$. But an invariant is a $G(y)$-equivariant map $\psi : M(x) \to N(y)$; so it maps any element of $M(x) = (M(x))^{G(y)}$ to $(N(y))^{G(y)} = 0$. Therefore y is the unique break of $\mathrm{Hom}_{\mathbf{Z}}(M(x), N(y))$ when $x < y$.

If $x > y$, then $G(x+)$ acts trivially on $\mathrm{Hom}_{\mathbf{Z}}(M(x), N(y))$, whereas $G(x)$ acts trivially on $N(y)$, so a $G(x)$-invariant in Hom is a $G(x)$-equivariant map $\psi : M(x) \to N(y)$, which necessarily factors through the covariants $(M(x))_{G(x)}$. But for any $\mathbf{Z}[1/p][G(y)]$ module, the natural map from invariant to covariants

$$(M(x))^{G(x)} \hookrightarrow M(x) \twoheadrightarrow (M(x))_{G(x)}$$

is an isomorphism, the inverse being the averaging operator $\pi(x)$. As $(M(x))^{G(x)} = 0$, we have $(Mx))_{G(x)} = 0$, so $\psi = 0$. Therefore x is the unique break of $\mathrm{Hom}_{\mathbf{Z}}(M(x), N(y))$ when $x > y$.

In the $x = y$ case, $G(x+)$ acts trivially on both $N(x)$ and $M(x)$, so on both $M(x) \otimes N(x)$ and on $\mathrm{Hom}_{\mathbf{Z}}(M(x), N(x))$. Therefore both of these have all breaks $\leq x$. ∎

1.4. Lemma. *If A is a $\mathbf{Z}[1/p]$-algebra, and M a left A-module on which P acts A-linearly through a finite discrete quotient, then in the break-decomposition*

$$M = \bigoplus_{x \geq 0} M(x),$$

each $M(x)$ is an A-submodule of M. For any A-algebra B, the break-decomposition of $B \otimes_A M$ is given by

$$B \otimes_A M = \bigoplus_{x} B \otimes_A M(x).$$

Proof. Applying 1.1(4) to left multiplication by an element $a \in A$, we see that each $M(x)$ is an A-module. The second assertion is clear from the construction of the break–decomposition by means of idempotents. ∎

1.5. Lemma. *Let A be noetherian local ring of residue characteristic $\neq p$, and M a free A-module of finite rank on which P acts A-linearly through a finite discrete quotient. Then each $M(x)$ is a free A-module of finite rank, whose rank is called the multiplicity of the break x. (Thus the total number of breaks of M, counted with multiplicity, is rank(M).) If $M^\vee$ denotes the A-linear dual of M (contragredient representation of P), then $(M^\vee(x) = (M(x))^\vee$; in particular M and $M^\vee$ have the same breaks and multiplicities. If $A \to B$ is a not-necessarily-local homomorphism to a second noetherian local ring (necessarily of residue characteristic $\neq p$) then M and $M \otimes_A B$ have the same breaks with the same multiplicities.*

Proof. For each $x \geq 0$, $M(x)$ is an A-module direct factor of M by the preceding corollary. For A noetherian local and M free of finite rank, we find that $M(x)$ is projective of finite type, so free of finite type. Because $M^\vee$ is a submodule of $\mathrm{Hom}_{\mathbf{Z}}(M, A)$, the second assertion follows from (1.3). The last assertion is obvious from the change of rings formula (1.4)

$$(M \otimes_A B)(x) = M(x) \otimes_A B. \quad ∎$$

1.6. Definition. For M as in 1.5 above, we define its Swan conductor to be the non-negative real number (it is in fact a rational number, cf. [Ka-5])

$$\mathrm{Swan}(M) = \sum_{x \geq 0} x \, \mathrm{rk}_A(M(x)).$$

1.7. Remarks. Clearly we have $\mathrm{Swan}(M) = 0$ if and only if $M = M(0) = M^P$ is trivial as a representation of P. For any ring homomorphism $A \to B$ with B noetherian local, we have

$$\mathrm{Swan}(M) = \mathrm{Swan}(M \otimes_A B).$$

The most useful cases of this are $B =$ residue field of A, and, if A is a domain, $B =$ fraction field of A. Notice also that M and $M^\vee = \mathrm{Hom}_A(M, A)$ have the same Swan conductor.

We now consider the situation where the given action of P extends to an action of I, or of D.

1.8. Lemma. *If M is a $\mathbf{Z}[1/p]$-module on which I (resp. D) acts, such that the action of P factors through a finite discrete quotient of P, then in the break-decomposition $M = \oplus M(x)$, each $M(x)$ is I-stable (resp. D-stable).*

Proof. The groups P and $I^{(x)}$ for $x > 0$ are all normal in I (resp. D), so the elements of I (resp. D) commute with all of the idempotents $\pi(x)$ and $\pi(x+)$ used to define the break-decomposition. ∎

Now let A be a complete noetherian local ring with finite residue field $\mathbf{F}_\lambda$ of characteristic $l \neq p$, and let M be a free A-module of finite rank on which D (resp. I) acts continuously. Because an open subgroup of finite index in $\mathrm{Aut}_A(M)$ is pro-l (e.g., the subgroup of elements which induce the identity on $M \otimes_A \mathbf{F}_\lambda$), while P is pro-p, the action of P on M automatically factors through a finite discrete quotient of P. Thus we may speak of the break-decomposition of M; it provides a canonical direct sum decomposition as $A[D]$ (resp. $A[I]$)-module

$$M = \bigoplus_{x \geq 0} M(x)$$

in which each $M(x)$ is a free A-module of finite rank, whose formation commutes with arbitrary extension of scalars $A \to B$.

1.9. Proposition. *Let A be a complete noetherian local ring with finite residue field $\mathbf{F}_\lambda$ of characteristic $l \neq p$, and M a free A-module of finite rank on which I acts continuously. Then for every $x \geq 0$, the product*

$x\,\mathrm{rk}_A(M(x))$ is an integer ≥ 0. In particular, the Swan conductor $\mathrm{Swan}(M)$ *is an integer ≥ 0, and* $\mathrm{Swan}(M) = 0$ *if and only if M is tame in the sense that $M = M^P$.*

Proof. In view of the break-decomposition $M = \bigoplus M(x)$, it suffices to prove universally that $\mathrm{Swan}(M)$ is an integer. Because $\mathrm{Swan}(M) = \mathrm{Swan}(M \otimes \mathbf{F}_\lambda)$, we are reduced to the case when A is a finite field $\mathbf{F}_\lambda$ of characteristic $l \neq p$. In this case, M is itself finite, so the representation of I factors through a finite quotient G of I. In this case, the compatibility ([Se-1], pp. 80-82) between upper and lower numbering shows that $\mathrm{Swan}(M)$ coincides with the integer "$b(M)$" of ([Se-2], 19.3). ∎

1.10. Remark. Suppose now that E_λ is a finite extension of $\mathbf{Q}_l$, $l \neq p$, with integer ring $\mathcal{O}_\lambda$ and residue field $\mathbf{F}_\lambda$. Let M be a finite-dimensional E_λ-vector space on which D (resp. I) operates continuously and E_λ-linearly. By compactness, there exists an $\mathcal{O}_\lambda$-lattice $\mathbf{M}$ in M, (i.e., a free $\mathcal{O}_\lambda$-module $\mathbf{M}$ of finite rank with $\mathbf{M} \underset{\mathcal{O}_\lambda}{\otimes} E_\lambda = M$) which is D-stable (resp. I-stable). Therefore P acts on $\mathbf{M}$, and hence on M, through a finite quotient. The break-decomposition of M is obtained from that of $\mathbf{M}$ by the extension of scalars $\mathcal{O}_\lambda \hookrightarrow E_\lambda$. Therefore the E_λ resp. $\mathcal{O}_\lambda$, resp. $\mathbf{F}_\lambda$-representations

$$\mathbf{M} \underset{\mathcal{O}_\lambda}{\otimes} E_\lambda = M, \qquad \mathbf{M}, \qquad \mathbf{M} \otimes \mathbf{F}_\lambda$$

all have the same breaks with the same multiplicities. In particular,

$$\mathrm{Swan}(M) = \mathrm{Swan}(\mathbf{M}) = \mathrm{Swan}(\mathbf{M} \otimes \mathbf{F}_\lambda)$$

for any $\mathcal{O}_\lambda$-form $\mathbf{M}$ of M.

1.11. Lemma. *Let l be a prime number $l \neq p$, E_λ a finite extension of $\mathbf{Q}_l$, $\mathcal{O}_\lambda$ its integer ring, $\mathbf{F}_\lambda$ its residue field. Let M be a non-zero finite dimensional E_λ (resp. $\mathbf{F}_\lambda$)-vector space which is a continuous representation of I. Suppose that $M^P = 0$ and* $\mathrm{Swan}(M) = 1$. *Then*

(1) *The unique break of M is $x = 1/\dim(M)$, and its multiplicity is* $\dim(M)$.
(2) *As representation of I, M is absolutely irreducible.*
(3) *If M is a quasi-unipotent representation of I (e.g., if M is the restriction to I of a continuous representation of D, and if no finite extension of the residue field of K contains all l-power roots of unity) then the image of I in* $\mathrm{Aut}(M)$ *is finite.*

(4) *If M as in (3) above is the restriction to I of a continuous representation of D, then an open subgroup of D acts by scalars. In particular, the image of D in* $\mathrm{Aut}(M)$ *is finite if and only if* $\det(M)$ *is a character of finite order of D.*

Proof. We have $\mathrm{Swan}(M) = \sum x \dim M(x) = 1$ and $M(0) = 0$. Since each term $x \dim M(x)$ with $x > 0$ is a non-negative integer, there is exactly one such term which is non-zero, say $x_0 \dim M(x_0) = 1$. Because $M(0) = 0$ by hypothesis, the decomposition of M as $\bigoplus M(x)$ shows $M = M(x_0)$, whence $\dim M(x_0) = \dim M$, $x_0 = 1/\dim M$. If $M' \subset M$ is a non-zero I-sub-representation, then $M'(x) \subset M(x)$ for every $x \geq 0$, so the only possible break of M' is $1/\dim M$, whence $\mathrm{Swan}(M') = \dim M'/\dim M$; as $\mathrm{Swan}(M')$ is an integer, we must have $M = M'$. As this argument is equally valid after extending scalars to any finite extension of E_λ (resp. $\mathbf{F}_\lambda$), we get the absolute irreducibility. For (3) and (4), only the E_λ case is not obvious. For (3), the condition on the residue field guarantees that the local monodromy theorem applies: there exists an open subgroup of I on which the representation is unipotent. The associated nilpotent endomorphism N is I-equivariant (cf. 7.0.5), hence a nilpotent scalar (I acts absolutely irreducibly), hence zero, whence the representation is trivial on an open subgroup of I. For (4), every element of D normalizes the finite (by (3)) image of I in $\mathrm{Aut}(M)$. Therefore an open subgroup of D commutes with the image of I. By the absolute irreducibility of I, this open subgroup of D acts by scalars in $\mathrm{Aut}(M)$. ∎

1.12. Lemma. *Let E_λ be a finite extension of $\mathbf{Q}_l$, $l \neq p$, with residue field $\mathbf{F}_\lambda$. Let M be a non-zero finite dimensional E_λ (resp. $\mathbf{F}_\lambda$)-vector space on which I acts continuously and irreducibly. Then the unique break of M is $x = \mathrm{Swan}(M)/\dim(M)$, and its multiplicity is* $\dim(M)$.

Proof. This is obvious from the fact that in the break-decomposition $M = \bigoplus M(x)$, each $M(x)$ is an I-submodule; by irreducibility M must be a single $M(x)$. ∎

1.13. Before continuing, we must recall that for every integer $N \geq 1$ prime to p, the inertia group I has a unique open subgroup $I(N)$ of index N. In terms of the short exact sequence

$$1 \to P \to I \to \prod_{l \neq p} \mathbf{Z}_l(1) \to 1,$$

$I(N)$ is the kernel of the projection of I onto the $\boldsymbol{\mu}_N$ quotient of $\prod_{l \neq p} \mathbf{Z}_l(1)$. If we think of I as being the absolute galois group of the maximal unramified extension K^{nr} of K inside K^{sep}, then $I(N)$ corresponds to the $\boldsymbol{\mu}_N$-extension of K^{nr} obtained by adjoining the N-th root of any uniformizing parameter π of K^{nr}.

The wild inertia subgroup P of I is also the p-Sylow subgroup of $I(N)$, but the upper numbering filtration on it changes; if we think of $I(N)$ as the absolute galois group of $K^{nr}(\pi^{1/N})$, its upper numbering filtration is related to that of I by the simple change of scale

$$I(N)^{(x)} = I^{(x/N)} \text{ for all } x \geq 0.$$

From this it follows that if E_λ is a finite extension of $\mathbf{Q}_l$, $l \neq p$, with residue field $\mathbf{F}_\lambda$, then we have the following behavior of breaks and multiplicities under the operations of restriction and induction of E_λ (resp. $\mathbf{F}_\lambda$)-representations.

1.13.1. If M is a finite-dimensional continuous E_λ (resp. $\mathbf{F}_\lambda$)-representation of I with breaks x_i of multiplicity n_i, then its restriction to $I(N)$, which we denote by $[N]^*(M)$, has breaks Nx_i with the same multiplicity n_i; more precisely, we have $([N]^*(M))(x) = M(x/N)$ for each $x \geq 0$, and consequently

$$\text{Swan}([N]^*(M)) = N\,\text{Swan}(M).$$

1.13.2. If M is induced from a finite dimensional E_λ (resp. $\mathbf{F}_\lambda$)-representation V of $I(N)$, written $M = [N]_*(V)$, we have

$$[N]^*(M) = \bigoplus \gamma V$$

(for γ running over a set of coset representatives of $I/I(N)$), whence

$$M(x/N) = ([N]^*(M))(x) = \bigoplus \gamma(V(x)) = \bigoplus (\gamma V)(x).$$

Thus if V has breaks x_i of multiplicity n_i, its induction $[N]_*(V)$ has breaks x_i/N of multiplicity Nn_i, and consequently

$$\text{Swan}([N]_*(V)) = \text{Swan}(V).$$

1.14. Proposition. *Let E_λ be a finite extension of $\mathbf{Q}_l, l \neq p$, with residue field $\mathbf{F}_\lambda$. Let M be a non-zero finite dimensional E_λ (resp. $\mathbf{F}_\lambda$)-vector space on which I acts continuously. Suppose that*

$$\text{Swan}(M) = a, \quad \dim(M) = n, \quad (a, n) = 1,$$

and that the unique break of M is a/n with multiplicity n. Then:

(1) *M is absolutely irreducible.*

(2) *Write $n = n_0 p^v$ with n_0 prime to p. Then over a finite extension of E_λ (resp. $\mathbf{F}_\lambda$), M is induced from a p^v-dimensional representation V of $I(n_0)$, and the restriction of V to P is absolutely irreducible. As $I(n_0)$-representation, all breaks of V are a/p^v.*

(3) *Over a finite extension of E_λ (resp. $\mathbf{F}_\lambda$), the restriction of M to P is the direct sum of n_0 pairwise-inequivalent absolutely irreducible p^v-dimensional representations of P, whose isomorphism classes are fixed by $I(n_0)$ and cyclically permuted by $I/I(n_0)$.*

Proof. If M' is a non-zero sub-representation of M, then its unique break is a/n. Therefore $\text{Swan}(M') = \dim(M')(a/n)$. Because $\text{Swan}(M')$ is an integer, but $(a, n) = 1$, we find that n divides $\dim(M')$. As $n = \dim(M)$, we infer that $\dim(M') = \dim(M)$, whence $M' = M$. Repeating the argument over a finite extension field, we see that M is absolutely irreducible.

To prove (2) and (3), we argue as follows. Because P acts on M through a finite p-group quotient G, M is P-semisimple. Extending scalars, we may and will assume that every irreducible representation of P which factors through G is absolutely irreducible. Let $M = \bigoplus_\alpha M_\alpha$ be the P-isotypical decomposition of M. Because P is normal in I, I permutes the non-zero M_α; because I operates irreducibly it permutes them transitively. The stabilizer of one of these blocks is open in I, and it contains P. Therefore it has some index N prime to p, so is equal to $I(N)$ by unicity. By unicity $I(N)$ is normal in I, so $I(N)$ is the stabilizer of each non-zero P-isotypical block M_α. If we pick one of them, say M_{α_0}, we have an $I(N)$-direct sum decomposition

$$M = \bigoplus \gamma M_{\alpha_0}$$

over a set of representatives γ of $I/I(N)$. This means that M is induced from the representation M_{α_0} of $I(N)$. Because M as I-representation has unique break a/n, it follows that M_{α_0} as $I(N)$-representation has unique break $Na/n = a(N/n)$, with multiplicity n/N, and it has $\text{Swan} = a$. In particular, n/N is an integer, and it is prime to a (because $(a, n) = 1$ by hypothesis). Thus we may replace I by $I(N)$ and M by M_{α_0}, and begin all over again with the *additional* hypothesis

M is isotypical as a representation of P.

Let V be a non-zero P-irreducible subspace of M. By our preliminary extension of scalars, we have insured that V is absolutely irreducible as a

P-representation. Therefore its dimension is a power of p: $\dim V = p^v$ for some $v \geq 0$. Thus it suffers to show that in fact $V = M$.

For this we argue as follows. Because M is P-isotypical, every transform γV of V by an element $\gamma \in I$ is isomorphic to V as a P-representation. If we denote by

$$\rho : P \to \mathrm{Aut}(V)$$

the action of P on V, this means that for any $\gamma \in I$, the representation $\rho^\gamma : P \to \mathrm{Aut}(V)$ defined by

$$P \ni g \mapsto \rho(\gamma g \gamma^{-1}),$$

is equivalent to ρ. We will show that ρ may be extended to a continuous representation of I on V, from which it follows that Swan(V) is an integer. Granting this for a moment, we notice that as V is a P-submodule of M, its unique break is a/n, with multiplicity dim(V). Therefore Swan(V) is given by

$$\mathrm{Swan}(V) = \dim(V)(a/n).$$

As Swan(V) is an integer, but $(a, n) = 1$, we have $\dim V \equiv 0 \bmod n$, whence $n = \dim V$, whence $V = M$, as required.

It remains to show that ρ extends to I. For this, we choose an element $\gamma \in I$ which maps onto a topological generator of $\prod_{l \neq p} \mathbf{Z}_l(1)$ in the short exact sequence

$$1 \to P \to I \to \prod_{l \neq p} \mathbf{Z}_l(1) \to 1.$$

Taking a cluster point in I of the sequence of $\{\gamma^{p^n}\}_{n \geq 0}$, we may further suppose γ has its pro-finite order prime to p. Then γ defines an isomorphism

$$P \ltimes \hat{\mathbf{Z}}_{\text{not p}} \xrightarrow{\sim} I.$$

Let $\mathbf{V}$ be an $\mathcal{O}_\lambda$-form of the P-module V (in the case of E_λ). Recall that because P acts irreducibly on V through a finite p-group, the $\mathbf{F}_\lambda[P]$-module $\mathbf{V} \otimes \mathbf{F}_\lambda$ is irreducible.

Now consider the representations ρ and ρ^γ of P on $\mathbf{V}$ (resp. on V). They are isomorphic over E_λ (resp. $\mathbf{F}_\lambda$), so we can certainly write down an E_λ (resp. $\mathbf{F}_\lambda$)-equivalence A between them, which by suitable scaling may be assumed in the E_λ-case to map $\mathbf{V}$ to $\mathbf{V}$ and to induce a non-zero map of $\mathbf{V} \otimes \mathbf{F}_\lambda$ to itself. Because $\mathbf{V} \otimes \mathbf{F}_\lambda$ (resp. V) is P-irreducible, the non-zero map $A \otimes \mathbf{F}_\lambda$ (resp. A) must be an isomorphism. Therefore we may choose an isomorphism A between ρ and ρ^γ on $\mathbf{V}$ (resp. on V):

$$\rho(\gamma g \gamma^{-1}) = A\rho(g)A^{-1} \text{ for all } g \in P,$$

with $A \in \mathrm{Aut}_{\mathcal{O}_\lambda}(\mathbf{V})$ (resp. $\mathrm{Aut}_{\mathbf{F}_\lambda}(V)$). Replacing A by a p^n-th power of A, and γ by the same p^n-th power of γ, we may further assume that A has pro-finite order prime to p, and that γ still defines an isomorphism

$$P \ltimes \hat{\mathbf{Z}}_{\text{not p}} \xrightarrow{\sim} I.$$

Now we can write down an explicit continuous extension $\tilde{\rho}$ of ρ on all of I, by defining, for $g \in P$ and $n \in \hat{\mathbf{Z}}_{\text{not p}}$,

$$\tilde{\rho}(g\gamma^n) = \rho(g)A^n. \quad ∎$$

1.14.1. Remark. In 1.14, the hypothesis on the breaks serves only to guarantee that M is absolutely irreducible:

1.14.2. Variant. *Let E_λ be a finite extension of $\mathbf{Q}_l$, $l \neq p$, with residue field $\mathbf{F}_\lambda$. Let M be a non-zero finite dimensional E_λ (resp. $\mathbf{F}_\lambda$)-vector space on which I acts continuously and absolutely irreducibly. Write* $\dim(M) =$ *$n_0 p^v$, with n_0 prime to p. Then over a finite extension of E_λ (resp. $\mathbf{F}_\lambda$), M is induced from a p^v-dimensional representation V of $I(n_0)$, and the restriction of V to P is absolutely irreducible.*

Furthermore, the restriction of M to P is the direct sum of n_0 pairwise inequivalent absolutely irreducible p^v-dimensional representations of P, whose isomorphism classes are fixed by $I(n_0)$ and cyclically permuted by $I/I(n_0)$.

Proof. Exactly as in the proof of 1.14, we reduce easily to the case when M is isotypical as a P-representation. If V is an absolutely irreducible P-submodule of M, the argument of 1.14 shows that V *extends* to a representation $\mathbf{V}_1$ of I. If $M|P \simeq nV$, then $\mathrm{Hom}_P(V_1, M)$ is an n-dimensional representation of I/P. Extending scalars if necessary, this representation of I/P contains a one-dimensional character, say χ^{-1}, of I/P, simply because I/P is abelian. Therefore we have

$$\mathrm{Hom}_I(V_1, M \otimes \chi) \neq 0.$$

Because both V_1 and $M \otimes \chi$ are I-irreducible, this non-zero I-morphism must be an isomorphism of V_1 with $M \otimes \chi$. Restricting to P, we obtain $V \simeq M|P$, as required. ∎

1.15. Corollary. *Hypotheses as in 1.14 above, if* $\dim(M)$ *is a power of p, then M is absolutely irreducible as a P-representation.*

Proof. We have $n_0 = 1$, whence over an extension field $M = V$ is absolutely irreducible as P-representation. ∎

1.16. Corollary. *Hypotheses and notations being as in Proposition 1.14, let $\chi : I \to E_\lambda^\times$ (resp. $\mathbf{F}_\lambda^\times$) be a continuous character of order dividing n_0 (recall n_0 = the "prime-to-p part" of* $\dim(M)$*). Then M is isomorphic to $M \otimes \chi$ as $E_\lambda[I]$ (resp, $\mathbf{F}_\lambda[I]$)-module.*

Proof. Both M and $M \otimes \chi$ are $E_\lambda[I]$ (resp. $\mathbf{F}_\lambda[I]$)-modules, so it suffices to show they become isomorphic after a finite field extension ([C-R], 29.7). But after such an extension of fields, M is induced from the subgroup $I(n_0) \subset \mathrm{Ker}(\chi)$, so the result follows from the "projection formula"

$$[n_0]_*(V) \otimes \chi \simeq [n_0]_*(V \otimes [n_0]^*(\chi)). \quad \blacksquare$$

1.17. Corollary. *Hypotheses and notations as in 1.14 above, suppose further that the given representation M of I is the restriction to I of a continuous action of D on M, and that the residue field k of K contains all the n_0-th roots of unity. Let $D(n_0) \subset D$ be the normal open subgroup of index n_0 corresponding to the cyclic ($\boldsymbol{\mu}_{n_0}$) extension of K obtained by adjoining the n_0-th root of any uniformizing parameter of K. Then over a finite extension of E_λ (resp. $\mathbf{F}_\lambda$), M as representation of D is induced from a p^v-dimensional representation V of $D(n_0)$, whose restriction to P is absolutely irreducible.*

Proof. We have already seen that over a finite extension of E_λ (resp. $\mathbf{F}_\lambda$), the P-isotypical decomposition of M is $M = \bigoplus_{\alpha=1}^{n_0} V_\alpha$ where $V_1, \ldots, V_{n_0}$ are pairwise inequivalent absolutely irreducible p^v-dimensional representations of P. The action of D on M permutes the V_α transitively (I already did so), so the stabilizer in D of any given V_α is an open subgroup D_α of D of index n_0. We must show that $D_\alpha = D(n_0)$.

We have already seen that the stabilizer of V_α in I is the subgroup $I(n_0)$, whence we have $D_\alpha \cap I = I(n_0)$. This means that if we denote by K_α the fixed field of D_α inside K^{sep}, then K_α is linearly disjoint from K^{nr}, the maximal unramified extension of K in K^{sep}. This means that K_α is a fully ramified extension of K of degree n_0. Because the residue field k of K contains n_0 distinct n_0-th roots of unity, the unique such extension of K is $K(\pi^{1/n_0})$ for π any uniformizing parameter of K. Therefore $D_\alpha = D(n_0)$. As $D(n_0)$ is normal in D, each V_α is $D(n_0)$-stable and M is induced from any V_α. $\blacksquare$

1.18. Detailed Analysis when Swan = 1

Throughout this section, we fix a prime number $l \neq p$, and a finite extension E_λ of $\mathbf{Q}_l$ with residue field $\mathbf{F}_\lambda$.

We suppose given a finite-dimensional E_λ (resp. $\mathbf{F}_\lambda$)-vector space M, together with a continuous linear representation $\rho : I \to \mathrm{Aut}_{E_\lambda}(M)$ (resp. $\rho : I \to \mathrm{Aut}_{\mathbf{F}_\lambda}(M)$). We assume that $\dim M \geq 2$, $M^P = 0$ and $\mathrm{Swan}(M) = 1$, and we write $\dim(M) = n = n_0 p^v$ with n_0 prime to p.

We denote by G the *Zariski closure* of the subgroup $\rho(I)$ in $\mathrm{Aut}(M)$, which we view as a linear algebraic group over E_λ (resp. $\mathbf{F}_\lambda$), given with a faithful representation on M. (Recall (1.11) that if ρ is quasi-unipotent, then $\rho(I)$ is finite, so in this case G is just the finite group $\rho(I)$, viewed as an algebraic group. For example, this is automatic over $\mathbf{F}_\lambda$.)

1.19. Lemma. *Hypotheses and notations as in 1.18 above, we have*

(1) *The given representation of G on M is absolutely irreducible; in particular, as $\dim M \geq 2, G$ is not abelian.*
(2) *Over an finite extension of E_λ (resp. of $\mathbf{F}_\lambda$), any linear representation of G of dimension $< n = \dim M$ is abelian. In particular, G has no faithful representations of dimension $< n$, and every absolutely irreducible representation of G of dimension $< n$ is of dimension one.*
(3) *Any linear representation of G of dimension $< n$ is trivial on the finite subgroup $\rho(P)$.*
(4) *If $\rho(I)$ is finite, then the index of $\rho(P)$ in $\rho(I)$ is prime to p and $\geq n$, and the quotient $\rho(I)/\rho(P)$ is cyclic.*

Proof. Assertion (1) holds because the subgroup $\rho(I)$ of the rational points of G already acts absolutely irreducibly on M (cf. 1.11.(2)). For (2), let $\psi : G \to GL(V)$ be a linear representation of G of dimension $< n$, and consider the composite

$$I \xrightarrow{\rho} \rho(I) \subset G \xrightarrow{\psi} GL(V).$$

Because ρ has all breaks $1/n$, i.e. is trivial on $I^{(x)}$ for $x > 1/n$, we see that this composite is also trivial on $I^{(x)}$ for $x > 1/n$, so all the breaks of $\psi \circ \rho$ are $\leq 1/n$. But $\dim(V) < n$, so $\mathrm{Swan}(V) < 1$, whence $\mathrm{Swan}(V) = 0$. Therefore $\psi \circ \rho$ is trivial on P, thus proving assertion (3); I acts on V through its abelian quotient I^{tame}. So for any α, $\beta \in I$, we have $[\psi(\rho(\alpha)), \psi(\rho(\beta))] = 0$ in $\mathrm{End}(V)$. Fixing β, we see by Zariski density that $[\psi(g), \psi(\rho(\beta))] = 0$ for all $g \in G$. Now fixing $g \in G$, we see by Zariski density that $[\psi(g), \psi(g')] = 0$ for all $g, g' \in G$, whence ψ is abelian.

For (4), the quotient $\rho(I)/\rho(P)$ is a finite quotient of I^{tame}, so must be cyclic of some order $N \geq 1$ prime to p. Consider the restriction of ρ to

$I(N)$. All of its breaks are N/n, and by construction $\rho(I(N)) = \rho(P)$ is a finite p-group. We must show that $N \geq n$, i.e., that the unique break of ρ restricted to $I(N)$ is ≥ 1. This is a special case of the following lemma, applied to $I = I(N)$ and to $V = M$ viewed as $I(N)$-representation. ∎

1.20. Lemma. *Let V be a finite-dimensional vector space over a field of characteristic $\neq p$, on which I acts linearly, and such that the action of I factors through a finite discrete p-group quotient of I. Then every non-zero break of V is ≥ 1.*

Proof. By the break-decomposition $V = \bigoplus V(x)$, we may reduce to the case when V has a single break, say x_0. We may suppose $x_0 > 0$, otherwise there is nothing to prove. Let G be a finite p-group quotient of I through which the action factors, and let

$$G = G_0 \supset G_1 \supset G_2 \supset \cdots$$

be the lower-numbering filtration. In terms of lower numbering the Swan conductor is given by

$$\mathrm{Swan}(V) = \sum_{i \geq 1} \frac{\#G_i}{\#G_0} \dim(V/V^{G_i}).$$

In our situation, G is a p-group, so $G_0 = G_1$, but all breaks of V are > 0, so $V^{G_1} = 0$, whence the term $i = 1$ of the formula gives the inequality $\mathrm{Swan}(V) \geq \dim V$. Because v has a single break, x_0, we have $\mathrm{Swan}(V) = x_0 \dim(V)$, whence $x_0 \geq 1$, as asserted. ∎

CHAPTER 2

Curves and Their Cohomology

2.0. Generalities. Let k be a perfect field, C/k a proper smooth geometrically connected curve over k, $K = k(C)$ the function field of C. For each closed point x of C, denote by $K_{\{x\}}$ the fraction field of the henselization of the discrete valuation ring in K which "is" x. Choose a separable closure $K_{\{x\}}^{\text{sep}}$ of $K_{\{x\}}$, and a K-linear field embedding $K^{\text{sep}} \hookrightarrow K_{\{x\}}^{\text{sep}}$. The local galois group

$$D_x \overset{\text{dfn}}{=} \text{Gal}(K_{\{x\}}^{\text{sep}}/K_{\{x\}})$$

is naturally identified to the decomposition subgroup of $\text{Gal}(K^{\text{sep}}/K)$ at the place $\bar{x}$ of K^{sep} defined by the chosen embedding of K^{sep} into $K_{\{x\}}^{\text{sep}}$. Similarly, the inertia subgroup $I_x \subset D_x$ is itself the inertia subgroup of $\text{Gal}(K^{\text{sep}}/K\, k^{\text{sep}})$, k^{sep} the separable closure of k in K^{sep}, for the same place $\bar{x}$ of K^{sep}.

Let $U \subset C$ be a non-empty open set. If we view the chosen K^{sep} as a geometric generic point $\bar{\eta}$ of U, we have a natural identification

2.0.1 $\pi_1(U, \bar{\eta}) =$ the quotient of $\text{Gal}(K^{\text{sep}}/K)$ by the smallest closed normal subgroup containing I_x for all $x \in U$.

Fix a prime number $l \neq \text{char}(k)$, and denote by A an "l-adic coefficient ring," i.e., A is either a finite extension E_λ of $\mathbf{Q}_l$, or a complete noetherian local ring with finite residue field $\mathbf{F}_\lambda$ of characteristic l. For $\mathcal{F}$ a lisse sheaf of finitely generated A-modules on U, its fibre $\mathcal{F}_{\bar{\eta}}$ is a continuous representation of $\pi_1(U_{\bar{\eta}})$ on a finitely generated A-module, and the construction $\mathcal{F} \mapsto \mathcal{F}_{\bar{\eta}}$ defines an exact A-linear equivalence of abelian categories

2.0.2 $$\begin{pmatrix}\text{lisse sheaves of} \\ \text{finitely} \\ \text{generated} \\ A\text{-modules on } U\end{pmatrix} \xrightarrow{\sim} \begin{pmatrix}\text{finitely generated} \\ A\text{-modules given with a} \\ \text{continuous } A\text{-linear action} \\ \text{of } \pi_1(U, \bar{\eta})\end{pmatrix}.$$

For a closed point $x \in C$, the decomposition group $D_x \subset \text{Gal}(K^{\text{sep}}/K)$ maps to $\pi_1(U, \bar{\eta})$, so acts on $\mathcal{F}_{\bar{\eta}}$. In particular, the inertia groups I_x at all points of $C - U$ act on $\mathcal{F}_{\bar{\eta}}$ (the I_x for $x \in U$ operate trivially by hypothesis).

2.0.3. For $x \in C - U$, we may speak of the break-decomposition of $\mathcal{F}_{\bar{\eta}}$ as P_x-representation. The breaks which occur in it are called "the breaks of $\mathcal{F}$ at x." We say that $\mathcal{F}$ is tame at x if P_x acts trivially on $\mathcal{F}_{\bar{\eta}}$, i.e., if 0 is the only break of $\mathcal{F}$ at x. We say that $\mathcal{F}$ is totally wild at x if $(\mathcal{F}_{\bar{\eta}})^{P_x} = 0$, i.e., if 0 is *not* a break of $\mathcal{F}$ at x.

2.0.4. In view of the exactness properties of the break-decomposition (cf. 1.1), each of the conditions "tame at x," "totally wild at x" defines a full sub-category, stable by sub-object, quotient, and extension, of the abelian category of all lisse sheaves of finitely generated A-modules on U. For $\mathcal{F}_1$ and $\mathcal{F}_2$ tame at x, and $\mathcal{G}$ totally wild at x, the tensor product $\mathcal{F}_1 \otimes_A \mathcal{F}_2$ is tame at x, while $\mathcal{F}_1 \otimes_A \mathcal{G}$ is totally wild at x.

For A an l-adic coefficient ring, and $\mathcal{F}$ a lisse sheaf of finitely generated A-modules on U, the compact and ordinary cohomology groups

$$H^i_c(U \otimes k^{\text{sep}}, \mathcal{F}) \quad \text{and} \quad H^i(U \otimes k^{\text{sep}}, \mathcal{F})$$

are finitely generated A-modules on which $\text{Gal}(k^{\text{sep}}/k)$ acts continuously, and both vanish for $i \neq 0, 1, 2$. In terms of the action of $\pi_1^{\text{geom}} = \pi_1(U \otimes k^{\text{sep}}, \bar{\eta})$ on $\mathcal{F}_{\bar{\eta}}$, we have, for $\mathcal{F}$ lisse on U:

2.0.5
$$\begin{aligned} H^0(U \otimes k^{\text{sep}}, \mathcal{F}) &= (\mathcal{F}_{\bar{\eta}})^{\pi_1^{\text{geom}}} \\ H^0_c(U \otimes k^{\text{sep}}, \mathcal{F}) &= 0 \quad \text{if } U \subsetneqq C \end{aligned}$$

and

2.0.6
$$\begin{aligned} H^2(U \otimes k^{\text{sep}}, \mathcal{F}) &= 0 \quad \text{if } U \subsetneqq C \\ H^2_c(U \otimes k^{\text{sep}}, \mathcal{F}) &= (\mathcal{F}_{\bar{\eta}})_{\pi_1^{\text{geom}}}(-1). \end{aligned}$$

(Of course if $U = C$ then $H^i_c = H^i$.) For $U \subsetneqq C$, there is a natural transformation $H^i_c \to H^i$, which sits in a $\text{Gal}(k^{\text{sep}}/k)$-equivariant exact sequence (the long cohomology sequence on $C \otimes k^{\text{sep}}$ attached to the cone, for $j : U \hookrightarrow C$ the inclusion, of $j_!\mathcal{F} \to Rj_*\mathcal{F}$)

2.0.7
$$\begin{aligned} 0 \to H^0(U \otimes k^{\text{sep}}, \mathcal{F}) \to & \bigoplus_{\substack{x \in C-U \\ \text{closed point}}} \bigoplus_{\bar{x} \in x(k^{\text{sep}})} (\mathcal{F}_{\bar{\eta}})^{I_x} \to \\ \to H^1_c(U \otimes k^{\text{sep}}, \mathcal{F}) \to & H^1(U \otimes k^{\text{sep}}, \mathcal{F}) \to \\ \to \bigoplus_{\substack{x \in C-U \\ \text{closed point}}} & \left(\bigoplus_{\bar{x} \in x(k^{\text{sep}})} (\mathcal{F}_{\bar{\eta}})_{I_x}(-1) \right) \to H^2_c(U \otimes k^{\text{sep}}, \mathcal{F}) \to 0. \end{aligned}$$

2.1. Cohomology of Wild Sheaves

2.1.1. Lemma. *For A as above, and x a chosen closed point of $C-U$, denote by $\mathcal{A}$ the abelian category of lisse sheaves of finitely generated A-modules on U which are totally wild at x.*

(1) $H^0(U \otimes k^{\text{sep}}, \mathcal{F}) = 0 = H^2_c(U \otimes k^{\text{sep}}, \mathcal{F})$ *for any* $\mathcal{F} \in \mathcal{A}$.

(2) *The functors* $\mathcal{F} \mapsto H^1_c(U \otimes k^{\text{sep}}, \mathcal{F})$ *and* $\mathcal{F} \mapsto H^1(U \otimes k^{\text{sep}}, \mathcal{F})$ *are exact functors from $\mathcal{A}$ to the category of finitely generated A-modules on which* $\text{Gal}(k^{\text{sep}}/k)$ *acts continuously.*

(3) *Both of the functors in (2) above carry A-flat $\mathcal{F}$'s in $\mathcal{A}$ to free A-modules of finite rank. Their formation is compatible with extensions of scalars $A \to A'$ of l-adic coefficient rings.*

Proof. (1) We have $H^0 = 0$ because $\mathcal{F}_{\bar{\eta}}$ has no non-zero P_x-invariants, so even fewer $\pi_1(U \otimes k^{\text{sep}}, \bar{\eta})$-invariants. Because $\mathcal{F}_{\bar{\eta}}$ is semisimple as a P_x-module, it has no non-zero P_x-coinvariants either, so even fewer (this time a quotient) coinvariants under $\pi_1(U \otimes k^{\text{sep}}, \bar{\eta})$. Therefore $H^2_c = 0$. (2) By (1), the H^1_c and H^1 are the only non-vanishing cohomology groups for $\mathcal{F} \in \mathcal{A}$, so each is exact. (3) If N is any finitely generated A-module, take a resolution $K. \to N$ by free finitely generated A-modules. For $\mathcal{F} \in \mathcal{A}, \mathcal{F} \otimes_A K.$ is a complex in $\mathcal{A}$, and by exactness either of our functors-generically denoted "H^1"-carries its homology objects to those of the complex $H^1(\mathcal{F} \otimes_A K.) = H^1(\mathcal{F}) \otimes_A K.$; concretely, we have

$$H^1(\text{Tor}_i^A(\mathcal{F}, N)) = \text{Tor}_i^A(H^1(\mathcal{F}), N).$$

Therefore if $\mathcal{F}$ is A-flat, then so is $H^1(\mathcal{F})$. Taking $i = 0$, we see that for any $\mathcal{F} \in \mathcal{A}$, and any finitely generated A-module N, we have

$$H^1(\mathcal{F}) \otimes_A N \xrightarrow{\sim} H^1(\mathcal{F} \otimes_A N).$$

Given an extension of scalars $A \to A'$, the desired formula

$$H^1(\mathcal{F}) \otimes_A A' \xrightarrow{\sim} H^1(\mathcal{F} \otimes_A A')$$

results from the above by writing A' as the A-module direct limit of its finitely generated A-submodules. ∎

2.2. Canonical Calculation of Cohomology

2.2.1. Lemma. *Suppose that $(C-U)^{\text{red}} = D_1 \coprod D_2$ is a decomposition of $C-U$ into two disjoint non-empty finite etale k-schemes (such decompositions exist precisely when $C-U$ contains ≥ 2 closed points of C). Denote* *by $j_1 : U \hookrightarrow C - D_2$ and $j_2 : U \hookrightarrow C - D_1$ the corresponding partial compactifications of $U = C - D_1 - D_2$. Then for A as above, and $\mathcal{F}$ any lisse sheaf of finitely generated A-modules on U, we have*

(1) $H^i((C - D_2) \otimes k^{\text{sep}}, (j_1)_!\mathcal{F}) = 0$ *for* $i \neq 1$.
(2) *The H^1 is an exact functor to finitely generated A-modules with a continuous action of* $\mathrm{Gal}(k^{\text{sep}}/k)$.
(3) *This functor carries A-flat $\mathcal{F}$'s to free A-modules of finite type. Its formation commutes with extension of coefficient rings $A \to A'$.*

Proof. (1) We have $H^0 = 0$ because $\mathcal{F}$ is lisse on U and then extended by 0, and we have $H^2 = 0$ because $(C - D) \otimes k^{\text{sep}}$ is an affine curve. (2) By exactness of $(j_1)_!$ and the cohomology sequence, the remaining H^1 is exact for variable $\mathcal{F}$; it is a finitely generated A-module because $(j_1)_!\mathcal{F}$ is a constructible sheaf of A-modules. Just as in the preceding lemma, (3) results formally from (1) and (2). ∎

2.2.2. Remark. If $\mathcal{F}$ is totally wild at every point of D_1, then $(j_1)_!\mathcal{F}$ is isomorphic to $R(j_1)_*(\mathcal{F})$, and so

$$H^1((C - D_2) \otimes k^{\text{sep}}, (j_1)_!\mathcal{F}) \xrightarrow{\sim} H^1(U \otimes k^{\text{sep}}, \mathcal{F}).$$

So taking "$D_1 = x$", we recover 2.1.1 as a special case of 2.2.1.

2.2.3. Remark. Consider the diagram of inclusions

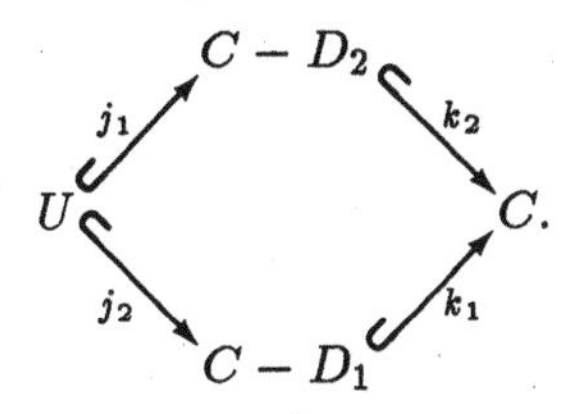

For $\mathcal{F}$ on U, we have

$$(Rk_2)_*(j_1)_!(\mathcal{F}) \simeq (k_1)_!(Rj_2)_*(\mathcal{F}),$$

simply because the closed subschemes D_1 and D_2 are disjoint. Thus we may also write describe our exotic cohomology as

$$H^i((C - D_2) \otimes k^{\text{sep}}, (j_1)_!\mathcal{F}) = H^i_c((C - D_1) \otimes k^{\text{sep}}, (Rj_2)_*\mathcal{F}).$$

2.2.4. Now let us consider the following situation: $U \subsetneq C$, $D \subset U$ is a closed non-empty subscheme, finite etale over k, and $j : U - D \hookrightarrow U$ is the inclusion. For $\mathcal{F}$ lisse on U, we have a tautological exact sequence

$$0 \to j_!(j^*\mathcal{F}) \to \mathcal{F} \to \mathcal{F}|D \to 0,$$

whence a long exact cohomology sequence

$$0 \to H^0(U \otimes k^{\text{sep}}, \mathcal{F}) \to H^0(D \otimes k^{\text{sep}}, \mathcal{F}|D) \xrightarrow{\delta}$$
$$\xrightarrow{\delta} H^1(U \otimes^{\text{sep}}, j_! j^* \mathcal{F}) \to H^1(U \otimes k^{\text{sep}}, \mathcal{F}) \to 0.$$

This means that we may calculate the cohomology groups $H^i(U \otimes k^{\text{sep}}, \mathcal{F})$ as the *cohomology* of the two-term complex

$$*(\mathcal{F}) : H^0(D \otimes k^{\text{sep}}, \mathcal{F}|D) \xrightarrow{\delta} H^1(U \otimes k^{\text{sep}}, j_! j^* \mathcal{F}).$$

2.2.5. Key Lemma.

(1) *For A as above, the construction $\mathcal{F} \mapsto *(\mathcal{F})$ is an exact functor from the category of lisse sheaves of finitely generated A-modules on U to the category of two-term complexes of finitely generated A-modules with a continuous action of* $\text{Gal}(k^{\text{sep}}/k)$.

(2) *If $\mathcal{F}$ is A-flat, then $*(\mathcal{F})$ is a two-term complex of free finitely generated A-modules.*

(3) *The formation of $*(\mathcal{F})$ commutes with extension coefficient rings $A \to A'$.*

Proof. Because $D \otimes k^{\text{sep}}$ is just a finite set of $\text{Spec}(k^{\text{sep}})$'s, the term

$$H^0(D \otimes k^{\text{sep}}, \mathcal{F}|D)$$

certainly has the asserted exactness properties for variable $\mathcal{F}$. That the term $H^1(U \otimes k^{\text{sep}}, j_!(j^* \mathcal{F}))$ does also is the special case $(U, D_1, D_2) = (U - D, D, C - U)$ of the previous lemma. ∎

2.2.6. We now consider how to "calculate" cohomology with compact supports. For A as above, and $\mathcal{F}$ a lisse sheaf of finitely generated A-modules on U, the complex $Rj_*(j^*\mathcal{F})$ has

$$R^0 j_*(j^*\mathcal{F}) \simeq \mathcal{F}$$
$$R^1 j_*(j^*\mathcal{F}) \simeq \mathcal{F}(-1)|D$$
$$R^i j_*(j^*\mathcal{F}) = 0 \quad \text{for } i \geq 2,$$

so we have a short exact sequence on U

$$0 \to \mathcal{F} \to Rj_* j^* \mathcal{F} \to (\mathcal{F}(-1)|D)[-1] \to 0.$$

The long exact cohomology sequence with compact supports reads

$$0 \to H^1_c(U \otimes k^{\text{sep}}, \mathcal{F}) \to H^1_c(U \otimes k^{\text{sep}}, Rj_*(j^*\mathcal{F})) \to$$
$$\to H^0(D \otimes k^{\text{sep}}, \mathcal{F}|D)(-1) \to H^2_c(U \otimes k^{\text{sep}}, \mathcal{F}) \to 0,$$

because $H^2_c(U \otimes k^{\text{sep}}, Rj_*(j^*\mathcal{F})) = 0$ by 2.2.1 (1) and 2.2.3. Indeed in terms of the diagram of inclusions

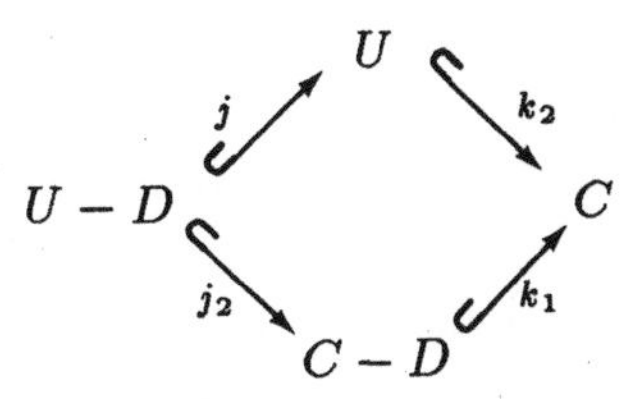

we have

$$\begin{aligned} H^i_c(U \otimes k^{\text{sep}}, Rj_*(j^*\mathcal{F})) &\overset{\text{dfn}}{=} H^i(C \otimes k^{\text{sep}}, (k_2)_! Rj_*(j^*\mathcal{F})) \\ &= H^i(C \otimes k^{\text{sep}}, (Rk_1)_*(j_2)_!(j^*\mathcal{F})) \\ &= H^i(C - D, (j_2)_!(j^*\mathcal{F})). \end{aligned}$$

Therefore the two-term complex, placed in degrees 1 and 2,

$$*_c(\mathcal{F}) : H^1_c(U \otimes k^{\text{sep}}, Rj_*(j^*\mathcal{F})) \to H^0(D \otimes k^{\text{sep}}, \mathcal{F}|D)(-1),$$

calculates the $H^i_c(U \otimes k^{\text{sep}}, \mathcal{F})$. The Key Lemma 2.2.5 is valid for the construction $\mathcal{F} \mapsto *_c(\mathcal{F})$, with the same proof.

2.2.7. Lemma. *Hypotheses as above, suppose that $\mathcal{F}$ is A-flat, and that $H^2_c(U \otimes k^{\text{sep}}, \mathcal{F}) = 0$. Then $H^1_c(U \otimes k^{\text{sep}}, \mathcal{F})$ is a free A-module of finite rank, and for any extension of scalars $A \to A'$ of l-adic coefficient rings, $H^i_c(U \otimes k^{\text{sep}}, \mathcal{F}) \otimes_A A' \xrightarrow{\sim} H^i_c(U \otimes^{\text{sep}}, \mathcal{F} \otimes_A A')$.*

Proof. Because $H^2_c = 0$, $*_c(\mathcal{F})$ is a two-term complex of free finitely generated A-modules of the form $N \twoheadrightarrow M$, so splits. ∎

2.3. The Euler-Poincaré and Lefschetz Trace Formulas.

For A as above, we now consider a lisse sheaf $\mathcal{F}$ on $U \subsetneq C$ of free finitely generated A-modules. As explained in the previous section, the choice of a non-empty finite subscheme $D \subset U$ allows us to construct functorial, exact-in-$\mathcal{F}$ two term complexes

$$*(\mathcal{F}), \quad *_c(\mathcal{F})$$

of free finitely generated A-modules given with a continuous action of $\text{Gal}(k^{\text{sep}}/k)$, whose cohomology groups are the H^i and H^i_c of $U \otimes k^{\text{sep}}$ with coefficients in $\mathcal{F}$. Those complexes allow us to define

(1) the euler characteristics $\chi(U \otimes k^{\text{sep}}, \mathcal{F})$ and $\chi_c(U \otimes k^{\text{sep}}, \mathcal{F})$, as the alternating sum of the A-ranks of the components of $*(\mathcal{F})$ and of $*_c(\mathcal{F})$ respectively.

(2) for any element $\sigma \in \mathrm{Gal}(k^{\mathrm{sep}}/k)$, the "alternating sum of the traces of σ on the $H^i(U \otimes k^{\mathrm{sep}}, \mathcal{F})$," or on the $H^i_c(U \otimes k^{\mathrm{sep}}, \mathcal{F})$, as equal to the alternating sum of its traces on the components of $*(\mathcal{F})$ and of $*_c(\mathcal{F})$ respectively.

Of course if *each* of the cohomology groups $H^i(U \otimes k^{\mathrm{sep}}, \mathcal{F})$, $i = 0, 1$ (resp.$H^i_c(U \otimes k^{\mathrm{sep}}, \mathcal{F})$, $i = 1, 2$) is itself a free A-module of finite rank, then the alternating sums defined in 1) and 2) above by means of the resolutions $*(\mathcal{F})$ (resp., $*_c(\mathcal{F})$) are equal to the literal alternating sums on the cohomology groups themselves.

Because $\mathcal{F}$ is itself a sheaf of free finitely generated A-modules, for each $x \in C - U$, we may speak not only of the break-decomposition of P_x acting on $\mathcal{F}_{\bar{\eta}}$, but also of the multiplicities of the breaks and of the Swan conductor, which we denote $\mathrm{Swan}_x(\mathcal{F})$.

2.3.1. Euler-Poincaré Formula

$$\begin{aligned}\chi(U \otimes k^{\mathrm{sep}}, \mathcal{F}) &= \chi_c(U \otimes k^{\mathrm{sep}}, \mathcal{F}) \\ &= rk(\mathcal{F})\chi_c(U \otimes^{\mathrm{sep}}) - \sum_{x \in C-U} \deg_k(x)\, \mathrm{Swan}_x(\mathcal{F})\end{aligned}$$

where the sum is over the closed points x of $C - U$, each weighted by the degree of its residue field over k, and in which $\chi_c(U \otimes k^{\mathrm{sep}})$ is the topological Euler characteristic ($= 2 - 2g - \sum_{x \in C-U} \deg_k(x)$, where g is the genus of C).

2.3.2. Lefschetz Trace Formula. When k is a finite field $\mathbf{F}_q$ and F is the inverse of the standard generator $\alpha \mapsto \alpha^q$ of $\mathrm{Gal}(k^{\mathrm{sep}}/k)$, we have the equality (in A)

$$\text{“}\sum(-1)^i \operatorname{trace}(F | H^i_c(U \otimes k^{\mathrm{sep}}, \mathcal{F}))\text{”} = \sum_{x \in U(k)} \operatorname{trace}(F_x | \mathcal{F}_{\bar{\eta}})$$

where F_x denotes the image in $\pi_1(U, \bar{\eta})$ of F under the map

$$\begin{array}{ccccc} \mathrm{Gal}(k^{\mathrm{sep}}/k) & \xleftarrow{\sim} & D_x/I_x & \rightarrow & \pi_1(U, \bar{\eta}) \\ \cup & & \cup & & \cup \\ F & \longleftrightarrow & F_x & \rightarrow & F_x. \end{array}$$

2.3.3. Explicitation. Notations and hypotheses being as above, suppose that $H^2_c(U \otimes k^{\mathrm{sep}}, \mathcal{F}) = 0$. Then

(1) $H^1_c(U \otimes k^{\mathrm{sep}}, \mathcal{F})$ is a free A-module of rank

$$\sum_{x \in C-U} \deg_k(x)\, \mathrm{Swan}_x(\mathcal{F}) - \mathrm{rk}(\mathcal{F})\chi_c(U \otimes k^{\mathrm{sep}}).$$

(2) if k is a finite field,

$$-\operatorname{trace}(F|H^1_c(U\otimes k^{\mathrm{sep}},\mathcal{F})) = \sum_{x\in U(k)} \operatorname{trace}(F_x|\mathcal{F}_{\bar\eta}).$$

If we assume further that $\mathcal{F}$ is totally wild at some point of $C-U$ then we also have

(1)′ $H^1(U\otimes k^{\mathrm{sep}},\mathcal{F})$ is a free A-module, and it has the same rank as $H^1_c(U\otimes k^{\mathrm{sep}},\mathcal{F})$.

We recall Grothendieck's proofs of these basic results (cf. Raynaud [Ray] and Deligne [De-2] for more details). By hypothesis, A is either E_λ or a complete noetherian local ring with finite residue field $\mathbf{F}_\lambda$ of residue characteristic $l\neq p$. In the E_λ-case, we may replace $\mathcal{F}$ by an $\mathcal{O}_\lambda$-form to reduce to the case when A is complete noetherian local. Because the resolution functor $\mathcal{F}\mapsto *_c(\mathcal{F})$ commutes with change of coefficient ring, it suffices to prove the formulas for $\mathcal{F}/\max(A)^n\mathcal{F}$ over $A/\max(A)^n$ for all $n\geq 1$. Thus we are reduced to the case when A is a finite local ring of residue characteristic l. Any given $\mathcal{F}$ then becomes constant on some connected finite etale galois covering $E\xrightarrow{\pi}U$, say with galois group G. Fix such a covering. The construction $\mathcal{F}\mapsto H^0(E,\pi^*\mathcal{F})$ is an exact A-linear equivalence of categories

$$\begin{pmatrix}\text{lisse sheaves of finitely generated}\\ A\text{-modules on } U \text{ which become}\\ \text{constant on the covering } E\end{pmatrix} \xrightarrow{\sim} \begin{pmatrix}\text{finitely generated}\\ \text{left } A[G]\text{-modules}\end{pmatrix}.$$

By means of this equivalence, we may view each of the two terms of the resolution functor $\mathcal{F}\mapsto *_c(\mathcal{F})$ as an exact A-linear functor, say T

$$\begin{pmatrix}\text{finitely generated}\\ \text{left } A[G]\text{-modules}\end{pmatrix} \xrightarrow{T} \begin{pmatrix}\text{finitely generated}\\ A\text{-modules}\end{pmatrix}.$$

Because $A[G]$ is a finite ring, "finitely generated" is the same as "finitely presented." Therefore our functor is necessarily of the form

$$M\mapsto T(A[G])\bigotimes_{A[G]} M$$

where $T(A[G])$ is viewed as a right $A[G]$-module through the right action of $A[G]$ on itself (one checks this formula by writing M as the cokernel of a map $(A[G])^n\to(A[G])^m$ of free left $A[G]$-modules, and applying the exact functor T). Because the functor T is exact, it follows that $T(A[G])$ is flat as a right $A[G]$-module. Because $T(A[G])$ is finite, this flatness implies that $T(A[G])$ is right $A[G]$-projective, i.e., a direct factor of a free right $A[G]$-module of finite rank.

The particular functor T at hand commutes with extension of coefficient rings. As A is a $\mathbf{Z}/l^n\mathbf{Z}$-algebra for $n \gg 0$, we see that

$$T(A[G]) = T(\mathbf{Z}_l[G]) \bigotimes_{\mathbf{Z}_l} A, \quad .$$

that $T(\mathbf{Z}_l[G])$ is a projective right $\mathbf{Z}_l[G]$-module of finite type, and that our original functor T on finitely generated left $A[G]$-modules is

$$M \mapsto T(\mathbf{Z}_l[G]) \bigotimes_{\mathbf{Z}_l[G]} M.$$

The endomorphisms of the functor T given by $\phi =$ identity and, in the Lefschetz case, by $\phi =$ Frobenius, are A-linear, and their formation is compatible with extension of coefficient rings. Any such ϕ is uniquely of the form $\phi_0 \otimes id$ where ϕ_0 is the right $\mathbf{Z}_l[G]$-linear endomorphism of $T(\mathbf{Z}_l[G])$ induced by ϕ (one checks this by calculating $T(M)$ by means of a free presentation of M).

Because $T(\mathbf{Z}_l[G])$ is right $\mathbf{Z}_l[G]$–projective of finite type, there exists a right $\mathbf{Z}_l[G]$-module Q such that $T(\mathbf{Z}_l[G]) \oplus Q$ is a free right $\mathbf{Z}_l[G]$-module of finite rank. Write the matrix of $\phi_0 \oplus 0$ in terms of a basis $e_1, \ldots, e_r$ of this free $\mathbf{Z}_l[G]$-module, say

$$(\phi_0 \oplus 0)(e_i) = \sum_j f_{ji} e_j, \quad f_{ij} \in \mathbf{Z}_l[G].$$

Then

$$T(M)) \oplus (Q \bigotimes_{\mathbf{Z}_l[G]} M) \simeq M^r,$$

$\phi \oplus 0$ operates on M^r by

$$(\phi \oplus 0) \begin{pmatrix} m_1 \\ \vdots \\ m_r \end{pmatrix} = \begin{pmatrix} f_{ij} \end{pmatrix} \begin{pmatrix} m_1 \\ \vdots \\ m_r \end{pmatrix}.$$

Suppose that M is A-flat. Then $T(M)$ is a free A-module of finite rank, so we may speak of the trace $\mathrm{trace}_A(\phi | T(M)) \in A$. Calculating this trace as the trace of $\phi \oplus 0$ on M^r, we see that

$$\operatorname{trace}_A(\phi|T(M)) = \operatorname{trace}_A((f_{ij})|M^r) = \sum_i \operatorname{trace}_A(f_{ii}|M)$$
$$= \operatorname{trace}_A\Bigl(\sum_i f_{ii}|M\Bigr).$$

The important thing is that there exists a *single* element (i.e., $\sum_i f_{ii}$)

$$\langle\phi_0\rangle \in \mathbf{Z}_l[G]$$

such that for *any* finite local $\mathbf{Z}_l$-algebra A, and *any* finitely generated left $A[G]$-module M which is A-flat, we have

$$\operatorname{trace}_A(\phi|T(M)) = \operatorname{trace}_A(\langle\phi_0\rangle|M).$$

Because G is a finite group and $\mathbf{Z}_l$ is flat over $\mathbf{Z}$, this property of $\langle\phi_0\rangle$ determines the image $\langle\phi_0\rangle^\natural$ of $\langle\phi_0\rangle$ under the map

$$\mathbf{Z}_l[G] \to \mathbf{Z}_l\text{-valued functions on the set of conjugacy classes in } G$$
$$\sum a_g[g] \mapsto \text{the function on conjugacy classes } C \to \sum_{g\in C} a_g.$$

Furthermore, given that $\langle\phi_0\rangle^\natural$ exists, it is already uniquely determined by knowing its trace on M's which are $\mathcal{O}_\lambda$-forms of the finitely many absolutely irreducible $\mathbf{Q}_l$-representations of G, all realizable over some finite extension E_λ of $\mathbf{Q}_l$. Reinterpreting M as a sheaf, we are thus reduced to proving that the Euler-Poincaré and Lefschetz formulas hold for $\mathcal{F}$ a lisse E_λ-sheaf on U which becomes constant on a finite etale connected galois covering $E \to U$, with galois group G. In this case, the formulas to be proven amount to Weil's "classical" Lefschetz trace formula with constant coefficients E_λ on the complete non-singular model C' of the open curve E, for the following endomorphisms of C':

(a) (for the Euler-Poincaré formula) the elements $g \neq \text{id}$ in G

(b) (for the Lefschetz trace formula) the endomorphisms $F \circ g^{-1}$ for all $g \in G$, where F denotes the relative-to-k Frobenius endomorphism of E. ∎

CHAPTER 3

Equidistribution in Equal Characteristic

3.0. For any "abstract" group Γ, we denote by $\Gamma^\natural$ = the set of conjugacy classes in Γ.

3.1. Let us fix

- a prime number p,
- a finite field $\mathbf{F}_q$ of characteristic p,
- a smooth, geometrically connected curve $U/\mathbf{F}_q$, the complement of a finite set of closed points in a proper smooth geometrically connected curve $C/\mathbf{F}_q$,
- a geometric point $\bar{x} : \mathrm{Spec}(\Omega) \to U$,
- a prime number $l \neq p$,
- a lisse $\overline{\mathbf{Q}}_l$-sheaf $\mathcal{F}$ on U of rank $n \geq 1$ (i.e., $\mathcal{F}$ is a lisse E_λ-sheaf for E_λ an unspecified finite extension of $\mathbf{Q}_l$), and
- an embedding $\overline{\mathbf{Q}}_l \to \mathbf{C}$.

Set

$$\pi_1^{\mathrm{arith}} \overset{\mathrm{dfn}}{=} \pi_1(U, \bar{x}), \qquad \pi_1^{\mathrm{geom}} \overset{\mathrm{dfn}}{=} \pi_1(U \underset{\mathbf{F}_q}{\otimes} \overline{\mathbf{F}}_q, \bar{x}),$$

let

$$\rho : \pi_1^{\mathrm{arith}} \to \mathrm{Aut}_{\overline{\mathbf{Q}}_l}(\mathcal{F}_{\bar{x}}) \simeq \mathrm{GL}(\bar{n}, \overline{\mathbf{Q}}_l)$$

be the monodromy representation of $\mathcal{F}$, and let G denote the Zariski closure of $\rho(\pi_1^{\mathrm{geom}})$ in $GL(n) \otimes \overline{\mathbf{Q}}_l$. Moreover, suppose that, with respect to the given complex embedding of $\overline{\mathbf{Q}}_l$, $\mathcal{F}$ is pure of weight zero.

3.2. By a fundamental result of Deligne [De-5], the purity of $\mathcal{F}$ implies that the identity component G^0 of the $\overline{\mathbf{Q}}_l$-algebraic group G is semi-simple. By means of the complex embedding $\overline{\mathbf{Q}}_l \hookrightarrow \mathbf{C}$, we may view $\mathbf{C}$ as a $\overline{\mathbf{Q}}_l$-algebra and so speak of the groups

$$\begin{array}{ccc} G^0(\mathbf{C}) & \subset & G(\mathbf{C}) \\ \cup & & \cup \\ G^0(\overline{\mathbf{Q}}_l) & \subset & G(\overline{\mathbf{Q}}_l). \end{array}$$

In the classical topology $G(\mathbf{C})$ is a complex semisimple Lie group, $G^0(\mathbf{C})$ its identity component. Let us denote by $K \subset G(\mathbf{C})$ a maximal compact subgroup of the complex Lie group $G(\mathbf{C})$. (Construction: take K^0 a compact form of $G^0(\mathbf{C})$ à la Weyl, then take for K the normalizer of K^0 in $G(\mathbf{C})$.) One knows that every compact subgroup of $G(\mathbf{C})$ is $G(\mathbf{C})$-conjugate to a subgroup of K, and that the functors

$$\begin{array}{c}
\left(\text{finite-dimensional } \overline{\mathbf{Q}}_l\text{-representations of } G \text{ as } \overline{\mathbf{Q}}_l\text{-algebraic group}\right) \\
\int\!\!\downarrow \text{extension of scalars } \overline{\mathbf{Q}}_l \to \mathbf{C} \\
\left(\begin{array}{l}\text{finite-dimensional } \mathbf{C}\text{-representations of } G \otimes_{\overline{\mathbf{Q}}_l} \mathbf{C} \text{ as } \mathbf{C}\text{-algebraic} \\ \text{group}\end{array}\right) \\
\int\!\!\downarrow \text{evaluation on } \mathbf{C}\text{-valued joints} \\
\left(\begin{array}{l}\text{finite-dimensional holomorphic representations of } G(\mathbf{C}) \text{ as} \\ \text{complex Lie group}\end{array}\right) \\
\int\!\!\downarrow \text{restriction to } K \\
\left(\begin{array}{l}\text{finite-dimensional continuous representations of } K \text{ as compact} \\ \text{group}\end{array}\right)
\end{array}$$

are all equivalences of categories.

3.3. We now make the further hypothesis that $\rho(\pi_1^{\text{arith}}) \subset G(\overline{\mathbf{Q}}_l)$ (inside $\mathrm{GL}(n, \overline{\mathbf{Q}}_l)$). For every closed point $u \in U$, the Frobenius conjugacy class $F_u \in (\pi_1^{\text{arith}})^\natural$ then defines a conjugacy class

$$\rho(F_u) \in G(\overline{\mathbf{Q}}_l)^\natural \subset G(\mathbf{C})^\natural,$$

whose semi-simple part, in the sense of Jordan decomposition, lies in a compact subgroup of $G(\mathbf{C})$ (because all its eigenvalues have absolute value one).

Therefore $\rho(F_u)^{ss}$ is $G(\mathbf{C})$-conjugate to an element of K. Because the traces of the finite dimensional continuous representations of K separate K-conjugacy classes, and all such representations are the restrictions to K of representations of $G(\mathbf{C})$, the element of K to which $\rho(F_u)^{ss}$ is $G(\mathbf{C})$-conjugate is well defined up to K-conjugacy. Thus we obtain a well-defined element of $K^\natural$,

$$\theta(u) \overset{\text{dfn}}{=} \rho(F_u)^{ss} \in K^\natural,$$

which we think of as the generalized "angle of Frobenius" at u. (For example, if $G = \mathrm{SL}(2)$, then in a suitable basis

$$\rho(F_u)^{\mathrm{ss}} = \begin{pmatrix} e^{i\theta(u)} & 0 \\ 0 & e^{-i\theta(u)} \end{pmatrix},$$

for the unique choice of real $\theta(u) \in [0, \pi]$ which satisfies $\mathrm{trace}(\rho(F_u)) = 2\cos(\theta(u))$; with $K = \mathrm{SU}(2)$, this identifies $K^\natural$ with the interval $[0, \pi]$.)

3.4. Viewing $K^\natural$ as a quotient space of K, it acquires a quotient topology for which it is compact, and for which the continuous functions on $K^\natural$ "are" precisely the continuous *central* functions on K. We denote by $\mu^\natural$ the direct image on $K^\natural$ of normalized (total mass 1) Haar measure on K.

Concretely, if f is any continuous $\mathbf{C}$-valued function on $K^\natural$, and if f_{central} denotes its associated (inverse image) continuous central function on K, then

$$\int_{K^\natural} f d\mu^\natural \overset{\mathrm{dfn}}{=} \int_K f_{\mathrm{central}} d\mu_{\mathrm{Haar}}.$$

(For example, with $G = \mathrm{SL}(2)$, we have $K = \mathrm{SU}(2)$, $K^\natural = [0, \pi]$, and $\mu^\natural$ is the "Sato-Tate measure" $(\frac{2}{\pi}) \sin^2 \theta d\theta$.)

3.5. By evaluating at the "generalized angles" $\theta(u) \in K^\natural$, one constructs three more or less natural sequences of positive measures of mass one on $K^\natural$, each indexed by integers n sufficiently large that U has a closed point of degree n:

$$X_n = \left(\frac{1}{\#U(\mathbf{F}_{q^n})}\right) \sum_{\deg(u)|n} \deg(u)\delta(\theta(u)^{n/\deg(u)})$$

$$Y_n = \left(\frac{1}{\#\{u \text{ of } \deg = n\}}\right) \sum_{\deg(u)=n} \delta(\theta(u))$$

$$Z_n = \left(\frac{1}{\#\{u \text{ of } \deg \leq n\}}\right) \sum_{\deg(u)\leq n} \delta(\theta(u)).$$

In the above formulas, $\delta(x)$ denotes the Dirac delta measure supported at x.

3.6. Theorem (Deligne [De-5]). *The sequences of measures $\{X_n\}$, $\{Y_n\}$ and $\{Z_n\}$ on $K^\natural$ all tend weak-$*$ to $\mu^\natural$; for any continuous $\mathbf{C}$-valued function*

f on $K^\natural$, we have

$$\int_{K^\natural} f d\mu^\natural = \lim_n \int_{K^\natural} f dX_n$$

$$= \lim_n \int_{K^\natural} f dY_n = \lim_n \int_{K^\natural} f dZ_n.$$

Proof. Because the $\mathbf{C}$-span of the characters of continuous finite-dimensional representations of K is uniformly dense in the space of all continuous functions on $K^\natural$, while all measures involved are positive and of total mass one, an obvious limiting argument reduces us to the case where f is the character of a finite-dimensional irreducible representation of K.

Because all measures involved have total mass one, the constant function $f = 1$ "works" even without passing to a limit over n, so we may assume the representation is non-trivial. Thus f is the trace of the restriction of ψ to $K \subset G(\mathbf{C})$, where

$$\psi : G \to \mathrm{GL}(M) \otimes \overline{\mathbf{Q}}_l$$

is an *irreducible* and *non-trivial* representation of G.

For such a representation, we have

$$0 = \int_K \operatorname{trace}(\psi) d\mu_{\mathrm{Haar}},$$

so what must be proven is that for such ψ the integrals $\int_{K^\natural} \operatorname{trace}(\psi) dX_n$, $\int_{K^\natural} \operatorname{trace}(\psi) dY_n$ and $\int_{K^\natural} \operatorname{trace}(\psi) dZ_n$ tend to 0 as n tends to ∞ (i.e., the Weyl Criterion for equidistribution).

In fact, we will establish a much more precise estimate, for ψ irreducible and non-trivial, as follows:

$$\left|\int_{K^\natural} \operatorname{trace}(\psi) dX_n\right| \leq O\left(\frac{\dim(\psi)}{q^{n/2}}\right)$$

$$\left|\int_{K^\natural} \operatorname{trace}(\psi) dY_n\right| \leq O\left(\frac{n \dim(\psi)}{q^{n/2}}\right)$$

$$\left|\int_{K^\natural} \operatorname{trace}(\psi) dZ_n\right| \leq O\left(\frac{n^2 \dim(\psi)}{q^{n/2}}\right),$$

where the constants in the big O are easily made explicit.

Let us denote by $\mathcal{F}(\psi)$ the lisse $\overline{\mathbf{Q}}_l$-sheaf of rank M on U whose monodromy representation is the composite

$$\psi \circ \rho : \pi_1^{\text{arith}} \xrightarrow{\rho} G(\overline{\mathbf{Q}}_l) \xrightarrow{\psi} \mathrm{GL}(M, \overline{\mathbf{Q}}_l).$$

Then $\mathcal{F}(\psi)$ is again pure of weight zero (for example, because it is a subquotient of some $\mathcal{F}^{\otimes n} \otimes (\mathcal{F}^\vee)^{\otimes m}$), and for every closed point u

$$((\psi \circ \rho)(F_u))^{\text{ss}} = \psi(\rho(F_u)^{\text{ss}}) = \psi(\theta(u)).$$

Because ψ is irreducible non-trivial, it has no non-zero invariants or coinvariants. Because $\rho(\pi_1^{\text{geom}})$ is Zariski dense in $G(\overline{\mathbf{Q}}_l)$, we have

$$H_c^2(U \underset{\mathbf{F}_q}{\otimes} \overline{\mathbf{F}}_q, \mathcal{F}(\psi)) = 0 = H^0(U \underset{\mathbf{F}_q}{\otimes} \overline{\mathbf{F}}_q, \mathcal{F}(\psi)).$$

The Lefschetz trace formula gives

$$(\#U(\mathbf{F}_{q^n})) \int_{K^\natural} \operatorname{trace}(\psi) dX_n = -\operatorname{trace}(F^n | H_c^1(U \underset{\mathbf{F}_q}{\otimes} \overline{\mathbf{F}}_q, \mathcal{F}(\psi)).$$

Because $\mathcal{F}(\psi)$ is lisse and pure of weight zero on $U \otimes \overline{\mathbf{F}}_q$, its H_c^1 is mixed of weight ≤ 1, so we get

3.6.1
$$\left| \int_{K^\natural} \operatorname{trace}(\psi) dX_n \right| \leq \frac{|\chi_c(U \otimes \overline{\mathbf{F}}_q, \mathcal{F}(\psi))| \sqrt{q}^n}{\#U(\mathbf{F}_q n)}.$$

The analogous consideration of H_c with constant coefficients gives

3.6.2
$$\#U(\mathbf{F}_{q^n}) \geq q^n - h_c^1(U \otimes \overline{\mathbf{F}}_q, \overline{\mathbf{Q}}_l) \sqrt{q}^n.$$

The Euler characteristic with coefficients in $\mathcal{F}(\psi)$ can be easily estimated by the Euler-Poincaré formula if we notice that at every "missing" point $y \in (C - U)(\overline{\mathbf{F}}_q)$, we have

$$\begin{pmatrix} \text{biggest break of} \\ \mathcal{F}(\psi) \text{ at } y \end{pmatrix} \leq \begin{pmatrix} \text{biggest break of} \\ \mathcal{F} \text{ at } y \end{pmatrix}$$

(simply because I_y acts on $\mathcal{F}$ through ρ, but on $\mathcal{F}(\psi)$ through $\psi \circ \rho$). Therefore we have, by the Euler-Poincaré formula,

$$\chi(U \otimes \overline{\mathbf{F}}_q, \mathcal{F}(\psi)) = \chi(U \otimes \overline{\mathbf{F}}_q, \overline{\mathbf{Q}}_l) \dim(\psi) - \sum_{y \in (C-U)(\overline{\mathbf{F}}_q)} \operatorname{Swan}_y(\mathcal{F}(\psi))$$

with

$$\begin{aligned} 0 \leq \operatorname{Swan}_y(\mathcal{F}(\psi)) &\leq \dim(\psi)(\text{biggest break of } \mathcal{F}(\psi) \text{ at } y) \\ &\leq \dim(\psi)(\text{biggest break of } \mathcal{F} \text{ at } y). \end{aligned}$$

So in terms of

$$\begin{cases} g = \text{genus of } C \\ N = \#(C-U)(\overline{\mathbf{F}}_q) \\ r_1, \ldots, r_N = \text{ the biggest breaks of } \mathcal{F} \text{ at the } N \text{ points of } C-U, \end{cases}$$

we have

$$|\chi(U \otimes \overline{\mathbf{F}}_q, \mathcal{F}(\psi))| \le \left(2g - 2 + N + \sum_1^N r_i\right) \dim(\psi), \tag{3.6.2.1}$$

whence

3.6.3 $$\left|\int_{K^\natural} \operatorname{trace}(\psi) dX_n\right| \le (2g-2+N+\textstyle\sum r_i)\dfrac{\dim(\psi)\sqrt{q}^n}{\#U(\mathbf{F}_{q^n})} \le O\left(\dfrac{\dim(\psi)}{q^{n/2}}\right).$$

We now explain briefly how to deduce from this estimate for Y_n, Z_n. Let us write

$$\begin{aligned} A_n &= \#U(\mathbf{F}_{q^n}) = O(q^n) \\ B_n &= \#\{u \text{ of } \deg = n\} \\ C_n &= \#\{u \text{ of } \deg \le n\}. \end{aligned}$$

Then

$$A_n = \sum_{r|n} rB_r,$$

and

$$A_n X_n - nB_n Y_n = \sum_{\substack{\deg(u)|n \\ \deg(u)<n}} \deg(u)\delta(\theta(u)^{n/\deg(u)})$$

is the sum of

$$\sum_{\substack{r|n \\ r<n}} rB_r \le \sum_{m \le [n/2]} A_m = O\left(\frac{n}{2} q^{n/2}\right)$$

δ measures supported at points of $K^\natural$. By compactness of K, for any $k \in K$ the matrix $\psi(k)$ has all eigenvalues of absolute value one, so

$$|\operatorname{trace}(\psi(k))| \le \dim \psi \text{ for all } k \in K.$$

Therefore

$$\left|A_n \int_{K^\natural} \operatorname{trace}(\psi) dX_n - nB_n \int_{K^\natural} \operatorname{trace}(\psi) dY_n\right| = O(n \dim(\psi) q^{n/2}).$$

Now both A_n and nB_n are equal to $q^n + O(nq^{n/2})$, so dividing by nB_n we find

$$\left|\int_{K^\natural} \operatorname{trace}(\psi) dY_n\right|$$

$$\leq (2g-2+N+\textstyle\sum r_i)\dim(\psi)\left(\frac{\sqrt{q}^n}{q^n+O(nq^{n/2})}\right)+O\left(\frac{n\dim(\psi)}{q^{n/2}+O(n)}\right)$$

$$\leq O\left(\frac{n\dim(\psi)}{q^{n/2}}\right).$$

From this estimate for Y_n's we get an estimate for Z_n, for we have

$$Z_n = \sum_{k=1}^{n}(B_k/C_n)Y_k$$

$$\left|\int_{K^\natural} \operatorname{trace}(\psi) dZ_n\right| \leq O\left(\sum_{k=1}^{n}\frac{kB_k\dim(\psi)}{C_n q^{k/2}}\right).$$

Now $nC_n \geq nB_n = q^n + O(nq^{n/2})$, so for some $\alpha > 0$ we have $C_n \geq \alpha q^n$ for all $n \gg 0$, while for suitable $\beta > 0$ we trivially get $B_k \leq \beta q^k$ for all k. Thus for $n \gg 0$, we find

$$\left|\int_{K^\natural} \operatorname{trace}(\psi) dZ_n\right| \leq O\left(\sum_{k=1}^{n}\frac{\beta k q^k\dim(\psi)}{\alpha q^n q^{k/2}}\right) \leq O\left(\frac{n^2\beta\dim(\psi)}{\alpha q^{n/2}}\right)$$

$$\leq O\left(\frac{n^2\dim(\psi)}{q^{n/2}}\right). \blacksquare$$

3.7. Remark. Let f be a continuous central function on K, whose representation-theoretic "Fourier series"

$$f = \sum_{\psi \text{ irred}} a(\psi)\operatorname{trace}(\psi)$$

has

$$\|f\|^\natural \overset{\text{dfn}}{=} \sum_{\psi \text{ irred}} |a(\psi)|\dim(\psi) < \infty.$$

Then our estimates show that, say for the X_n, we have

$$\left|\int_{K^\natural} f d\mu^\natural - \int_{K^\natural} f dX_n\right| \leq O\left(\frac{\|f\|^\natural}{q^{n/2}}\right).$$

For this reason, we will discuss $\|f\|^\natural$ in some detail.

We will suppose that $G = G^0$ and that G^0 is non-trivial. Because $G = G^0$ is connected, K is connected. Once we pick a maximal torus $T \subset K$, we may view f as a continuous function on T which is invariant under the Weyl group $W = N(T)/T$. One knows that "restriction to T" is a bijection

$$\begin{pmatrix}\text{continuous central}\\ \text{function on } K\end{pmatrix} \xrightarrow{\sim} \begin{pmatrix}\text{continuous } W\text{-invariant}\\ \text{functions on } T\end{pmatrix}.$$

Viewing f as a continuous function on T which happens to be W-invariant, we may also write its "usual" Fourier series on T

$$f(t) = \sum_{r \in \mathrm{Hom}(T,S^1)} b(r)t^r.$$

Now given any continuous f on T, with Fourier series as above, let us define

$$\|f\|_1 \stackrel{\mathrm{dfn}}{=} \sum_{r \neq 0} |b(r)|;$$

if $T \simeq (S^1)^l$, this is easily seen to be a Sobolev semi-norm of class $C^{(l/2)+\varepsilon}$.

We will express $\|f\|^{\natural}$ as the $\|g\|_1$ for a function g which is obtained from f by applying to f a (rather fundamental) differential operator. To do this, we will apply the Weyl character and dimension formulas. Let us pick an ordering in $\mathrm{Hom}(T,S^1) \otimes_{\mathbf{Z}} \mathbf{R}$, so that we can speak of the positive roots R_+. At the possible expense of passing to a connected double covering of K, we may and will assume that $\rho = \frac{1}{2}\sum_{\alpha \in R_+} \alpha$ lies in $\mathrm{Hom}(T,S^1)$. Let us also pick a positive definite W-invariant scalar product $\langle , \rangle$ on $\mathrm{Hom}(T,S^1) \otimes \mathbf{R}$. Then we have

$$\langle \rho, \alpha \rangle > 0 \quad \text{for } \alpha \in R_+.$$

For each $\alpha \in R_+$, we denote by D_α the invariant derivation of T characterized by

$$D_\alpha(t^\beta) = \left(\frac{\langle \beta, \alpha \rangle}{\langle \rho, \alpha \rangle}\right) t^\beta$$

for all $\beta \in \mathrm{Hom}(T,S^1)$. These D_α mutually commute, and we define

$$D_{\mathrm{Weyl}} = \prod_{\alpha \in R_+} D_\alpha,$$

an invariant differential operator on T of order $\#(R_+)$. As customary, we define

$$J(\rho)(t) = \sum_{w \in W} \mathrm{sgn}(w) t^{\rho \circ w},$$

where sgn is the determinant of the homomorphism $W \to \mathrm{GL}(l, \mathbf{Z})$ given by the action of W on $\mathrm{Hom}(T,S^1) \simeq \mathbf{Z}^l$; this is the "denominator" of the

Weyl character formula, that says that if ψ is an irreducible representation of K with highest weight λ, then

$$J(\rho)(t).\operatorname{trace}(\psi)(t) = \sum_{w\in W} \operatorname{sgn}(w)t^{(\rho+\lambda)\circ w}.$$

One verifies easily that for any $\beta \in \operatorname{Hom}(T, S^1)$, and any $w \in W$,

$$\frac{D_{\mathrm{Weyl}}(t^{\beta\circ w})}{t^{\beta\circ w}} = (\operatorname{sgn}(w))\frac{D_{\mathrm{Weyl}}(t^{\beta})}{t^{\beta}}.$$

This being the case, the Weyl dimension formula

$$\dim(\psi) = \prod_{\alpha\in R_+} \frac{\langle \rho+\lambda, \alpha\rangle}{\langle \rho, \alpha\rangle}$$

may be rewritten

$$D_{\mathrm{Weyl}}(J(\rho)\operatorname{trace}(\psi)) = \dim(\psi) \sum_{w\in W} t^{(\rho+\lambda)\circ w}.$$

Therefore, if $f = \sum a(\psi)\operatorname{trace}(\psi)$, we have

$$D_{\mathrm{Weyl}}(J(\rho)f) = \sum_{\psi} a(\psi)\dim(\psi) \sum_{w\in W} t^{(\rho+\lambda)\circ w},$$

where in the inner sum the λ is the highest weight of ψ. Thus we find

$$\|D_{\mathrm{Weyl}}(J(\rho)f)\|_1 = \#(W)\sum_{\psi} |a(\psi)|\dim(\psi) = \#(W)\|f\|^{\natural}.$$

In particular, we see that $f \mapsto \|f\|^{\natural}$ is given by a Sobolev seminorm on T of class C^k, with

$$k = l/2 + \varepsilon + \#(R_+) < \dim G.$$

For example, if $G = G^0 = \mathrm{SL}(2)$, then $K = \mathrm{SU}(2)$ and $T = S^1$ via

$$\theta \mapsto \begin{pmatrix} e^{i\theta} & 0 \\ 0 & e^{-i\theta}. \end{pmatrix}$$

The action of $W = \{\pm 1\}$ on T is $\theta \mapsto \pm\theta$. The irreducible representations of k are $\operatorname{Symm}^n(\mathrm{std})$ for $n = 0, 1, 2, \ldots$, and we have

$$\dim(\operatorname{Symm}^n(\mathrm{std})) = n+1$$
$$\operatorname{trace}(\operatorname{Symm}^n(\mathrm{std}))(\theta) = \frac{\sin((n+1)\theta)}{\sin\theta}.$$

Thus the representation-theoretic Fourier series of f on $K^{\natural}$ is

$$f \sim \sum_{n\geq 0} a(n)\frac{\sin((n+1)\theta)}{\sin\theta},$$

while its "naive" Fourier series, as even function on T, is

$$\begin{aligned} f &\sim \sum_n b(n)e^{in\theta} \qquad (b(n) = b(-n)) \\ &= b(0) + 2\sum_{n\geq 1} b(n)\cos(n\theta). \end{aligned}$$

The function $J(p)$ is just $2i\sin(\theta)$, the differential operator D_{Weyl} is

$$D_{\mathrm{Weyl}} = -i\frac{d}{d\theta},$$

and the previous discussion boils down to the identity

$$-i\frac{d}{d\theta}(2i\sin(\theta)f(\theta)) = 2\sum_{n\geq 0}(n+1)a(n)\cos((n+1)\theta).$$

CHAPTER 4

Gauss Sums and Kloosterman Sums: Kloosterman Sheaves

4.0. Let p be a prime number, $\mathbf{F}_q$ a finite field of characteristic p, $\psi : (\mathbf{F}_q, +) \to \mathbf{Q}(\zeta_p)^\times$ a *non-trivial* additive character of $\mathbf{F}_q$, and $\chi : \mathbf{F}_q^\times \to \mathbf{Q}(\zeta_{q-1})^\times$ a (possibly trivial) multiplicative character of $\mathbf{F}_q^\times$. The gauss sum $g(\psi, \chi) \in \mathbf{Q}(\zeta_p, \zeta_{q-1})$ is defined by

$$g(\psi, \chi) = \sum_{a \in \mathbf{F}_q^\times} \psi(a)\chi(a).$$

For χ trivial, $g(\psi, \mathbf{1}) = -1$, while for non-trivial χ we have $|g(\psi, \chi)| = \sqrt{q}$ for any embedding $\mathbf{Q}(\zeta_p, \zeta_{q-1}) \to \mathbf{C}$.

For a fixed choice of non-trivial ψ, it is sometimes convenient to view the gauss sum $g(\psi, \chi)$ as a function of χ. This amounts to viewing $\chi \mapsto g(\psi, \chi)$ as the multiplicative Fourier transform of the function on $\mathbf{F}_q^\times$ defined by $a \mapsto \psi(a)$. For any function f on $\mathbf{F}_q^\times$, say with values in an overfield E of $\mathbf{Q}(\zeta_{q-1})$, we define its multiplicative Fourier transform $\hat{f}$ to be the E-valued function on characters given by

$$\hat{f}(\chi) = \sum_{a \in \mathbf{F}_q^\times} f(a)\chi(a).$$

The Fourier inversion formula,

$$f(a) = \frac{1}{q-1} \sum_{\chi} \overline{\chi}(a)\hat{f}(\chi),$$

allows us to recover f from $\hat{f}$.

Given two functions f, g on $\mathbf{F}_q^\times$, their convolution $f * g$ is the function on $\mathbf{F}_q^\times$ defined by

$$(f * g)(a) = \sum_{xy=a} f(x)g(y).$$

The Fourier transform of the convolution is given by the product of the transforms: $(\widehat{f * g})(\chi) = \hat{f}(\chi)\hat{g}(\chi)$. Summing this relation over χ and dividing by $q-1$, we obtain

$$(f*g)(1)=\frac{1}{q-1}\sum_{\chi}\hat{f}(\chi)\hat{g}(\chi),$$

that is,

$$\sum_{a\in\mathbf{F}_q^\times} f(a)g(a^{-1})=\frac{1}{q-1}\sum_{\chi}\hat{f}(\chi)\hat{g}(\chi).$$

If $e\mapsto\bar{e}$ is any automorphism of E which induces complex conjugation on $\mathbf{Q}(\zeta_{q-1})$, then by applying the above equality with g replaced by its "conjugate" function $g^*(a)=\overline{g(a^{-1})}$, and noting that

$$\widehat{g^*}(\chi)=\sum_a g^*(a)\chi(a)=\sum_a \overline{g(a^{-1})}\overline{\chi}(a^{-1})=\sum_a\overline{g(a)\chi(a)}=\overline{\hat{g}(\chi)},$$

we obtain the Parseval identity

$$\sum_{a\in\mathbf{F}_q^\times} f(a)\overline{g(a)}=\frac{q}{q-1}\sum_{\chi}\hat{f}(\chi)\overline{\hat{g}(\chi)}.$$

From this point of view, Kloosterman sums occur naturally as the inverse Fourier transforms of monomials in gauss sums. More precisely, we have the following table of functions f on $\mathbf{F}_q^\times$ and of their multiplicative Fourier transforms $\hat{f}$, in which n is an integer ≥ 1, $\chi_1,\ldots,\chi_n$ are fixed multiplicative characters of $\mathbf{F}_q^\times$, and $b_1,\ldots,b_n$ are strictly positive integers:

	$f(a)$	$\hat{f}(\chi)$
1.	$\psi(a)$	$g(\psi,\chi)$
2.	$\sum_{\substack{x_1\ldots x_n=a\\ \text{all } x_i\in\mathbf{F}_q^\times}}\psi(x_1+\cdots+x_n)$	$g(\psi,\chi)^n$
3.	$\psi(a)\chi_1(a)$	$g(\psi,\chi\chi_1)$
4.	$\sum_{\substack{x_1\ldots x_n=a\\ \text{all } x_i\in\mathbf{F}_q^\times}}\psi(x_1+\cdots+x_n)\chi_1(x_1)\ldots\chi_n(x_n)$	$\prod_{i=1}^n g(\psi,\chi\chi_i)$
5.	$\sum_{\substack{x^{b_1}=a\\ \text{all } x\in\mathbf{F}_q^\times}}\psi(x)\chi_1(x)$	$g(\psi,\chi^{b_1}\chi_1)$
6.	$\sum_{\substack{x_1^{b_1}\ldots x_n^{b_n}=a\\ \text{all } x_i\in\mathbf{F}_q^\times}}\psi(x_1+\cdots+x_n)\chi_1(x_1)\ldots\chi_n(x_n)$	$\prod_{i=1}^n g(\psi,\chi^{b_i}\chi_i)$.

(The odd-numbered lines are checked by applying the definition of $\hat{f}$, the even-numbered lines by observing that the function f in question is an n-fold convolution of the sort of function in the odd-numbered preceding line.)

Given ψ, $n \geq 1$, $\chi_1, \ldots, \chi_n$, $b_1, \ldots, b_n$ as above, and an element $a \in \mathbf{F}_q^\times$, we denote by

$$\mathrm{Kl}(\psi; \chi_1, \ldots, \chi_n; b_1, \ldots, b_n)(\mathbf{F}_q, a)$$

the sum

$$\sum_{\substack{x_1^{b_1} \ldots x_n^{b_n} = a \\ \text{all } x_i \in \mathbf{F}_q^\times}} \psi(\sum x_j)\chi_1(x_1) \ldots \chi_n(x_n).$$

Given a finite extension k of $\mathbf{F}_q$, and an element $a \in k^\times$, we denote by

$$\text{trace} = \text{trace}_{k/\mathbf{F}_q} : k \to \mathbf{F}_q$$
$$\mathrm{N} = \mathrm{Norm}_k / \mathbf{F}_q : k^\times \to \mathbf{F}_q^\times$$

the trace and norm maps, and we denote by

$$\mathrm{Kl}(\psi; \chi_1, \ldots, \chi_n; b_1, \ldots, b_n)(k, a)$$

the sum

$$\sum_{\substack{x_1^{b_1} \ldots x_n^{b_n} = a \\ \text{all } x_j \in k^\times}} \psi(\mathrm{Trace}(\sum x_i))\chi_1(\mathrm{N}(x_1)) \cdots \chi_n(\mathrm{N}(x_n)),$$

viewed as lying in the subring $\mathbf{Z}$[values of ψ, χ_i's] of $\mathbf{Q}(\zeta_p, \zeta_{q-1})$.

4.0.1. Scholium. For any finite extension k of $\mathbf{F}_a^\times$, the multiplicative Fourier transform of the function on $k^\times$ defined by

$$a \mapsto \mathrm{Kl}(\psi; \chi_1, \ldots, \chi_n; b_1, \ldots b_n)(k, a)$$

is the function on the character group of $k^\times$ defined by

$$\chi \mapsto \prod_{i=1}^{n} g(\psi \circ \text{trace}, \ \chi^{b_i}(\chi_i \circ \mathrm{N})).$$

4.1. The Existence Theorem for Kloosterman sheaves. Let p be a prime number, $\mathbf{F}_q$ a finite field of characteristic p, E a finite extension of $\mathbf{Q}$, $\psi : (\mathbf{F}_q, +) \to E^\times$ a non-trivial additive character, $n \geq 1$ an integer, $\chi_1, \ldots, \chi_n$ multiplicative characters $\chi_i : \mathbf{F}_q^\times \to E^\times$, and $b_1, \ldots, b_n$ strictly positive integers. Write each b_i in the form $b_i = b_i' p^{n_i}$, with b_i' prime to p.

Let l be a prime number $l \neq p$, λ an l-adic place of E, E_λ the λ-adic completion of E, $\mathcal{O}_\lambda$ the ring of integers in E_λ and $\mathbf{F}_\lambda$ the residue field of $\mathcal{O}_\lambda$.

4.1.1. Theorem (cf. Deligne [De-3], Thm. 7.8). *There exists a lisse sheaf of free $\mathcal{O}_\lambda$-modules of finite rank on $\mathbf{G}_m \otimes \mathbf{F}_q$, denoted*

$$\mathrm{Kl}(\psi; \chi_1, \ldots, \chi_n;\ b_1, \ldots, b_n),$$

or simply Kl, *with the following properties:*

1. Kl *is lisse of rank $\sum b_i'$, and pure of weight $n-1$.*
2. *For any finite extension k of $\mathbf{F}_q$, and any element $a \in k^\times = \mathbf{G}_m(k)$, denoting by $\bar{a}$ a geometric point lying over a and by $F_{k,a}$ the inverse of the standard generator $x \mapsto X^{\#(k)}$ of $\mathrm{Gal}(\bar{k}/k) \simeq \pi_1(\mathrm{Spec}(k), \bar{a})$, we have the identity in $\mathcal{O}_\lambda$*

   $$\mathrm{trace}(F_{a,k} | \mathrm{Kl}_{\bar{a}}) = (-1)^{n-1}\,\mathrm{Kl}(\psi; \chi_1, \ldots, \chi_n; b_1, \ldots, b_n)(k, a).$$

3. *At ∞,* Kl *is totally wild, and* $\mathrm{Swan}_\infty(\mathrm{Kl}) = 1$.
4. *At* 0, Kl *is tame.*

Before discussing the proof of this theorem, let us deduce from it some rather striking corollaries.

4.1.2. Corollary (Rigidity).

1. *The sheaves $\mathrm{Kl} \otimes \mathbf{F}_\lambda$ and $\mathrm{Kl} \otimes E_\lambda$ are absolutely irreducible as representations of I_∞. A fortiori, they are absolutely irreducible as representations of the larger groups $\pi_1^{\mathrm{geom}} = \pi_1(\mathbf{G}_m \otimes \overline{\mathbf{F}}_q, \bar{\eta})$ and $\pi_1 = \pi_1(\mathbf{G}_m \otimes \mathbf{F}_q, \bar{\eta})$.*
2. *Let $\mathcal{F}$ be a lisse sheaf of free $\mathcal{O}_\lambda$-modules of finite rank on $\mathbf{G}_m \otimes \mathbf{F}_q$ whose local traces are given by (2). Then*

   (a) There exists an isomorphism $\alpha : \mathcal{F} \xrightarrow{\sim} \mathrm{Kl}$ as lisse $\mathcal{O}_\lambda$ sheaves on $\mathbf{G}_m \otimes \mathbf{F}_q$ i.e., as $\mathcal{O}_\lambda[\pi_1(\mathbf{G}_m \otimes \mathbf{F}_q), \bar{\eta}]$-modules.

   (b) If G is any of the three groups

   $$\pi_1(\mathbf{G}_m \otimes \mathbf{F}, \bar{\eta}), \qquad \pi_1(\mathbf{G}_m \otimes \overline{\mathbf{F}}_q, \bar{\eta}), \qquad I_\infty$$

   and R is any of the three rings

   $$\mathcal{O}_\lambda,\ E_\lambda,\ \mathbf{F}_\lambda,$$

   $\mathrm{Hom}_{R[G]}(\mathcal{F} \otimes R, \mathrm{Kl} \otimes R)$ *is a free R-module of rank 1 with basis α, and* $\mathrm{Isom}_{R[G]}(\mathcal{F} \otimes R, \mathrm{Kl} \otimes R)$ *consists precisely of the $R^\times$-multiples of α.*

(3) *Let $\mathcal{F}$ be a lisse sheaf of $\mathbf{F}_\lambda$-modules of rank $< l + \sum b_i'$ on $\mathbf{G}_m \otimes \mathbf{F}_q$ whose local traces are given by (2)$\otimes \mathbf{F}_\lambda$. Then $\mathcal{F}$ is isomorphic to* $\mathrm{Kl} \otimes \mathbf{F}_\lambda$ *as lisse $\mathbf{F}_\lambda$-sheaf on $\mathbf{G}_m \otimes \mathbf{F}_q$.*

Proof. Assertion (1) results from the fact that Kl is totally wild at ∞ with $\mathrm{Swan}_\infty(\mathrm{Kl}) = 1$, cf. 1.11. For (2), Cebataroff shows that $\mathcal{F} \otimes E_\lambda$'s semisimplification as $E_\lambda[\pi_1(\mathbf{G}_m \otimes \mathbf{F}_q, \bar{\eta})]$-module is isomorphic to that of $\mathrm{Kl} \otimes E_\lambda$. As $\mathrm{Kl} \otimes E_\lambda$ is absolutely irreducible, we may infer $\mathcal{F} \otimes E_\lambda$ is itself absolutely irreducible and isomorphic to $\mathrm{Kl} \otimes E_\lambda$ as lisse E_λ-sheaf on $\mathbf{G}_m \otimes \mathbf{F}_q$. Therefore $\mathcal{F} \otimes E_\lambda$ is totally wild at ∞, and has $\mathrm{Swan}_\infty = 1$. By 1.10, it follows that first $\mathcal{F}$ and then $\mathcal{F} \otimes \mathbf{F}_\lambda$ have the same behavior at ∞. In particular, $\mathcal{F} \otimes \mathbf{F}_\lambda$ is absolutely irreducible as I_∞-representation. To get an isomorphism $\alpha : \mathcal{F} \to \mathrm{Kl}$ over $\mathcal{O}_\lambda$, begin with one over E_λ and multiply by a constant in $E_\lambda^\times$ until it maps $\mathcal{F}$ to Kl and is non-zero $\otimes \mathbf{F}_\lambda$; by the absolute irreducibility of $\mathcal{F} \otimes \mathbf{F}_\lambda$ and $\mathrm{Kl} \otimes \mathbf{F}_\lambda$, it must be an isomorphism $\otimes \mathbf{F}_\lambda$ because it is non-zero, whence an isomorphism over $\mathcal{O}_\lambda$. That (b) holds once we have α follows from all the absolute irreducibilities in a straightforward way.

Assertion (3) is slightly more delicate, but similar. It suffices to show that $\mathcal{F} \otimes \overline{\mathbf{F}}_\lambda$ and $\mathrm{Kl} \otimes \overline{\mathbf{F}}_\lambda$ are isomorphic (cf. [C-R], 29.7). Let us write the $\overline{\mathbf{F}}_\lambda[\pi_1]$-semisimplification of $\mathcal{F} \otimes \overline{\mathbf{F}}_\lambda$

$$(\mathcal{F} \otimes \overline{\mathbf{F}}_\lambda)^{\mathrm{ss}} = \sum n_i V_i$$

with integers $n_i \geq 1$, and V_i pairwise inequivalent finite-dimensional $\overline{\mathbf{F}}_\lambda$-representations of π_1. Taking trace functions, we have (again by Cebataroff)

$$\mathrm{trace}_{\mathrm{Kl} \otimes \overline{\mathbf{F}}_\lambda} = \sum n_i \, \mathrm{trace}_{V_i},$$

an identity of $\overline{\mathbf{F}}_\lambda$-valued functions on π_1. By the linear independence of traces of irreducible representations over $\overline{\mathbf{F}}_\lambda$ (cf. [C-R]) and the fact that $\mathrm{Kl} \otimes \overline{\mathbf{F}}_\lambda$ has been proven irreducible, we infer that all but one of the n_i is divisible by l, that the remaining one, say n_1, satisfies $n_1 \equiv 1 \bmod l$, and that V_1 has the same trace functions as $\mathrm{Kl} \otimes \overline{\mathbf{F}}_\lambda$, hence is isomorphic to it. Thus we find

$$(\mathcal{F} \otimes \overline{\mathbf{F}}_\lambda)^{\mathrm{ss}} = \mathrm{Kl} \otimes \overline{\mathbf{F}}_\lambda \bigoplus \begin{pmatrix} \text{some other representation} \\ \text{repeated } l \text{ times} \end{pmatrix}.$$

From the hypothesis $\mathrm{rk}(\mathcal{F}) < \mathrm{rk}(\mathrm{Kl}) + l$, we infer that $(\mathcal{F} \otimes \overline{\mathbf{F}}_\lambda)^{\mathrm{ss}}$ is isomorphic to $\mathrm{Kl} \otimes \overline{\mathbf{F}}_\lambda$; as the latter is irreducible, we find that $\mathcal{F} \otimes \overline{\mathbf{F}}_\lambda$ is irreducible, and is isomorphic to it. ∎

4.1.3. Corollary (Duality). *For given ψ, χ_i, b_i, and* $\mathrm{Kl} = \mathrm{Kl}(\psi, \chi_i, b_i)$, *let us denote by* $\overline{\mathrm{Kl}}$ *the "complex conjugate" Kloosterman sheaf*

$$\overline{\mathrm{Kl}} = \mathrm{Kl}(\overline{\psi}, \overline{\chi_1}, \ldots, \overline{\chi_n}; b_1, \ldots, b_n).$$

Then there exists an isomorphism of lisse $\mathcal{O}_\lambda$-sheaves on $\mathbf{G}_m \otimes \mathbf{F}_q$

$$\overline{\mathrm{Kl}} \xrightarrow{\sim} \underline{\mathrm{Hom}}_{\mathcal{O}_\lambda}(\mathrm{Kl}, \mathcal{O}_\lambda(1-n)) = \mathrm{Kl}^\vee(1-n).$$

Proof. Because Kl is pure of weight $n-1$, $\mathrm{Kl}^\vee(1-n)$ has the same local traces as $\overline{\mathrm{Kl}}$. ∎

4.1.4. Corollary. *Suppose that the pairs $(b_i, \overline{\chi_i})$, $i = 1, \ldots, n$, are just a permutation of the pairs (b_i, χ_i), $i = 1, \ldots, n$, and suppose that the character $\prod_1^n \chi_i$ of $\mathbf{F}_q^\times$ satisfies $(\prod \chi_i)(-1) = 1$ (this last condition always holds if we replace $\mathbf{F}_q$ by its quadratic extension). Then*

(1) *If $(-1)^{\sum b_i} = 1$ in $\mathbf{F}_q$, there exists an isomorphism of lisse $\mathcal{O}_\lambda$-sheaves on $\mathbf{G}_m \otimes \mathbf{F}_q$*

$$\overline{\mathrm{Kl}} \xrightarrow{\sim} \mathrm{Kl}.$$

(2) *If $(-1)^{\sum b_i} = -1$ and $-1 \neq 1$ in $\mathbf{F}_q$, then there exists an isomorphism of lisse $\mathcal{O}_\lambda$-sheaves on $\mathbf{G}_m \otimes \mathbf{F}_q$*

$$[t \mapsto -1]^*(\mathrm{Kl}) \xrightarrow{\sim} \overline{\mathrm{Kl}}.$$

Proof. In both cases one verifies easily that both sides have the same local traces. ∎

We will now examine the consequences of the last two corollaries on the possible autoduality of Kloosterman sheaves.

4.1.5. Proposition *For any Kloosterman sheaf* Kl, *and any $\zeta \in \mathbf{F}_q^\times$ with $\zeta \neq 1$, the sheaves* Kl *and*

$$\mathrm{Trans}_\zeta^*(\mathrm{Kl}) \stackrel{\mathrm{dfn}}{=} [t \mapsto \zeta t]^*(\mathrm{Kl})$$

are not isomorphic. More precisely, for R any of the rings $\mathcal{O}_\lambda, E_\lambda, \mathbf{F}_\lambda$, and for G any of the groups $\pi_1(\mathbf{G}_m \otimes \mathbf{F}_q, \bar{\eta})$, $\pi_1(\mathbf{G}_m \otimes \overline{\mathbf{F}}_q, \bar{\eta})$, I_∞, we have

$$\mathrm{Hom}_{R[G]}(\mathrm{Trans}_\zeta^*(\mathrm{Kl} \otimes R),\ \mathrm{Kl} \otimes R) = 0.$$

Proof. Clearly it suffices to prove this with the smallest G, namely I_∞, and with $R = E_\lambda$ or $\mathbf{F}_\lambda$. This is a special case of part (3) of the following, applied to the completion at ∞ of the function field of $\mathbf{G}_m \otimes \overline{\mathbf{F}}_q$, and to $\mathcal{F} = \mathrm{Kl} \otimes R$. ∎

4.1.6. Proposition. *Let k be an algebraically closed field of characteristic $p > 0$, K the field $k((x))$ of Laurent series in one variable over k, $I = \mathrm{Gal}(K^{\mathrm{sep}}/K)$. Let l be a prime $l \neq p$, E_λ a finite extension of $\mathbf{Q}_l$, $\mathbf{F}_\lambda$ its residue field. For $R = \mathbf{F}_\lambda$ or E_λ, let $\mathcal{F}$ be a lisse R-sheaf of finite rank on $\mathrm{Spec}(K)$, viewed as finite-dimensional R-representation of I. For $\alpha \in k^\times$, denote by $\mathrm{Trans}_\alpha : \mathrm{Spec}(K) \to \mathrm{Spec}(K)$ the spec of the continuous k-linear automorphism of $K = k((x))$ given by $x \mapsto \alpha x$. Then*

(1) *If there exists an isomorphism of $R[I]$-modules $\mathrm{Trans}_\alpha^*(\mathcal{F}) \simeq \mathcal{F}$ with α not a root of unity, then $\mathcal{F}$ is tame.*
(2) *If $\mathcal{F}$ is absolutely irreducible as R-linear representation of I, and if $\zeta \in k^\times$ is a root of unity for which there exists an isomorphism of $R[I]$-modules $\mathrm{Trans}_\zeta^*(\mathcal{F}) \to \mathcal{F}$, then denoting by $N \geq 1$ the exact order of the root of unity ζ, we have the congruence $\mathrm{Swan}(\mathcal{F}) \equiv 0 \mod N$.*
(3) *In particular, if $\mathcal{F}$ is totally wild and $\mathrm{Swan}(\mathcal{F}) = 1$, then*

$$\mathrm{Hom}_{R[I]}(\mathrm{Trans}_\alpha^*(\mathcal{F}), \mathcal{F}) = 0$$

for any $\alpha \neq 1$ in $k^\times$.

Proof. Assertion (1) is an extremely useful result of Verdier ([Ver], Prop. 1.1). Assertion (3) is a special case of (2). To prove assertion (2), we observe that because $\mathcal{F}$ is absolutely irreducible, the "theta group" G of pairs (ζ, A_ζ) with $\zeta \in \boldsymbol{\mu}_N(k)$ and $A_\zeta \in \mathrm{Isom}_{R[I]}(\mathrm{Trans}_\zeta^*(\mathcal{F}), \mathcal{F})$ sits in a central extension

$$1 \to R^\times \to G \to \boldsymbol{\mu}_N(k) \to 1.$$

At the expense of extending the field R by adjoining the N'th root of g^N for some $g \in G$ whose image in $\boldsymbol{\mu}_N(k)$ is a generator, we may split this extension. But such a splitting is just a *descent* of $\mathcal{F}$ through the finite etale $\boldsymbol{\mu}_N(k)$-covering

$$\begin{array}{c} \mathrm{Spec}(k((x))) \\ {\scriptstyle [N]}\downarrow \circlearrowright {\scriptstyle \boldsymbol{\mu}_N(k)} \\ \mathrm{Spec}(k((x^N))). \end{array}$$

Writing $\mathcal{F} \simeq [N]^*\mathcal{G}$ for some lisse $\mathcal{G}$ on $\mathrm{Spec}(k((x^N)))$, we have $\mathrm{Swan}(\mathcal{G}) = N\,\mathrm{Swan}(\mathcal{F})$ (cf. 1.13 (1)). ∎

4.1.7. Corollary. *Suppose that the pairs $(b_i, \overline{\chi_i})$, $i = 1, \ldots, n$, are a permutation of the pairs (b_i, χ_i), $i = 1, \ldots, n$, and that $(-1)^{\sum b_i} \neq 1$ in $\mathbf{F}_q$.*

Then for R any of the rings $\mathcal{O}_\lambda, E_\lambda, \mathbf{F}_\lambda$, any R-bilinear I_∞-equivariant pairing $\langle , \rangle : \mathrm{Kl}\otimes R \times \mathrm{Kl}\otimes R \to R(1-n)$, with I_∞ acting trivially on $R(1-n)$, vanishes.

Proof. Such a pairing is just an element of

$$\begin{aligned}\mathrm{Hom}_{R[I]}(\mathrm{Kl}\otimes R, \mathrm{Kl}^\vee(1-n)\otimes R) &\simeq \mathrm{Hom}_{R[I]}(\mathrm{Kl}\otimes R, \overline{\mathrm{Kl}}\otimes R) \\ &\simeq \mathrm{Hom}_{R[I]}(\mathrm{Trans}^*_{-1}(\overline{\mathrm{Kl}}\otimes R), \overline{\mathrm{Kl}}\otimes R) \\ &= 0. \ \blacksquare\end{aligned}$$

4.1.8. Remark. Here is an elementary argument to prove that for any Kloosterman sheaf Kl, any $\zeta \neq 1$ in $\mathbf{F}_q^\times$, and R any of the rings $\mathcal{O}_\lambda, E_\lambda, \mathbf{F}_\lambda$, the sheaves $\mathrm{Kl}\otimes R$ and $\mathrm{Trans}^*_\zeta(\mathrm{Kl}\otimes R)$ are not isomorphic: they don't even have the same local traces at all rational points $a \in \mathbf{F}_q^\times = \mathbf{G}_m(\mathbf{F}_q)$! To prove this it suffices to treat the case $\mathbf{F}_\lambda$ (because the traces in question begin life in $\mathcal{O}_\lambda$). We may extend scalars to assume that $\mathbf{F}_\lambda$ contains all the $q-1$'st roots of unity. Let us denote by $f : \mathbf{F}_q^\times \to \mathbf{F}_\lambda$ the trace function of $\mathrm{Kl}\otimes\mathbf{F}_\lambda$:

$$f(a) = \mathrm{trace}(F_a | \mathrm{Kl}_{\overline{a}}) \mod \lambda.$$

Then the trace function g of $\mathrm{Trans}^*_\zeta(\mathrm{Kl}\otimes\mathbf{F}_\lambda)$ is related to f by

$$g(a) = f(\zeta a).$$

Passing to multiplicative Fourier transforms, we find that for every character $\chi : \mathbf{F}_q^\times \to \mathbf{F}_\lambda^\times$, we have

$$\hat{g}(\chi) = \sum_a \chi(a)g(a) = \sum_a \chi(a)f(\zeta a) = \overline{\chi}(\zeta)\hat{f}(\chi).$$

But $\hat{f}(\chi)$ is a unit in $\mathbf{F}_\lambda$, because it is the reduction mod λ of a product of gauss sums for $\mathbf{F}_q$, each of which is an algebraic integer which divides q. Therefore, if $f = g$, we find $\hat{f}(\chi) = \overline{\chi}(\zeta)\hat{f}(\chi)$, whence $\overline{\chi}(\zeta) = 1$ for every characrter χ, whence $\zeta = 1$.

4.1.9. Lemma. *Let $\mathrm{Kl}(\psi; \chi_1, \ldots, \chi_n; b_1, \ldots, b_n)$ be an $\mathcal{O}_\lambda$-Kloosterman sheaf on $\mathbf{G}_m\otimes\mathbf{F}_q$. For $\xi \in \mathbf{F}_q^\times$, denote by ψ_ξ the additive character $x \mapsto \psi(\xi x)$ of $\mathbf{F}_q$, and let r denote the exact order of the $(q-1)$-st root of unity $(\prod \chi_i)(\xi)$. Then the $\mathcal{O}_\lambda$-sheaves $\mathrm{Kl}(\psi_\xi; \chi_1, \ldots, \chi_n; b_1, \ldots, b_n)$ and*

$$(\mathrm{Trans}_{\xi^{\Sigma b_i}})^*(\mathrm{Kl}(\psi; \chi_1, \ldots, \chi_n; b_1, \ldots, b_n))$$

on $\mathbf{G}_m \otimes \mathbf{F}_q$ become $\mathcal{O}_\lambda$-isomorphic when pulled back to $\mathbf{G}_m \otimes \mathbf{F}_{q^r}$.

Proof. Both have the same local traces when pulled back to $\mathbf{G}_m \otimes \mathbf{F}_{q^r}$. $\blacksquare$

4.1.10. Corollary. *Notations as in the above lemma, suppose that* $\xi^{\sum b_i} \neq 1$. *Then for R any of the rings* $E_\lambda, \mathcal{O}_\lambda, \mathbf{F}_\lambda$, *and G any of the groups* $\pi_1(\mathbf{G}_m \otimes \mathbf{F}_q, \bar{\eta})$, $\pi_1(\mathbf{G}_m \otimes \overline{\mathbf{F}}_q, \bar{\eta})$, I_∞, *we have*

$$\operatorname{Hom}_{R[G]}(\mathrm{Kl}(\psi_\xi; \chi's, b's) \otimes R, \mathrm{Kl}(\psi; \chi's, b's) \otimes R) = 0.$$

Proof. For fixed R, it suffices to prove this in the hardest case $G = I_\infty$, which we obtain by combining the above lemma with Proposition 4.1.5. ∎

We now turn to the *autodual* Kloosterman sheaves.

4.1.11. Corollary. Suppose that the pairs $(b_i, \overline{\chi}_i)$, $i = 1, \ldots, n$, are a permutation of the pairs (b_i, χ_i), $i = 1, \ldots, n$, that $(\prod \chi_i)(-1) = 1$, and that

$$(-1)^{\sum b_i} = 1 \quad \text{in} \quad \mathbf{F}_q.$$

Then there exists a perfect autoduality pairing of lisse sheaves of free $\mathcal{O}_\lambda$-modules of finite rank on $\mathbf{G}_m \otimes \mathbf{F}_q$,

$$\langle , \rangle : \mathrm{Kl} \times \mathrm{Kl} \rightarrow \mathcal{O}_\lambda(1-n).$$

For R any of the rings $\mathcal{O}_\lambda, E_\lambda, \mathbf{F}_\lambda$, and G any of the groups

$$\pi_1(\mathbf{G}_m \otimes \mathbf{F}_q, \bar{\eta}), \qquad \pi_1(\mathbf{G}_m \otimes \overline{\mathbf{F}}_q, \bar{\eta}), \qquad I_\infty,$$

any G-equivariant R-bilinear pairing

$$\langle , \rangle' : (\mathrm{Kl} \otimes R) \times (\mathrm{Kl} \otimes R \rightarrow R(1-n)$$

is an R-multiple of $\langle , \rangle \otimes R$.

Proof. We have $\mathrm{Kl}^\vee(1-n) \simeq \overline{\mathrm{Kl}} \simeq \mathrm{Kl}$, whence such a pairing is just an element of $\operatorname{Hom}_{R[G]}(\mathrm{Kl} \otimes R, \mathrm{Kl} \otimes R)$, and we apply part (2) of the rigidity Corollary 4.1.2. ∎

4.2. Signs. By the uniqueness up to a scalar of the above autoduality pairing on Kl, it is either symmetric or it is anti-symmetric (for if we denote the pairing $\langle x, y \rangle$, then $\langle y, x \rangle$ is another one, so an $\mathcal{O}_\lambda^\times$-multiple of it, say $\langle y, x \rangle = \epsilon \langle x, y \rangle$, whence $\epsilon^2 = 1$ in $\mathcal{O}_\lambda$, i.e. $\epsilon = \pm 1$). Although our ultimate *construction* of the Kloosterman sheaves by cohomological methods will make transparent in principle what the sign is, we will give here an *arithmetic* determination of the sign, due to Ekedahl, which works in fair generality. (We do not know how to treat the most general case by this method.) We first explain the idea behind the method (compare [Ka-6]).

Suppose we are given a lisse E_λ-sheaf $\mathcal{F}$ on a smooth geometrically connected curve C over $\mathbf{F}_q$, which is pure of some weight $n-1$, and which is absolutely irreducible as a representation of $\pi_1^{\text{geom}} = \pi_1(C \otimes \overline{\mathbf{F}}_q, \bar{\eta})$. Suppose further that we are given a perfect pairing of lisse E_λ-sheaves on C, i.e. equivariant for $\pi_1(C, \bar{\eta})$,

$$\langle , \rangle : \mathcal{F} \times \mathcal{F} \to E_\lambda(1-n).$$

As above, it is either symmetric or anti-symmetric, and it is the unique (up to an E_λ-factor) $\pi_1(C \otimes \overline{\mathbf{F}}_q, \bar{\eta})$-invariant morphism

$$(\mathcal{F} \otimes \mathcal{F})(n-1) \to E_\lambda,$$

or, equivalently, it is the unique (up to an E_λ-factor) *coinvariant* of π_1^{geom} on $(\mathcal{F}\otimes\mathcal{F})(n-1)$. In view of the description of $H_c^2(C\otimes\overline{\mathbf{F}}_q, \mathcal{G})$ as the π_1^{geom}-coinvariants of $\mathcal{G}(-1)$, for any lisse $\mathcal{G}$ on C, we see that $H_c^2(C\otimes\overline{\mathbf{F}}_q, \mathcal{F}\otimes\mathcal{F})(n)$ is a one-dimensional E_λ-space with basis our pairing $\langle , \rangle$, and that $F \in \mathrm{Gal}(\overline{\mathbf{F}}_q/\mathbf{F}_q)$ operator trivially on this space. By means of the decomposition

$$\mathcal{F} \otimes \mathcal{F} = \Lambda^2(\mathcal{F}) \oplus \mathrm{Symm}^2(\mathcal{F}),$$

we see that if the pairing is *symmetric*, then $H_c^2(C \otimes \overline{\mathbf{F}}_q, \Lambda^2(\mathcal{F}))(n) = 0$, $H_c^2(C\otimes\overline{\mathbf{F}}_q, \mathrm{Symm}^2(\mathcal{F}))(n)$ is one-dimensional, and F acts as 1, while if the pairing is alternating, the situation is reversed.

We now use the fact that $\mathcal{F}$ is pure of weight $n-1$. Then $\Lambda^2(\mathcal{F})$ and $\mathrm{Symm}^2(\mathcal{F})$ are pure of weight $2n-2$, and it follows from Deligne [De-5] that, for $i \leq 1$, the two cohomology groups $H_c^i(C \otimes \overline{\mathbf{F}}_q, \Lambda^2(\mathcal{F}))(n)$ and $H_c^i(C \otimes \overline{\mathbf{F}}_q, \mathrm{Symm}^2(\mathcal{F}))(n)$ are mixed of weight ≤ -1.

Therefore, for a suitable constant, we have for every integer $r \geq 1$:

$$\sum_i (-1)^i \,\mathrm{trace}\big(F^r | H_c^i(C \otimes \overline{\mathbf{F}}_q, \mathrm{Symm}^2(\mathcal{F}))(n)\big)$$
$$- \sum_i (-1)^i \,\mathrm{trace}\big(F^r | H_c^i(C \otimes \overline{\mathbf{F}}_q, \Lambda^2(\mathcal{F}))(n)\big)$$
$$= \begin{cases} 1 + O(1/\sqrt{q^r}) & \text{if the pairing is symmetric} \\ -1 + O(1/\sqrt{q^r}) & \text{if the pairing is alternating.} \end{cases}$$

Now using the Lefschetz trace formula, and simplifying each local term by the linear algebra identity

$$\mathrm{trace}(A^2) = \mathrm{trace}(\mathrm{Symm}^2(A)) - \mathrm{trace}(\Lambda^2(A)),$$

we find that for $r \geq 1$, we have

$$\frac{1}{q^{nr}} \sum_{a \in C(\mathbf{F}_{q^r})} \mathrm{trace}\big(F_a^2 | \mathcal{F}_{\bar{a}}\big) = \epsilon + O(1/\sqrt{q^r}),$$

where $\epsilon = \pm 1$ is the sign of the pairing.

4.2.1. Proposition (Ekedahl). *Suppose that we have $\chi_i^2 = 1$ for $i = 1, \ldots, n$, that $(\prod \chi_i)(-1) = 1$, and that*

$$(-1)^{\sum b_i} = 1 \text{ in } \mathbf{F}_q.$$

Then the sign of the pairing

$$\langle , \rangle : \mathrm{Kl} \times \mathrm{Kl} \to \mathcal{O}_\lambda(1-n).$$

is given by

$$\epsilon = (-1)^{n-1+\delta}$$

where δ is the number of indices i for which $\chi_i^2 = 1$ but $\chi_i \neq 1$.

Proof. We will apply the method explained above. For $a \in \mathbf{F}_q^\times$, we have

$$\text{(4.2.1.1)} \quad \mathrm{trace}(F_a^2 | \mathrm{Kl}_{\bar{a}}) = (-1)^{n-1} \mathrm{Kl}(\psi; \chi_1, \ldots, \chi_n; b_1, \ldots, b_n)(\mathbf{F}_{q^2}, a)$$

by property (2) of the Kloosterman sheaves. Explicitly, this is

$$\text{(4.2.1.2)} \quad \mathrm{trace}(F_a^2 | \mathrm{Kl}_{\bar{a}}) = (-1)^{n-1} \sum_{\substack{\prod x_i^{b_i} = a \\ x_i \in \mathbf{F}_{q^2}}} \psi(\mathrm{trace}(\sum x_i)) \prod \chi_i(\mathrm{N}(x_i)),$$

where trace and norm are relative to the quadratic extension $\mathbf{F}_{q^2}/\mathbf{F}_q$. Summing over all $a \in \mathbf{F}_q^\times$, we obtain

(4.2.1.3)

$$(-1)^{n-1} \sum_{a \in \mathbf{F}_q^\times} \mathrm{tr}(F_a^2 | \mathrm{Kl}_{\bar{a}}) = \sum_{\substack{x_1, \ldots, x_n \in \mathbf{F}_{q^2}^\times \\ \text{with} \prod x_1^{b_i} \in \mathbf{F}_q^\times}} \psi(\mathrm{trace}(\sum x_1)) \prod \chi_i(\mathrm{N}(x_i))$$

$$= \sum_{\substack{x_1, \ldots, x_n \in \mathbf{F}_{q^2}^\times \\ \prod x_1^{b_i(q-1)} = 1}} \psi(\mathrm{trace}(\sum x_1)) \prod \chi_i(\mathrm{N}(x_i)),$$

which is itself a Kloosterman sum over $\mathbf{F}_{q^2}$, namely

$$\text{(4.2.1.4)} \quad \mathrm{Kl}(\psi \circ \mathrm{trace}, \chi_1 \circ \mathbf{N}, \ldots, \chi_n \circ N; b_1(q-1), \ldots, b_n(q-1))(\mathbf{F}_{q^2}, 1).$$

By inverse multiplicative Fourier transform on $\mathbf{F}_{q^2}^\times$, this is equal to the sum

$$\text{(4.2.1.5)} \qquad \frac{1}{q^2-1} \sum_{\substack{\text{chars } \rho \\ \text{of } \mathbf{F}_{q^2}^\times}} \prod_{i=1}^{n} g(\psi \circ \mathrm{trace}, (\chi_i \circ \mathbf{N}) \rho^{b_i(q-1)}).$$

We next turn to the individual gauss sums. It is in analyzing these that the hypothesis $\chi_i^2 = 1$ will enter. For fixed i and fixed ρ, we have

$$(4.2.1.6)\quad g(\psi \circ \text{trace}, (\chi_i \circ \mathbf{N})\rho^{b_i(q-1)}) = \sum_{x \in \mathbf{F}_{q^2}^\times} \psi(\text{trace}(x))\chi_i(x^{q+1})\rho^{b_i}(x^{q-1}).$$

If we take $y \in \mathbf{F}_q^\times$, then $y^{q-1} = 1$ and $y^{q+1} = y^2$, so that if we replace x by xy (y fixed in $\mathbf{F}_q^\times$) in the summation, this same sum is equal to (using $\chi_i^2 = 1$)

$$(4.2.1.7)\quad \sum_{x \in \mathbf{F}_{q^2}} \psi(y\ \text{trace}(x))\chi_i(x^{q+1})\rho^{b_i}(x^{q-1}).$$

Averaging over $y \in \mathbf{F}_q^\times$, and using the orthogonality relation for ψ in the form

$$(4.2.1.8)\quad \frac{1}{q-1}\sum_{y \in \mathbf{F}_q^\times} \psi(y\ \text{trace}(x)) = \begin{cases} 1 & \text{if trace}(x) = 0 \\ \dfrac{-1}{q-1} & \text{if trace}(x) \neq 0, \end{cases}$$

our sum becomes

$$(4.2.1.9)\quad \sum_{\substack{x \in \mathbf{F}_{q^2}^\times \\ \text{trace}(x)=0}} \chi_i(x^{q+1})\rho^{b_i}(x^{q-1}) - \frac{1}{q-1}\sum_{\substack{x \in \mathbf{F}_{q^2}^\times \\ \text{trace}(x)\neq 0}} \chi_i(x^{q+1})\rho^{b_i}(x^{q-1})$$

which we reassemble as

$$(4.2.1.10)\quad \frac{q}{q-1}\sum_{\text{trace}=0} - \frac{1}{q-1}\sum_{\text{all } x}.$$

The elements $x \in \mathbf{F}_{q^2}^\times$ of trace zero are the non-zero solutions of $x^q + x = 0$, i.e., they are the solutions of $x^{q-1} = -1$. If we fix one solution $x_0 \in \mathbf{F}_{q^2}^\times$ of $\text{trace}(x_0) = 0$, any other solution is an $\mathbf{F}_q^\times$-multiple, so again using $\chi_i^2 = 1$ we find

$$(4.2.1.11)\quad \frac{q}{q-1}\sum_{\substack{x \in \mathbf{F}_{q^2}^\times \\ \text{trace}(x)=0}} \chi_i(x^{q+1})\rho^{b_i}(x^{q-1}) = q\chi_i(x_0^{q+1})\rho^{b_i}(-1).$$

Now the second term is given explicitly by

(4.2.1.12)

$$\frac{-1}{q-1}\sum_{x \in \mathbf{F}_{q^2}^\times} \chi_i(x^{q+1})\rho^{b_i}(x^{q-1}) = \begin{cases} 0 & \text{if } (\chi_i \circ \mathbf{N})\rho^{b_i(q-1)} \neq 1 \\ -(q+1) & \text{if } (\chi_i \circ \mathbf{N})\rho^{b_i(q-1)} = 1. \end{cases}$$

whence we find

(4.2.1.13)

$$g(\psi\circ\text{trace}, (\chi_i\circ\mathbf{N})\rho^{b_i(q-1)}) = \begin{cases} g\chi_i(x_0^{q+1})\rho^{b_i}(-1) & \text{if } (\chi_i\circ\mathbf{N})\rho^{b_i(q-1)} \neq 1 \\ -1 & \text{if } (\chi_i\circ\mathbf{N})\rho^{b_i(q-1)} = 1. \end{cases}$$

For fixed i and χ_i, the number of characters ρ for which $\rho^{b_i(q-1)} = (\chi_i^{-1}\circ\mathbf{N})$ is certainly bounded by $b_i(q-1)$, so the number of ρ for which this happens for at least one $i = 1, 2, \ldots, n$ is bounded by $(q-1)\sum b_i$. Thus we find that

(4.2.1.14)
$$\frac{1}{q^2-1}\sum_{\substack{\text{chars } \rho \\ \text{of } \mathbf{F}_{q^2}^\times}} \left[\prod_i g(\psi\circ\text{trace}, (\chi_i\circ\mathbf{N})\rho^{b_i(q-1)}) - \prod_i q\chi_i(x_0^{q+1})\rho^{b_i}(-1)\right]$$

has absolute value bounded by $\dfrac{1}{q^2-1}(\sum b_i)(q-1)q^n$.

For fixed ρ, the second product is equal to

$$\prod_{i=1}^n (q\chi_i(x_0^{q+1})\rho^{b_i}(-1)) = q^n(\prod\chi_i)(x_0^{q+1})\rho((-1)^{\sum b_i}) = q^n(\prod\chi_i)(x_0^{q+1}),$$

independent of ρ, whence we find, substituting via 4.2.1.5,

(4.2.1.15)
$$\left|(-1)^{n-1}\sum_{a\in\mathbf{F}_q^\times}\text{tr}(F_a^2|\,\text{Kl}_{\bar{a}}) - q^n(\prod\chi_i)(x_0^{q+1})\right| \leq \frac{(\sum b_i)q^n}{q+1}.$$

We next claim that $(\prod\chi_i)(x_0^{q+1}) = (-1)^\delta$, where δ is the number of i for which χ_i, assumed to be of order 1 or 2, has exact order 2. If δ is even, then $\prod\chi_i$ is the trivial character, so there is nothing to prove. If δ is odd, then the characteristic p must be odd, and $\prod\chi_i$ is *the* quadratic character of $\mathbf{F}_q^\times$. The hypothesis $(\prod\chi_i)(-1) = 1$ means that -1 is a square in $\mathbf{F}_q^\times$. The value of $(\prod\chi_i)(x_0^{q+1})$ is the unique choice of ± 1 in $\mathbf{Z}$ whose image in $\mathbf{F}_q^\times$ is equal to $(x_0^{q+1})^{\frac{q-1}{2}} = (x_0^{q-1})^{\frac{q+1}{2}} = (-1)^{\frac{q+1}{2}}$. Because -1 is a square in $\mathbf{F}_q^\times$, we have $(-1)^{\frac{q-1}{2}} = 1$, whence $(-1)^{\frac{q+1}{2}} = -1 = (-1)^\delta$, as required.

We thus find, upon dividing the above inequality by q^n, an estimate

$$\left|\frac{1}{q^n}\sum_{a\in\mathbf{G}_m(\mathbf{F}_q)}\text{tr}(F_a^2|\,\text{Kl}_{\bar{a}}) - (-1)^{n-1+\delta}\right| \leq \frac{\sum b_i}{q+1}.$$

Replacing $\mathbf{F}_q$ by $\mathbf{F}_{q^r}$ where $r \uparrow \infty$, we find that $\epsilon = (-1)^{n-1+\delta}$. ∎

4.3. The Existence Theorem for $n = 1$, via the sheaves $\mathcal{L}_\psi$ and $\mathcal{L}_\chi$. Let $\mathbf{F}_q$ be a finite field, and G a smooth commutative group-scheme over $\mathbf{F}_q$ with geometrically connected fibres. Denoting by $F : G \to G$ the Frobenius endomorphism of G relative to $\mathbf{F}_q$ (for any $\mathbf{F}_q$-algebra R, F is the group endomorphism of $G(R)$ "raise all coordinates to the q-th power"). The Lang isogeny $1-F : G \to G$ is an $\mathbf{F}_q$-endomorphism of group-schemes, which sits in a short exact sequence (as abelian etale sheaves on $(\mathrm{Sch}/\mathbf{F}_q)$)

$$0 \longrightarrow G(\mathbf{F}_q) \longrightarrow G \xrightarrow{1-F} G \longrightarrow 0.$$

For any ring A, and any homomorphism of abstract groups

$$\rho : G(\mathbf{F}_q) \to A^\times,$$

we obtain by "push out" a short exact sequence

$$0 \to A^\times \to E_\rho \to G \to 0,$$

whose middle term E_ρ is thus an etale $A^\times$-torsor over G. The associated locally constant sheaf of free A-modules of rank one on G is denoted $\mathcal{L}_\rho$. (Note for the specialist: because we have taken $1-F$ rather than $F-1$ as the Lang isogeny, the usual need to replace ρ by its inverse in the pushout defining $\mathcal{L}_\rho$ is avoided.) The basic properties of $\mathcal{L}_\rho$ are the following (cf. [De-3]).

(1) For k any finite extension of $\mathbf{F}_q$, $a \in G(k)$ any k-valued point of G, and $\bar{a}$ any geometric point of G lying over a, the inverse $F_{a,k}$ of the standard generator $x \mapsto x^{\#(k)}$ of $\mathrm{Gal}(\bar{k}/k) = \pi_1(\mathrm{Spec}(k), \bar{a})$ acts A-linearly on the free rank one A-module $(\mathcal{L}_\rho)_{\bar{a}}$ with

$$\mathrm{trace}\big(F_{a,k} | (\mathcal{L}_\rho)_{\bar{a}}\big) = \rho\big(\mathrm{trace}_{k/\mathbf{F}_q}(a)\big),$$

where the inner "trace" is in the sense of the group G.

(2) For k any finite extension of $\mathbf{F}_q$, denote by $\pi : G \underset{\mathbf{F}_q}{\otimes} k \to G$ the canonical projection, and by $\rho \circ \mathrm{trace}$ the homomorphism

$$\begin{array}{ccc} (G \otimes k)(k) = G(k) & \xrightarrow{\mathrm{trace}_{k/\mathbf{F}_q}} & G(\mathbf{F}_q) \\ & \searrow^{\rho \circ \mathrm{trace}} & \downarrow \rho \\ & & A^\times. \end{array}$$

We have a canonical isomorphism of sheaves of A-modules on $G \otimes k$

$$\pi^*(\mathcal{L}_\rho) \simeq \mathcal{L}_{\rho \circ \mathrm{trace}}.$$

(3) Under the "sum" map $G \times G \to G$, we have a canonical isomorphism of sheaves of A-modules on $G \times G$

$$\mathrm{sum}^*(\mathcal{L}_\rho) \simeq pr_1^*(\mathcal{L}_\rho) \underset{A}{\otimes} pr_2^*(\mathcal{L}_\rho).$$

We will make particular use of this theory in the two cases $G = \mathbf{G}_a = \mathbf{A}^1$ and $G = \mathbf{G}_m$. To distinguish them we will always use the notation $\mathcal{L}_\psi$ in the $\mathbf{G}_a$ case, and $\mathcal{L}_\chi$ in the $\mathbf{G}_m$ case.

Let A be complete noetherian local ring with finite residue field $\mathbf{F}_\lambda$ of characteristic $l \neq p$, or a finite extension E_λ of $\mathbf{Q}_1$ with $l \neq p$. One sees easily that

(1) If $\psi : (\mathbf{F}_q, +) \to A^\times$ is a *non-trivial* additive character, then $\mathcal{L}_\psi$ is a lisse sheaf of free rank-one A-modules on $\mathbf{A}^1 \otimes \mathbf{F}_q$, satisfying $\mathrm{Swan}_\infty(\mathcal{L}_\psi) = 1$. (Reduce to $A = \mathbf{F}_\lambda$, and calculate using lower numbering.)

(2) If $\chi : \mathbf{F}_q^\times \to A^\times$ is any multiplicative character, then $\mathcal{L}_\chi$ is a lisse sheaf of free rank-one A-modules on $\mathbf{G}_m \otimes \mathbf{F}_q$, which is tame at both 0 and ∞. (Indeed for $\mathbf{G}_m$, $1 - F : \mathbf{G}_m \to \mathbf{G}_m$ is an $\mathbf{F}_q^\times$-torsor, and $\mathbf{F}_q^\times$ has order prime to p.)

With these preliminaries out of the way, the construction of Kloosterman sheaves with $n = 1$ is absolutely trivial. We have

$$\mathrm{Kl}(\psi, \chi, b) = [b]_*(\mathcal{L}_\psi \otimes \mathcal{L}_\chi),$$

where $[b]$ denotes the endomorphism $x \mapsto x^b$ of $\mathbf{G}_m$, and where $\mathcal{L}_\psi \otimes \mathcal{L}_\chi$ means the lisse sheaf on $\mathbf{G}_m$ obtained by tensoring $\mathcal{L}_\chi$ with the restriction to $\mathbf{G}_m$ of $\mathcal{L}_\psi$.

Let us check that this construction has all the required properties. First, $\mathcal{L}_\psi \otimes \mathcal{L}_\chi$ is lisse of rank one and pure of weight zero on $\mathbf{G}_m$, tame at 0 and with break one at ∞. Writing $b = b'p^n$ with b' prime to p, we have

$$[b]_* = [b']_*[p^n]_* = [b']_* F_*^n.$$

Because $(F^n)_*$ is an equivalence with quasi-inverse $(F^n)^*$, we have

$$(F^n)_*(\mathcal{L}_\psi \otimes \mathcal{L}_\chi) = \mathcal{L}_{\psi'} \otimes \mathcal{L}_{\chi'}$$

where ψ' and χ' are the unique additive and multiplicative characters of $\mathbf{F}_q$ satisfying

$$\psi'(a^{p^n}) = \psi(a) \quad \text{for } a \in \mathbf{F}_q$$
$$\chi'(a^{p^n}) = \chi(a) \quad \text{for } a \in \mathbf{F}_q^\times.$$

Thus we have

$$[b]_*(\mathcal{L}_\psi \otimes \mathcal{L}_\chi) = [b'_*(\mathcal{L}_{\psi'} \otimes \mathcal{L}_{\chi'}).$$

Because b' is prime to p, the map $[b']$ is finite etale of degree b', tame and fully ramified at both 0 and ∞. Therefore $[b']_*(\mathcal{L}_{\psi'} \otimes \mathcal{L}_{\chi'})$ is lisse of rank b', pure of weight zero, tame at zero, and with all its breaks at ∞ equal to $1/b'$. That it has the correct local traces is obvious.

The trace functions on $\mathbf{G}_m(\mathbf{F}_q) = \mathbf{F}_q^\times$ of n-variable Kloosterman sheaves are simply n-fold *convolutions* of the trace functions of Kloosterman sheaves for $n = 1$. In the next chapter, we will show that the natural notion of *convolution of sheaves* on $\mathbf{G}_m \otimes \mathbf{F}_q$ permits us to construct the Kloosterman sheaves for general n by successively convolving those constructed above for $n = 1$.

CHAPTER 5

Convolution of Sheaves on $\mathbf{G}_m$

5.0. Let k be a perfect field of characteristic $p > 0$, l a prime number $l \neq p$, and A an "l-adic coefficient ring," i.e., a complete noetherian local ring with finite residue field $\mathbf{F}_\lambda$, of residue characteristic l, or a finite extension E_λ of $\mathbf{Q}_l$. Denote by $\pi : (\mathbf{G}_m \otimes k) \underset{k}{\times} (\mathbf{G}_m \otimes k) \to \mathbf{G}_m \otimes k$ the group operation $\pi(x, y) = xy$. For $\mathcal{F}, \mathcal{G}$ lisse sheaves of finitely generated A-modules on $\mathbf{G}_m \otimes k$, denote by $\mathcal{F} \boxtimes \mathcal{G}$ the lisse sheaf $\mathrm{pr}_1^*(\mathcal{F}) \otimes_A \mathrm{pr}_2^*(\mathcal{G})$ on $(\mathbf{G}_m \otimes k) \underset{k}{\times} (\mathbf{G}_m \otimes k)$. Finally, denote by $\mathcal{C} = \mathcal{C}_{A,k}$ the category of all lisse sheaves of free A-modules of finite rank on $\mathbf{G}_m \otimes k$ which are both tame at zero and totally wild at infinity.

5.0.1. Lemma.

(1) *For $\mathcal{F}$ in $\mathcal{C}$, its A-linear dual $\mathcal{F}^\vee = \underline{\mathrm{Hom}}(\mathcal{F}, A)$ lies in $\mathcal{C}$, and $\mathrm{Swan}_\infty(\mathcal{F}) = \mathrm{Swan}_\infty(\mathcal{F}^\vee)$ (in fact, $\mathcal{F}$ and $\mathcal{F}^\vee$ have the same breaks with the same multiplicities).*

(2) *For $\mathcal{F}$ in $\mathcal{C}$, and $\mathcal{L}$ a lisse sheaf of free A-modules of finite rank on $\mathbf{G}_m \otimes k$ which is tame at both 0 and ∞, $\mathcal{F} \otimes_A \mathcal{L}$ lies in $\mathcal{C}$, and $\mathrm{Swan}_\infty(\mathcal{F} \otimes \mathcal{L}) = \mathrm{Swan}_\infty(\mathcal{F}) \,\mathrm{rank}(\mathcal{L})$.*

(3) *For $\mathcal{F} \in \mathcal{C}$, the cohomology groups $H_c^i(\mathbf{G}_m \otimes \bar{k}, \mathcal{F})$ and $H^i(\mathbf{G}_m \otimes \bar{k}, \mathcal{F})$ vanish for $i \neq 1$. For $i = 1$, each is a free A-module of rank $\mathrm{Swan}_\infty(\mathcal{F})$, whose formation, for variable $\mathcal{F}$ in $\mathcal{C}$, is exact.*

(4) *For $A \to A'$ a homomorphism of rings of the type considered (i.e., either E_λ or complete noetherian local with finite residue field $\mathbf{F}_\lambda$ of characteristic $l \neq p$), the extension of scalars functor $\mathcal{F} \mapsto \mathcal{F} \bigotimes_A A'$ is an exact functor $\mathcal{C}_{A,k} \to \mathcal{C}_{A',k}$, and we have canonical isomorphisms*

$$H_c^1(\mathbf{G}_m \otimes \bar{k}, \mathcal{F}) \underset{A}{\otimes} A' \xrightarrow{\sim} H_c^1(\mathbf{G}_m \otimes \bar{k}, \mathcal{F} \underset{A}{\otimes} A'),$$

$$H^1(\mathbf{G}_m \otimes \bar{k}, \mathcal{F}) \underset{A}{\otimes} A' \xrightarrow{\sim} H_0^1(\mathbf{G}_m \otimes \bar{k}, \mathcal{F} \underset{A}{\otimes} A').$$

(5) *For any integer $N \geq 1$, denote by $[N] : \mathbf{G}_m \to \mathbf{G}_m$ the N-th power map $x \mapsto x^N$. The constructions $\mathcal{F} \mapsto [N]_*(\mathcal{F})$ and $\mathcal{F} \mapsto [N]^*(\mathcal{F})$ are exact functors carrying $\mathcal{C}$ to $\mathcal{C}$.*

(6) *For $a \in k^\times = \mathbf{G}_m(k)$, denote by $\mathrm{Trans}_a : \mathbf{G}_m \otimes k \to \mathbf{G}_m \otimes k$ the "translation by a" map $x \mapsto ax$. Then the constructions $\mathcal{F} \mapsto (\mathrm{Trans}_a)_*(\mathcal{F})$ and $\mathcal{F} \mapsto (\mathrm{Trans}_a)^*(\mathcal{F})$ are quasi-inverse equivalences of categories $\mathcal{C} \to \mathcal{C}$.*

Proof. This is just "mise pour memoire," cf. 1.5, 1.3, 2.1.1, 2.3.3, 1.13. ∎

5.1. Convolution Theorem. *For $\mathcal{F}, \mathcal{G}$ objects in $\mathcal{C}$, we have*

(1) *$R^i\pi_!(\mathcal{F} \boxtimes \mathcal{G}) = 0$ for $i \neq 1$, and $R^1\pi_!(\mathcal{F} \boxtimes \mathcal{G})$ lies in $\mathcal{C}$; we denote it $\mathcal{F} * \mathcal{G}$. Formation of $\mathcal{F} * \mathcal{G}$ is bi-exact. For any homomorphism $A \to A'$ of l-adic coefficient rings, we have a canonical isomorphism*

$$(\mathcal{F} \otimes_A A') * (\mathcal{G} \otimes_A A') \xrightarrow{\sim} (\mathcal{F} * \mathcal{G}) \otimes_A A'.$$

(2) *The formation of $R\pi_*(\mathcal{F} \boxtimes \mathcal{G})$ commutes with passage to fibres, and the natural "forget supports" map $R\pi_!(\mathcal{F} \boxtimes \mathcal{G}) \to R\pi_*(\mathcal{F} \boxtimes \mathcal{G})$ is an isomorphism.*

(3) *The A-linear duals $\mathcal{F}^\vee = \underline{\mathrm{Hom}}_A(\mathcal{F}, A)$ and $\mathcal{G}^\vee = \underline{\mathrm{Hom}}_A(\mathcal{G}, A)$ lie in $\mathcal{C}$, and the pairing*

$$(\mathcal{F} * \mathcal{G}) \times (\mathcal{F}^\vee * \mathcal{G}^\vee) \to A(-1)$$

defined by cup product

$$\begin{array}{c} R^1\pi_!(\mathcal{F} \boxtimes \mathcal{G}) \times R^1\pi_!(\mathcal{F}^\vee \boxtimes \mathcal{G}^\vee) \\ \downarrow \text{cup} \\ R^2\pi_!((\mathcal{F} \boxtimes \mathcal{G}) \otimes (\mathcal{F}^\vee \boxtimes \mathcal{G}^\vee)) \\ \downarrow \text{contraction} \\ R^2\pi_!(A) \\ \wr \downarrow \text{trace} \\ A(-1) \end{array}$$

is a perfect duality of lisse sheaves of free A-modules of finite rank on $\mathbf{G}_m \otimes k$.

(4) *The rank of $\mathcal{F} * \mathcal{G}$ is given by*

$$\mathrm{rank}(\mathcal{F} * \mathcal{G}) = \mathrm{rank}(\mathcal{F})\,\mathrm{Swan}_\infty(\mathcal{G}) + \mathrm{rank}(\mathcal{G})\,\mathrm{Swan}_\infty(\mathcal{F}).$$

(5) *The* Swan_∞ *of* $\mathcal{F} * \mathcal{G}$ *is given by*

$$\mathrm{Swan}_\infty(\mathcal{F} * \mathcal{G}) = \mathrm{Swan}_\infty(\mathcal{F})\,\mathrm{Swan}_\infty(\mathcal{G}).$$

(6) *The* H^1_c *and* H^1 *on* $\mathbf{G}_m \otimes \bar{k}$ *of* $\mathcal{F}$, $\mathcal{G}$, *and* $\mathcal{F} * \mathcal{G}$ *sit in a commutative diagram*

$$\begin{array}{ccc} H^1_c(\mathcal{F} * \mathcal{G}) & \xrightarrow{\sim} & H^1_c(\mathcal{F}) \otimes_A H^1_c(\mathcal{G}) \\ \text{forget supports}\downarrow & & \downarrow\text{forget supports} \\ H^1(\mathcal{F} * \mathcal{G}) & \xrightarrow{\sim} & H^1(\mathcal{F}) \otimes_A H^1(\mathcal{G}). \end{array}$$

(Remember that by 5.0.1(3) we have $H^i_c = H^i = 0$ *for* $i \neq 1$.*)*

(7) *If* k *is a finite field* $\mathbf{F}_q$, *if* $A = E_\lambda$ *or* $\mathcal{O}_\lambda$, *and if* $\mathcal{F}$ *and* $\mathcal{G}$ *are pure, of weights* $w(\mathcal{F})$ *and* $w(\mathcal{G})$ *respectively, then* $\mathcal{F} * \mathcal{G}$ *is pure, of weight*

$$w(\mathcal{F} * \mathcal{G}) = 1 + w(\mathcal{F}) + w(\mathcal{G}).$$

(8) *If* k *is a finite field* $\mathbf{F}_q$, *denote by* $\mathrm{trace}_{\mathcal{F}} : \mathbf{F}_q^\times \to A$ *the function*

$$\mathrm{trace}_{\mathcal{F}}(a) = \mathrm{trace}(F_a | \mathcal{F}_{\bar{a}}).$$

Then the trace functions of $\mathcal{F}, \mathcal{G}$ *and* $\mathcal{F} * \mathcal{G}$ *are related by*

$$\mathrm{trace}_{\mathcal{F} * \mathcal{G}} = -(\mathrm{trace}_{\mathcal{F}}) * (\mathrm{trace}_{\mathcal{G}}).$$

If $\chi : \mathbf{F}_q^\times \to A^\times$ *is any multiplicative character, then the multiplicative Fourier transform of* $\mathrm{trace}_{\mathcal{F}}$ *at* χ *is given by*

$$\widehat{\mathrm{trace}_{\mathcal{F}}}(\chi) = -\,\mathrm{trace}(F | H^1_c(\mathbf{G}_m \otimes \overline{\mathbf{F}}_q, \mathcal{F} \otimes \mathcal{L}_\chi).$$

(9) *If* k *contains the finite field* $\mathbf{F}_q$, *and if* $\chi : \mathbf{F}_q^\times \to A^\times$ *is a multiplicative character, then denoting by* $\mathcal{L}_\chi$ *the inverse image on* $\mathbf{G}_m \otimes k$ *of the sheaf* $\mathcal{L}_\chi$ *on* $\mathbf{G}_m \otimes \mathbf{F}_q$ *(cf. 4.3) we have a canonical isomorphism*

$$(\mathcal{F} \otimes \mathcal{L}_\chi) * (\mathcal{G} \otimes \mathcal{L}_\chi) \xrightarrow{\sim} (\mathcal{F} * \mathcal{G}) \otimes \mathcal{L}_\chi.$$

(10) *For any integer* $N \geq 1$, *denoting by* $[N] : \mathbf{G}_m \otimes k \to \mathbf{G}_m \otimes k$ *the* N*-th power map* $X \mapsto X^N$, *we have a canonical isomorphism*

$$([N]_*((\mathcal{F})) * ([N]_*(\mathcal{G})) \simeq [N]_*(\mathcal{F} * \mathcal{G}).$$

(11) *For any element* $a \in k^\times = \mathbf{G}_m(k)$, *denote by* $\mathrm{Trans}_a : \mathbf{G}_m \otimes k \to \mathbf{G}_m \otimes k$ *the "translation by* a*" map* $x \mapsto ax$. *Then we have canonical isomorphisms, for* $a, b \in k^\times$,

$$([\mathrm{Trans}_a]_*(\mathcal{F})) * ([\mathrm{Trans}_b]_*(\mathcal{G})) \xrightarrow{\sim} [\mathrm{Trans}_{ab}]_*(\mathcal{F} * \mathcal{G}),$$
$$([\mathrm{Trans}_a]^*(\mathcal{F})) * ([\mathrm{Trans}_b]^*(\mathcal{G})) \xrightarrow{\sim} [\mathrm{Trans}_{ab}]^*(\mathcal{F} * \mathcal{G}).$$

(12) *For any integer $N \leq 1$, there is a canonical injection*

$$[N]^*(\mathcal{F} * \mathcal{G}) \hookrightarrow [N]^*(\mathcal{F}) * [N]^*(\mathcal{G}),$$

with A-flat cokernel.

5.2. Proof of the convolution theorem and variations. For technical convenience, we introduce three auxiliary categories of sheaves on $\mathbf{G}_m \otimes k$:

$\mathcal{T} =$ lisse sheaves of free A-modules of finite type on $\mathbf{G}_m \otimes k$ which are tame at zero

$\mathcal{T}_1 =$ lisse sheaves of (not-necessarily free) A-modules of finite type on $\mathbf{G}_m \otimes k$ which are tame at zero

$\mathcal{C}_1 =$ lisse sheaves of (not necessarily free) A-modules of finite type on $\mathbf{G}_m \otimes k$ which are tame at zero and totally wild at infinity.

5.2.1. Proposition. *For $\mathcal{F} \in \mathcal{T}$ and $\mathcal{G} \in \mathcal{T}_1$, we have*

(1) $R^i\pi_!(\mathcal{F} \boxtimes \mathcal{G}) = 0$ *for* $i \neq 1, 2$.
(2) *the sheaves $R^i\pi_!(\mathcal{F} \boxtimes \mathcal{G})$ are lisse on $\mathbf{G}_m \otimes k$, and their formation commutes with passage to fibres.*
(3) *the sheaves $R^i\pi_!(\mathcal{F} \boxtimes \mathcal{G})$ lie in $\mathcal{T}_1$, i.e., they are tame at 0.*
(4) $R^i\pi_*(\mathcal{F} \boxtimes \mathcal{G}) = 0$ *for* $i \neq 0, 1$.
(5) *the sheaves $R^i\pi_*(\mathcal{F} \boxtimes \mathcal{G})$ are lisse on $\mathbf{G}_m \otimes k$, and their formation commutes with passage to fibres.*
(6) *the sheaves $R^i\pi_*(\mathcal{F} \boxtimes \mathcal{G})$ lie in $\mathcal{T}_1$, i.e., they are tame at zero.*

Proof. The case $A = E_\lambda$ results from the case $A = \mathcal{O}_\lambda$ (every lisse E_λ-sheaf has an $\mathcal{O}_\lambda$-form). The case of noetherian local A with finite residue field $\mathbf{F}_\lambda$ results from the case A = finite local with residue field $\mathbf{F}_\lambda$ by a standard passage to the limit argument.

For fixed $\mathcal{F}$ in $\mathcal{T}$, the functor $\mathcal{G} \mapsto \mathcal{F} \boxtimes \mathcal{G}$ is exact, so the long exact cohomology sequences for the $R^i\pi_!$ and the $R^i\pi_*$, before and after base change, allow us to reduce to the case when $\mathcal{G}$ is an $\mathbf{F}_\lambda$-sheaf. For such $\mathcal{G}$'s, $\mathcal{F}$ acts through $\mathcal{F} \otimes_A \mathbf{F}_\lambda$, so we are reduced to the case $A = \mathbf{F}_\lambda$.

Thus let $\mathcal{F}, \mathcal{G}$ be two lisse $\mathbf{F}_\lambda$-sheaves on $\mathbf{G}_m \otimes k$, both tame at zero. The morphism $\pi : \mathbf{G}_m \times \mathbf{G}_m \to \mathbf{G}_m$, $\pi(x, y) = xy$, may be viewed in new coordinates $(x, t = xy)$ as the projection onto the second factor of

$\mathbf{G}_m \times \mathbf{G}_m$,

$$\begin{array}{ccc} \mathcal{F} \boxtimes \mathcal{G} : \mathbf{G}_m \times \mathbf{G}_m & \xrightarrow[(x,y)\mapsto(x,xy)]{\sim} & \mathbf{G}_m \times \mathbf{G}_m : \mathcal{F}_x \otimes \mathcal{G}_{t/x} \\ & {}_{\pi(x,y)=xy}\searrow \quad \swarrow_{\mathrm{pr}_2(x,t)=t} & \\ & \mathbf{G}_m & \end{array}$$

Viewed in this way, it is natural to compactify pr_2 by embedding the $\mathbf{G}_m$ of x's into the $\mathbf{P}^1$ of x's:

$$\begin{array}{ccc} \mathcal{F}_x \otimes \mathcal{G}_{t/x} : \mathbf{G}_m \times \mathbf{G}_m & \overset{j}{\hookrightarrow} & \mathbf{P}^1 \times \mathbf{G}_m : j_!(\mathcal{F}_x \otimes \mathcal{G}_{t/x}) \\ & {}_{\mathrm{pr}_2}\searrow \quad \swarrow_{\mathrm{pr}_2} & \\ & \mathbf{G}_m & \end{array}$$

The sheaf $j_!(\mathcal{F}_x \otimes \mathcal{G}_{t/x})$ on $\mathbf{P}^! \times \mathbf{G}_m$ is lisse outside the two sections $0 \times \mathbf{G}_m$ and $\infty \times \mathbf{G}_m$ of $\overline{\mathrm{pr}}_2$, and along these it vanishes. By Deligne's semi-continuity theorem (cf. [Lau-1]), the necessary and sufficient condition that all the sheaves $R^i\pi_!(\mathcal{F} \boxtimes \mathcal{G}) \simeq R^i(\overline{\mathrm{pr}}_2)_*(j_!(\mathcal{F}_x \otimes \mathcal{G}_{t/x}))$ be lisse on $\mathbf{G}_m \otimes k$ is that the function on variable geometric points $t_0 : \mathrm{Spec}(\Omega) \to \mathbf{G}_m \otimes k$ defined by

$$t_0 \mapsto \mathrm{Swan}_0(\mathcal{F}_x \otimes \mathcal{G}_{t_0/x}) + \mathrm{Swan}_\infty(\mathcal{F}_x \otimes \mathcal{G}_{t_0/x})$$

be *constant.* In our situation, the sheaves $\mathcal{F}$ and $\mathcal{G}$ are both tame at 0, so by inversion (both $\mathcal{F}_{t_0/x}$ and) $\mathcal{G}_{t_0/x}$ are tame at ∞, so we find that this function is equal to

$$\mathrm{rank}(\mathcal{F})\,\mathrm{Swan}_0(\mathcal{G}_{t_0/x}) + \mathrm{Swan}_\infty(\mathcal{F})\,\mathrm{rank}(\mathcal{G}).$$

But as $x \mapsto t_0/x$ is an automorphism of $\mathbf{G}_m \otimes \Omega$ which interchanges 0 and ∞, we have

$$\mathrm{Swan}_0(\mathcal{G}_{t_0/x}) = \mathrm{Swan}_\infty(\mathcal{G}),$$

whence we see that we have the required constancy. Therefore the $R^i\pi_!(\mathcal{F} \boxtimes \mathcal{G})$ are all lisse sheaves on $\mathbf{G}_m \otimes k$. By proper base-change, formation of the $R^i\pi_!(\mathcal{F} \boxtimes \mathcal{G})$ commutes with passage to fibres. Looking at fibres, we see $R^i\pi_!(\mathcal{F} \boxtimes \mathcal{G}) = 0$ for $i \neq 1, 2$.

By $\mathbf{F}_\lambda$-Poincaré duality applied to the smooth morphism π and the lisse sheaf $\mathcal{F} \boxtimes \mathcal{G}$, cf. ([De-4], 2.1), the lissité of all the $R^i\pi_!(\mathcal{F} \boxtimes \mathcal{G})$ guarantees that after any base change the sheaves $R^i\pi_*(\mathcal{F}^\vee \boxtimes \mathcal{G}^\vee)$ are also lisse, and that the cup-product pairings

$$R^i\pi_!(\mathcal{F}\boxtimes\mathcal{G})\times R^{2-i}\pi_*(\mathcal{F}^\vee\boxtimes\mathcal{G}^\vee)\to R^2\pi_!(\mathbf{F}_\lambda)\xrightarrow[\text{trace}]{\sim}\mathbf{F}_\lambda(-1)$$

are perfect pairings of lisse $\mathbf{F}_\lambda$-sheaves. Inverting the roles of $\mathcal{F}$ and $\mathcal{F}^\vee$, $\mathcal{G}$ and $\mathcal{G}^\vee$, we see that the $R^i\pi_*(\mathcal{F}\boxtimes\mathcal{G})$ are lisse on $\mathbf{G}_m\otimes k$, of formation compatible with passage to fibres, and that $R^i\pi_*(\mathcal{F}\boxtimes\mathcal{G})$ vanishes for $i\neq 0,1$.

It remains to show that the $R^i\pi_!(\mathcal{F}\boxtimes\mathcal{G})$ and the $R^i\pi_*(\mathcal{F}\boxtimes\mathcal{G})$ are tame at zero. By $\mathbf{F}_\lambda$-Poincaré duality, it suffices to treat the $R^i\pi_!$. We first reduce to the case when both $\mathcal{F}$ and $\mathcal{G}$ are lisse at zero. By hypothesis there exists an integer $N\geq 1$ prime to p such that under the N'th power map $[N]:\mathbf{G}_m\otimes k\to\mathbf{G}_m\otimes k$, both $[N]^*(\mathcal{F})$ and $[N]^*(\mathcal{G})$ extend to lisse $\mathbf{F}_\lambda$-sheaves on $\mathbf{A}^1$. Extending the field k, we may assume that k contains N distinct N'th roots of unity. To show that the $R^i\pi_!(\mathcal{F}\boxtimes\mathcal{G})$ are tame at zero, it is sufficient to show that the $[N]^*(R^i\pi_!(\mathcal{F}\boxtimes\mathcal{G}))$ are tame at zero, for some $N\geq 1$. But these are related to the $R^i\pi_!([N]^*(\mathcal{F})\boxtimes[N]^*(\mathcal{G}))$ by the following lemma.

5.2.1.1. Lemma. *Let $N\geq 1$ be an integer prime to p, and suppose that k contains N distinct N-th roots of unity. Then for $\mathcal{F}\in\mathcal{T}$ and $\mathcal{G}\in\mathcal{T}_1$, we have a spectral sequence of lisse sheaves on $\mathbf{G}_m\otimes k$*

$$E_2^{a,b}=H^a(\boldsymbol{\mu}_N(k),R^b\pi_!([N]^*(\mathcal{F})\boxtimes[N]^*(\mathcal{G}))$$
$$\Longrightarrow E_\infty^{a+b}=[N]^*(R^{a+b}\pi_!(\mathcal{F}\boxtimes\mathcal{G})),$$

which yields a canonical isomorphism

$$[N]^*(R^1\pi_!(\mathcal{F}\boxtimes\mathcal{G}))\xrightarrow{\sim}E_2^{0,1}=\left(R^1\pi_!([N]^*(\mathcal{F})\boxtimes[N]^*(\mathcal{G}))\right)^{\boldsymbol{\mu}_N}$$

and a four-term exact sequence

$$0\to E_2^{1,1}\to[N]^*(R^2\pi_!(\mathcal{F}\boxtimes\mathcal{G}))\to E_2^{0,2}\xrightarrow{d_2}E_2^{2,1}\to 0.$$

Proof. Because $[N]$ is a group-endomorphism, the diagram

$$\begin{array}{ccc}\mathbf{G}_m\times\mathbf{G}_m & \xleftarrow{[N]\times[N]} & \mathbf{G}_m\times\mathbf{G}_m\\ \downarrow\pi & & \downarrow\pi\\ \mathbf{G}_m & \xleftarrow{[N]} & \mathbf{G}_m\end{array}$$

is commutative, so we can factor its upper right corner through the fibre-product:

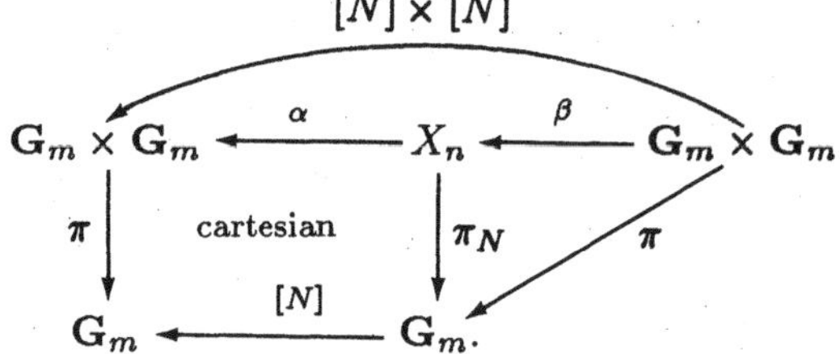

Because the morphism $[N] : \mathbf{G}_m \otimes k \to \mathbf{G}_m \otimes k$ is a finite etale $\boldsymbol{\mu}_N(k)$-torseur, it follows that α and β are each finite etale $\boldsymbol{\mu}_N(k)$-torseurs. By etale base change, we have

$$[N]^*(R^i\pi_!(\mathcal{F} \boxtimes \mathcal{G})) \xrightarrow{\sim} R^i\pi_N(\alpha^*(\mathcal{F} \boxtimes \mathcal{G})).$$

Temporarily writing $K = \alpha^*(\mathcal{F} \boxtimes \mathcal{G})$ on X_N, we have

$$\beta^*(K) = \beta^*\alpha^*(\mathcal{F} \boxtimes \mathcal{G}) = ([N] \times [N])^*(\mathcal{F} \boxtimes \mathcal{G}) \simeq [N]^*(\mathcal{F}) \boxtimes [N]^*(\mathcal{G}).$$

Because β is a finite etale $\boldsymbol{\mu}_N(k)$-torseur, we have, for any sheaf K on X_N, a spectral sequence of sheaves on $\mathbf{G}_m \otimes k$,

$$E_2^{a,b} = H^a(\boldsymbol{\mu}_N(k), R^b\pi_!(\beta^*K)) \Longrightarrow R^{a+b}(\pi_N)_!(K).$$

For $K = \alpha^*(\mathcal{F} \boxtimes \mathcal{G})$, this spectral sequence reads

$$E_2^{a,b} = H^a(\boldsymbol{\mu}_N(k), R^b\pi_!([N]^*(\mathcal{F}) \boxtimes [N]^*(\mathcal{F})) \Longrightarrow [N]^*(R^{a+b}\pi_!(\mathcal{F} \boxtimes \mathcal{G})).$$

By the vanishing of the $R^i\pi_!$ for $i = 1, 2$, we see that $E_2^{a,b} = 0$ for $b \neq 1, 2$ and $E_\infty^{a,b} = 0$ for $a + b \neq 1, 2$, whence the asserted isomorphism and four-term exact sequence. ∎

5.2.1.2. Variant Lemma. *Hypotheses and notations being as in the preceding lemma, we have a spectral sequence of lisse sheaves on* $\mathbf{G}_m \otimes k$

$$E_2^{a,b} = H^a(\boldsymbol{\mu}_n(k), R^b\pi_*([N]^*(\mathcal{F}) \boxtimes [N] * (\mathcal{G})) \\ \Longrightarrow E_\infty^{a+b} = [N]^*(R^{a+b}\pi_*(\mathcal{F} \boxtimes \mathcal{G})),$$

which yields a canonical isomorphism

$$[N]^*(\pi_*(\mathcal{F} \boxtimes \mathcal{G})) \simeq \left(\pi_*([N]^*(\mathcal{F}) \boxtimes [N]^*(\mathcal{G}))\right)^{\boldsymbol{\mu}_N(k)}$$

and a four-term exact sequence

$$0 \to E_2^{1,0} \to [N]^*(R^1\pi_*(\mathcal{F} \boxtimes \mathcal{G})) \to E_2^{0,1} \xrightarrow{d_2} E_2^{2,0} \to 0.$$

Proof. Entirely analogous to 5.2.1.1, the vanishing of $R^i\pi_*$ for $i \neq 0, 1$ forcing $E_2^{a,b} = 0$ for $b \neq 0, 1$, and $E_\infty^{a,b} = 0$ for $a + b \neq 0, 1$. ∎

By the spectral sequence of 5.2.1.1, we verify that if all the sheaves $R^i\pi_!([N]^*(\mathcal{F}) \boxtimes [N]^*(\mathcal{G}))$ are tame at zero, the same is true of all the $[N]^*(R^i\pi_!(\mathcal{F} \boxtimes \mathcal{G}))$, whence of all the $R^i\pi_!(\mathcal{F} \boxtimes \mathcal{G}))$. So to complete the proof of the proposition, we need only treat the case when $\mathcal{F}$ and $\mathcal{G}$ are lisse $\mathbf{F}_\lambda$-sheaves on all of $\mathbf{A}^1$.

5.2.1.3. Proposition. *Suppose that $\mathcal{F}$ and $\mathcal{G}$ extend to lisse $\mathbf{F}_\lambda$ (resp. E_λ)-sheaves on $\mathbf{A}^1 \otimes k$. Then the action of I_0 on $R^i\pi_!(\mathcal{F} \boxtimes \mathcal{G})$ is at most three-step unipotent for $i = 1$, and it is trivial for $i = 2$.*

Proof. Consider the coordinates $(1/x, t = xy)$ on $\mathbf{G}_m \times \mathbf{G}_m$. In terms of them, we have a commutative diagram

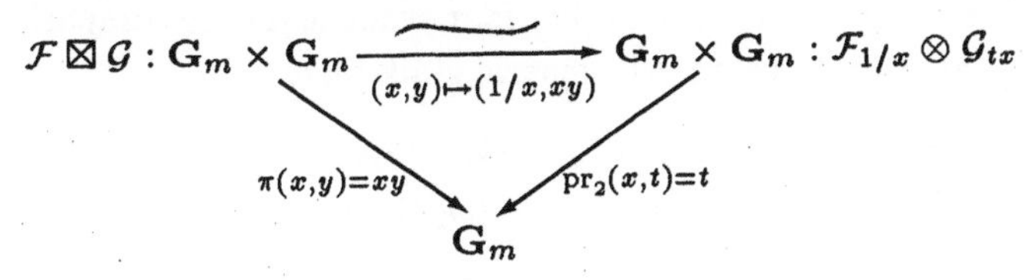

Denoting by $\mathcal{F}$ and $\mathcal{G}$ now the unique lisse sheaves on $\mathbf{A}^1$ which extend $\mathcal{F}$ and $\mathcal{G}$ on $\mathbf{G}_m$, we can embed the right half of this diagram in a cartesian diagram

$$\begin{array}{ccc} \mathbf{G}_m \times \mathbf{G}_m & \xrightarrow{(x,t)\mapsto(x,t)} & \mathbf{G}_m \times \mathbf{A}^1; \mathcal{F}_{1/x} \times \mathcal{G}_{tx} \\ \mathrm{pr}_2 \downarrow & & \downarrow \mathrm{pr}_2 \\ \mathbf{G}_m & \xrightarrow[t\mapsto t]{} & A^1, \end{array}$$

where $\mathcal{F}_{1/x} \otimes \mathcal{G}_{tx}$ makes sense as a lisse sheaf on all of $\mathbf{G}_m \times \mathbf{A}^1$. We now compactify the right half of this diagram by completing the fiber $\mathbf{G}_m$ to $\mathbf{P}^1$, and extending our lisse sheaf by 0. The picture is

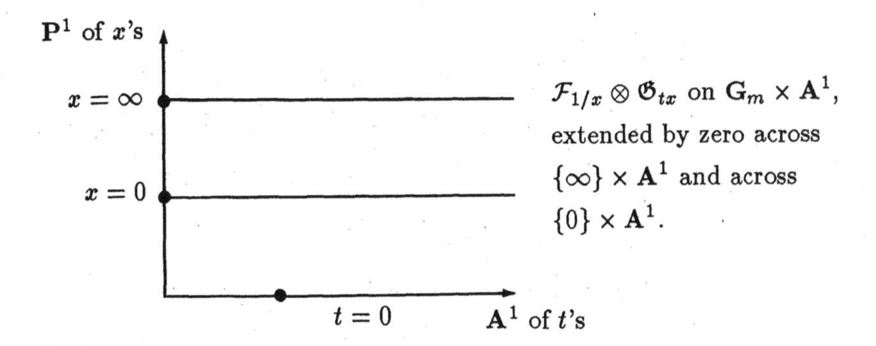

The sheaf $\mathcal{G}_{tx}$ is lisse on all of $\mathbf{A}^1 \times \mathbf{A}^1$; in particular it is lisse in a neighborhood of $\{0\} \times \mathbf{A}^1$. Therefore the Swan_0 of $\mathcal{F}_{1/x} \otimes \mathcal{G}_{tx}$ for any geometric point t of $\mathbf{A}^1$ is constant, equal to $\mathrm{rank}(\mathcal{G})\,\mathrm{Swan}_0(\mathcal{F}_{1/x}) = \mathrm{rank}(\mathcal{G})\,\mathrm{Swan}_\infty(\mathcal{F})$, and the morphism $\overline{\mathrm{pr}}_2 : \mathbf{P}^1 \times \mathbf{A}^1 \to \mathbf{A}^1$ is consequently universally locally acyclic for the sheaf "$\mathcal{F}_{1/x} \otimes \mathcal{G}_{tx}$, extended by 0," except possibly along the section $\{\infty\} \times \mathbf{A}^1$. But $\mathcal{F}_{1/x}$ is lisse in a neighborhood of $\{\infty\} \times A^1$, so the Swan_∞ of $\mathcal{F}_{1/x} \otimes \mathcal{G}_{tx}$, for t a geometric point of $\mathbf{A}^1$, is equal to $\mathrm{rank}(\mathcal{F})\,\mathrm{Swan}_\infty(\mathcal{G}_{tx})$. This Swan_∞ is therefore constant for $t \neq 0$ (because $x \mapsto tx$ is an automorphism of $\mathbf{G}_m$ fixing 0 and ∞, whence $\mathrm{Swan}_\infty(\mathcal{G}_{tx}) = \mathrm{Swan}_\infty(\mathcal{G})$). Therefore the morphism $\overline{\mathrm{pr}}_2 : \mathbf{P}^1 \times \mathbf{A}^1 \to \mathbf{A}^1$ is universally locally acyclic for the sheaf "$\mathcal{F}_{1/x} \otimes \mathcal{G}_{tx}$ extended by 0," except possibly at the single point $(x = \infty, t = 0)$. As explained in ([K-M], XIV), the only possibly non-vanishing group of vanishing cycles is the $R^1\phi_{(\infty,0)}$, and it sits in a five term D_0-equivariant exact sequence comparing the stalk of $R\,\overline{\mathrm{pr}}_{2!}$ at 0 to its geometric generic stalk at η:

$$\begin{aligned} 0 &\to H^1_c(\mathbf{G}_m \otimes \bar{k}, \mathcal{F}_{1/x} \otimes \mathcal{G}_0) \to H^1_c(\mathbf{G}_m \otimes k(\bar{\eta}), \mathcal{F}_{1/x} \otimes \mathcal{G}_{\eta x}) \to \\ &\to R^1\phi_{(\infty,0)}(\mathcal{F}_{1/x} \otimes \mathcal{G}_{tx} \text{ extended by } 0, \overline{\mathrm{pr}}_2) \to \\ &\to H^2_c(\mathbf{G}_m \otimes \bar{k}, \mathcal{F}_{1/x} \otimes \mathcal{G}_0) \to H^2_c(\mathbf{G}_m \otimes k(\bar{\eta}), \mathcal{F}_{1/x} \otimes \mathcal{G}_{\eta x}) \to 0. \end{aligned}$$

In this exact sequence, I_0 acts trivially on the first and fourth terms (the cohomology of the special fibre). Thus it suffices to show that I_0 also acts at worst two-step unipotently on the third term $R^1\phi$. Because of the fact that $\mathcal{F}_{1/x}$ is lisse near $x = \infty$, we have an isomorphism

$$R^1\phi_{(\infty,0)}(\mathcal{F}_{1/x} \otimes \mathcal{G}_{tx}, \overline{\mathrm{pr}}_2) \xleftarrow{\sim} \mathcal{F}_0 \otimes R^1\phi_{(\infty,0)}(\mathcal{G}_{tx}, \overline{\mathrm{pr}}_2).$$

Now it remains to analyze the action of I_0 on $R^1\phi_{(\infty,0)}(\mathcal{G}_{tx}, \overline{\mathrm{pr}}_2)$. This group sits the same vanishing cycle exact sequence as above, but now with $\mathcal{F}$ replaced by the constant sheaf:

$$\begin{aligned} 0 &\to H^1_c(\mathbf{G}_m \otimes \bar{k}, \mathcal{G}_0) \to H^1_0(\mathbf{G}_m \otimes k(\bar{\eta}), \mathcal{G}_{\eta x}) \to R^1\phi_{(\infty,0)}(\mathcal{G}_{tx}, \overline{\mathrm{pr}}_2) \to \\ &\to H^2_c(\mathbf{G}_m \otimes \bar{k}, \mathcal{G}_0) \to H^2_c(\mathbf{G}_m \otimes k(\bar{\eta}), \mathcal{F}_{\eta x}) \to 0. \end{aligned}$$

In view of this D_0-equivariant exact sequence, in which I_0 acts trivially on the first and fourth terms, it suffices to show that I_0 acts trivially on $H^1_c(\mathbf{G}_m \otimes k(\bar{\eta}), \mathcal{G}_{\eta x})$. In fact, the entire $\pi_1(\mathbf{G}_m \otimes \bar{k}, \bar{\eta})$ acts trivially, as follows from the following standard lemma, applied to $S = \mathrm{Spec}(k)$, $X = G = \mathbf{G}_m \otimes k$, G acting itself by translations.

5.2.1.4. Lemma. *Let S be a scheme, $f : X \to S$ an S-scheme, $\mathcal{G}$ a constructible torsion etale sheaf on X, $\phi : G \to S$ a group-scheme over S,* *and* action : $G \underset{S}{\times} X \to X$ *an S-action of G on X. Consider the situation*

$$\begin{array}{c}\text{action}^*(\mathcal{G}) \text{ on } G \times X \\ \big\downarrow \text{pr}_1 \\ G.\end{array}$$

Then

(1) *For every i we have a canonical isomorphism of sheaves on G*

$$\phi^*(R^i f_! \mathcal{G}) \to R^i(\text{pr}_1)_!(\text{action}^*(\mathcal{G})).$$

(2) *If $\phi : G \to S$ is a* <u>smooth</u> *group-scheme, then for every i we have a canonical isomorphism of sheaves on G*

$$\phi^*(R^i f_* \mathcal{G}) \to R^i(\text{pr}_1)_*(\text{action}^*(\mathcal{G})).$$

In particular, if $S = \text{Spec}(k)$ with k a separably closed field, then the $R^i(\text{pr}_1)_!(\text{action}^*(\mathcal{G}))$ are constant sheaves on G, and the same is true of the $R^i(\text{pr}_1)_*(\text{action}^*(\mathcal{G}))$ if G is smooth over k.

Proof. We have a commutative diagram of S-schemes

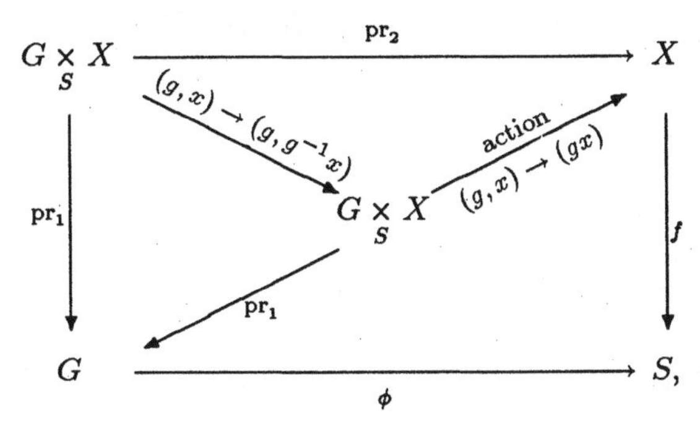

in which the outer square is cartesian. The left-hand triangle gives canonical isomorphisms

$$R^i(\text{pr}_1)_!(\text{action}^*(\mathcal{G})) \xrightarrow{\sim} R^i(\text{pr}_1)_!(\text{pr}_2^*(\mathcal{G})),$$
$$R^i(\text{pr}_1)_*(\text{action}^*(\mathcal{G})) \xrightarrow{\sim} R^i(\text{pr}_1)_*(\text{pr}_2^*(\mathcal{G})).$$

The assertions now result from the proper and smooth base-change theorems respectively. ∎

This concludes(!) the proof of Proposition 5.2.1. ∎

5.2.2. Proposition. *Let $\mathcal{F} \in \mathcal{C}$. For variable $\mathcal{G}$ in $\mathcal{T}_1$, we have*

(1) $R^i\pi_!(\mathcal{F} \boxtimes \mathcal{G}) = 0$ *for* $i \neq 1$, *its formation commutes with passage to fibres, and* $\mathcal{G} \mapsto R^1\pi_!(\mathcal{F} \boxtimes \mathcal{G})$ *is an exact functor* $\mathcal{T}_1 \to \mathcal{T}_1$. *We denote it* $\mathcal{F}_! * \mathcal{G}$. *By exactness, it carries* $\mathcal{T} \to \mathcal{T}$.

(2) $R^i\pi_*(\mathcal{F} \boxtimes \mathcal{G}) = 0$ *for* $i \neq 1$, *its formation commutes with passage to fibres, and* $\mathcal{G} \mapsto R^1\pi_*(\mathcal{F} \boxtimes \mathcal{G})$ *is an exact functor* $\mathcal{T}_1 \to \mathcal{T}_1$. *We denote it* $\mathcal{F}_* * \mathcal{G}$. *By exactness, it carries* $\mathcal{T} \to \mathcal{T}$.

(3) *For any homomorphism of l-adic coefficient rings* $A \to A'$, *we have canonical isomorphisms*

$$(\mathcal{F}_! * \mathcal{G}) \underset{A}{\otimes} A' \xrightarrow{\sim} (\mathcal{F} \underset{A}{\otimes} A')_! * (\mathcal{G} \underset{A}{\otimes} A'),$$

$$(\mathcal{F}_* * \mathcal{G}) \underset{A}{\otimes} A' \xrightarrow{\sim} (\mathcal{F} \underset{A}{\otimes} A')_* * (\mathcal{G} \underset{A}{\otimes} A').$$

(4) *For* $\mathcal{G} \in \mathcal{T}$, *the cup-product pairing* $(\mathcal{F}_! * \mathcal{G}) \times (\mathcal{F}^\vee{}_* * \mathcal{G}^\vee) \to A(-1)$ *is a perfect duality of objects of* $\mathcal{T}$.

(5) *For* $\mathcal{G} \in \mathcal{C}_1$, *the natural map* $\mathcal{F}_! * \mathcal{G} \to \mathcal{F}_* * \mathcal{G}$ *is an isomorphism.*

Proof. In view of Proposition 4.2.1 and the exact cohomology sequences for the $R^i\pi_!$ and the $R^i\pi_*$, assertions (1) and (2) result from the vanishings

$$R^2\pi_!(\mathcal{F} \boxtimes \mathcal{G}) = 0 = R^0\pi_*(\mathcal{F} \boxtimes \mathcal{G}),$$

for $\mathcal{F} \in \mathcal{C}$ and $\mathcal{G} \in \mathcal{T}_1$. Because both are lisse sheaves on $\mathbf{G}_m \otimes k$ whose formation commutes with passage to fibres, it suffices to verify that their fibres at a single geometric point, say at $t = 1$, vanish. These fibres are the H^2_c and the H^0 of $\mathbf{G}_m \otimes \bar{k}$ with coefficients in $\mathcal{F} \otimes (t \mapsto 1/t)^*(\mathcal{G})$. Because $\mathcal{F} \in \mathcal{C}$ and $\mathcal{G} \in \mathcal{T}_1$, $\mathcal{F}$ is totally wild at ∞, while $(t \mapsto 1/t)^*(\mathcal{G})$ is tame at ∞. Therefore $\mathcal{F} \otimes (t \to 1/t)^*(\mathcal{G})$ is totally wild at ∞, whence $H^2_c = H^0 = 0$ (cf. 2.1.1).

To prove (3), i.e., that the natural change-of-coefficient morphisms are isomorphisms, it once again suffices to check on a single geometric fibre, and there the assertion has already been proven (cf. 2.1.1).

To prove (4), we may by (3) reduce to the case $A = \mathbf{F}_\lambda$ (either by passing directly from A to its residue field, or by passing first from E_λ to $\mathcal{O}_\lambda$ and then to $\mathbf{F}_\lambda$); in this case, the assertion is the form of $\mathbf{F}_\lambda$-Poincaré duality already used in proving Proposition 5.2.1.

To prove (5), it again suffices to check at the fibre over, say, $t = 1$. Thus we must show that $H^1_c \xrightarrow{\sim} H^1$ on $\mathbf{G}_m \otimes \bar{k}$ with coefficients in the sheaf $\mathcal{F} \otimes (t \mapsto 1/t)^*(\mathcal{G})$. Because $\mathcal{F} \in \mathcal{C}$ and $\mathcal{G} \in \mathcal{C}_1$, this sheaf is totally wild at both 0 and ∞, and the assertion follows from 2.0.7. ∎

5.2.2.1. Corollary. *For $\mathcal{F} \in \mathcal{C}$ and $\mathcal{G} \in \mathcal{T}_1$, we have canonical isomorphisms for every $i \leq 0$*

$$H^i_c(\mathbf{G}_m \otimes \bar{k}, \mathcal{F}_! * \mathcal{G}) \xrightarrow{\sim} H^1_c(\mathbf{G}_m \otimes \bar{k}, \mathcal{F}) \underset{A}{\otimes} H^i_c(\mathbf{G}_m \otimes \bar{k}, \mathcal{G})$$

$$H^i(\mathbf{G}_m \otimes \bar{k}, \mathcal{F}_* * \mathcal{G}) \xrightarrow{\sim} H^1(\mathbf{G}_m \otimes \bar{k}, \mathcal{F}) \underset{A}{\otimes} H^i(\mathbf{G}_m \otimes \bar{k}, \mathcal{G})$$

which sit in a commutative diagram, whose horizontal arrows are all "oubli de support":

$$\begin{array}{ccc}
H^i_c(\mathbf{G}_m \otimes \bar{k}, \mathcal{F}_! * \mathcal{G}) & \xrightarrow{\sim} & H^1_c(\mathbf{G}_m \otimes \bar{k}, \mathcal{F}) \underset{A}{\otimes} H^i_c(\mathbf{G}_m \otimes \bar{k}, \mathcal{G}) \\
\downarrow & & \\
H^i_c(\mathbf{G}_m \otimes \bar{k}, \mathcal{F}_* * \mathcal{G}) & & \downarrow \\
\downarrow & & \\
H^i(\mathbf{G}_m \otimes \bar{k}, \mathcal{F}_* * \mathcal{G}) & \xrightarrow{\sim} & H^1(\mathbf{G}_m \otimes \bar{k}, \mathcal{F}) \underset{A}{\otimes} H^i(\mathbf{G}_m \otimes \bar{k}, \mathcal{G}).
\end{array}$$

In particular, if $\mathcal{F} \neq 0$ and if $\mathcal{F}_! * \mathcal{G}$ (resp. $\mathcal{F}_* * \mathcal{G}$) lies in $\mathcal{C}_1$, then for $i \neq 1$ we have $H^i_c(\mathbf{G}_m \otimes \bar{k}, \mathcal{G}) = 0$ (resp. $H^i(\mathbf{G}_m \otimes \bar{k}, \mathcal{G}) = 0$).

Proof. We know that $R^j\pi_!(\mathcal{F} \boxtimes \mathcal{G}) = 0$ for $j \neq 1$, so for every $i \leq 0$ we have

$$H^i_c(\mathbf{G}_m \otimes \bar{k}, R^1\pi_!(\mathcal{F} \boxtimes \mathcal{G})) \simeq H^{i+1}_c((\mathbf{G}_m \times \mathbf{G}_m) \otimes \bar{k}, \mathcal{F} \boxtimes \mathcal{G}).$$

Because $\mathcal{F} \in \mathcal{C}$, we know that $H^i_c(\mathbf{G}_m \otimes \bar{k}, \mathcal{F})$ is zero for $i \neq 1$ and a free A-module of rank $\mathrm{Swan}_\infty(\mathcal{F})$ for $i = 1$. Therefore the Kunneth formula gives, for all $i \geq 0$:

$$H^{i+1}_c(\mathbf{G}_m \times \mathbf{G}_m) \otimes \bar{k}, \mathcal{F} \boxtimes \mathcal{G}) \xleftarrow{\sim} H^1_c(\mathbf{G}_m \otimes \bar{k}, \mathcal{F}) \underset{A}{\otimes} H^i_c(\mathbf{G}_m \otimes \bar{k}, \mathcal{G}),$$

whence the asserted isomorphism for $H^i_c(\mathbf{G}_m \otimes \bar{k}, \mathcal{F}_! * \mathcal{G})$. The second isomorphism is obtained similarly, replacing H^i_c by H^i and $R\pi_!$ by $R\pi_*$.

If $\mathcal{F} \neq 0$ lies in $\mathcal{C}$, then $\mathrm{Swan}_\infty(\mathcal{F}) \neq 0$, so $H^1_c(\mathbf{G}_m \otimes \bar{k}, \mathcal{F})$ is a faithfully flat A-module, being A-free of rank $= \mathrm{Swan}_\infty(\mathcal{F})$. If $\mathcal{F}_! * \mathcal{G}$ lies in $\mathcal{C}_1$, then $H^i_c(\mathbf{G}_m \otimes \bar{k}, \mathcal{F}_! * \mathcal{G}) = 0$ for $i \neq 1$, so the first of the above isomorphisms forces $H^i_c(\mathbf{G}_m \otimes \bar{k}, \mathcal{G}) = 0$ for $i \neq 1$. Similarly for $\mathcal{F}_* * \mathcal{G}$ and $H^i(\mathbf{G}_m \otimes \bar{k}, \mathcal{G})$. ∎

5.2.3. Proposition. *For $\mathcal{F} \in \mathcal{C}$ and $\mathcal{G} \in \mathcal{C}_1$, $\mathcal{F}_! * \mathcal{G}$ lies in $\mathcal{C}_1$.*

Proof. We must show that $\mathcal{F}_! * \mathcal{G}$ is totally wild at ∞. We may reduce first to the case when A is finite, then, by filtering $\mathcal{G}$, to the case when $A = \mathbf{F}_\lambda$.

For any finite-dimensional continuous $\mathbf{F}_\lambda$-representation M of I_∞, we have a canonical direct sum decomposition

$$M = M^{P_\infty} \oplus \Big(\bigoplus_{\text{breaks } x>0} M(x) \Big) = M^{\text{tame}} \oplus M^{\text{wild}},$$

where M^{tame} is a representation of

$$I_\infty^{\text{tame}} \simeq \prod_{l\neq p} \mathbf{Z}_l(1) = \varprojlim_{p\nmid N} \boldsymbol{\mu}_N(\bar{k}) = \varprojlim_{\text{Norm}} \mathbf{G}_m(\mathbf{F}_{q^{n!}}),$$

and where $(M^{\text{wild}})^{P_\infty} = 0$. Therefore over a finite extension of $\mathbf{F}_\lambda$, M^{tame} is a successive extension, as representation of I_∞, of finitely many (inverse images on $\mathbf{G}_m \otimes k$ of) $\mathcal{L}_\chi$'s, for χ's characters of $(\mathbf{F}_{q^{n!}})^\times$, $n \gg 0$, with values in a finite extension of $\mathbf{F}_\lambda$. Therefore, we may conclude that

(5.2.3.0) if $M^{\text{tame}} \neq 0$ there exists $\mathcal{L}_\chi$ such that $(M \otimes \mathcal{L}_{\overline{\chi}})^{I_\infty} \neq 0$. ∎

5.2.3.1. Lemma. *For any finite subfield $\mathbf{F}_q$ of k, any multiplicative character $\chi : \mathbf{F}_q^\times \to A^\times$, and any $\mathcal{F} \in \mathcal{T}$, $\mathcal{G} \in \mathcal{T}_1$, we have canonical isomorphisms, for all i,*

$$\big(R^i\pi_!(\mathcal{F} \boxtimes \mathcal{G})\big) \otimes \mathcal{L}_\chi \xrightarrow{\sim} R^i\pi_!\big((\mathcal{F} \otimes \mathcal{L}_\chi) \boxtimes (\mathcal{G} \otimes \mathcal{L}_\chi)\big)$$
$$\big(R^i\pi_*(\mathcal{F} \boxtimes \mathcal{G})\big) \otimes \mathcal{L}_\chi \xrightarrow{\sim} R^i\pi_*\big((\mathcal{F} \times \mathcal{L}_\chi) \boxtimes (\mathcal{G} \times \mathcal{L}_\chi)\big).$$

Proof. For any situation $f : X \to S$, $\mathcal{H}$ a sheaf on X of A-modules, $\mathcal{L}$ on S a lisse sheaf of free A-modules of finite rank, we have canonical isomorphisms ("projection formula")

$$(R^i f_!(\mathcal{H})) \otimes \mathcal{L} \xrightarrow{\sim} R^i f_!(\mathcal{H} \underset{A}{\otimes} f^*(\mathcal{L}))$$
$$(R^i f_*(\mathcal{H})) \otimes \mathcal{L} \xrightarrow{\sim} R^i f_*(\mathcal{H} \otimes f^*(\mathcal{L})).$$

In our situation, $f = \pi$, $\mathcal{H} = \mathcal{F} \boxtimes \mathcal{G}$, $\mathcal{L} = \mathcal{L}_\chi$, and by (4.3) we have canonically $\pi^*(\mathcal{L}_\chi) \simeq \mathcal{L}_\chi \boxtimes \mathcal{L}_\chi$. ∎

In view of this lemma, and the above criterion (5.2.3.0) for total wildness at ∞, it suffices to show universally that if $\mathcal{F}, \mathcal{G}$ are $\mathbf{F}_\lambda$-sheaves in $\mathcal{C}$, then for all $\chi : I_\infty^{\text{tame}} \to \mathbf{F}_\lambda^\times$, we have $\big((\mathcal{F} \otimes \mathcal{L}_{\overline{\chi}})_! * (\mathcal{G} \otimes \mathcal{L}_{\overline{\chi}})\big)^{I_\infty} = 0$. But $\mathcal{F} \otimes \mathcal{L}_{\overline{\chi}}$ and $\mathcal{G} \otimes \mathcal{L}_{\overline{\chi}}$ are again in $\mathcal{C}$, so we are reduced to showing that for $\mathcal{F}, \mathcal{G}$ two $\mathbf{F}_\lambda$-sheaves in $\mathcal{C}$, we have $(\mathcal{F}_! * \mathcal{G})^{I_\infty} = 0$.

We do this by a trick of Deligne's (cf. [De-3], 7.10.4). Consider the inclusion $j : \mathbf{G}_m \otimes k \hookrightarrow \mathbf{A}^1 \otimes k$, and the inclusion $k : \mathbf{A}^1 \otimes k \to \mathbf{P}^1 \otimes k$, all sitting in the diagram

$$\begin{array}{ccccc} \mathcal{F}\boxtimes\mathcal{G} : \mathbf{G}_m \times \mathbf{G}_m & \xrightarrow{j\times j} & \mathbf{A}^1 \times \mathbf{A}^1 : (j_!\mathcal{F})\boxtimes(j_!\mathcal{G}) & & \\ {\scriptstyle\pi=xy}\downarrow & \text{cartesian} & \downarrow{\scriptstyle\tilde{\pi}=xy} & & \\ \mathbf{G}_m & \xrightarrow{j} & \mathbf{A}^1 & \xrightarrow{k} & \mathbf{P}^1. \end{array}$$

5.2.3.2. Lemma (cf. Deligne [De-3]). *For $\mathcal{F} \in \mathcal{C}$ and $\mathcal{G} \in \mathcal{C}_1$, the natural morphism in $D^b_c(\mathbf{P}^1 \otimes k, A)$*

$$k_! R\tilde{\pi}_!((j_!\mathcal{F}(\boxtimes(j_!\mathcal{G})) \to Rk_* R\tilde{\pi}_*((j_!\mathcal{F})\boxtimes(j_!\mathcal{F}))$$

is an isomorphism.

Proof. Over $\mathbf{G}_m \otimes k$, this reduces to $R\pi_!(\mathcal{F}\boxtimes\mathcal{G}) \xrightarrow{\sim} R\pi_*(\mathcal{F}\boxtimes\mathcal{G})$. Therefore the mapping cylinder has cohomology sheaves with punctual support, at 0 and ∞, and these are detected by taking cohomology on $\mathbf{P}^1 \otimes \bar{k}$. So we are reduced to checking that for every i, the induced map of cohomology groups are isomorphisms

$$\mathbf{H}^i(\mathbf{P}^1 \otimes \bar{k}, k_! R\tilde{\pi}_!((j_!\mathcal{F})\boxtimes(j_!\mathcal{G}))) \overset{?}{\xrightarrow{\sim}} \mathbf{H}(\mathbf{P}^1 \otimes \bar{k}, Rk_* R\tilde{\pi}_*((j_!\mathcal{F})\boxtimes(j_!\mathcal{G}))),$$

that is,

$$\mathbf{H}^i_c(\mathbf{A}^1 \otimes \bar{k}, R\tilde{\pi}_!((j_!\mathcal{F})\boxtimes(j_!\mathcal{G}))) \overset{?}{\xrightarrow{\sim}} \mathbf{H}(\mathbf{A}^1 \otimes \bar{k}, R\pi_*((j_!\mathcal{F})\boxtimes(j_!\mathcal{G}))),$$

or again

$$\mathbf{H}^i_c(\mathbf{A}^2 \otimes \bar{k}, (j_!\mathcal{F})\boxtimes(j_!\mathcal{G})) \overset{?}{\xrightarrow{\sim}} H^i(\mathbf{A}^2 \otimes \bar{k}, (j_!\mathcal{F})\boxtimes(j_!\mathcal{G})).$$

Now to check that this map is an isomorphism, we wish to invoke the Kunneth formula for both H^i_c and for H^i (cf. [De-4], 1.11). For $\mathcal{F}$ lisse on $\mathbf{G}_m$, the functors $\mathcal{F} \mapsto H^i(\mathbf{A}^1 \otimes \bar{k}, j_!\mathcal{F})$ vanish for $i \neq 1$, and the H^1 is thus exact, so transforms sheaves of flat A-modules into flat A-modules. For $\mathcal{F}$ totally wild at ∞, we have

$$H^i_c(\mathbf{A}^1 \otimes \bar{k}, j_!\mathcal{F}) \xrightarrow{\sim} H^i(A^1 \otimes \bar{k}, j_!\mathcal{F}).$$

Combining all this, we see that for $\mathcal{F} \in \mathcal{C}$ and $\mathcal{G} \in \mathcal{C}_1$, we have

$$H^i_c(\mathbf{A}^1 \otimes \bar{k}, j_!\mathcal{F}) \xrightarrow{\sim} H^i(\mathbf{A} \otimes \bar{k}, j_!\mathcal{F}) = \begin{cases} \text{free } A\text{-module for } i = 1 \\ 0 \text{ for } i \neq 1 \end{cases}$$

$$H^i_c(\mathbf{A}^1 \otimes \bar{k}, j_!\mathcal{G}) \xrightarrow{\sim} H^i(\mathbf{A} \otimes \bar{k}, j_!\mathcal{G}) = 0 \text{ for } i \neq 1.$$

By the Kunneth formulas, we find that for the map in question, both source and target vanish for $i \neq 2$, and for $i = 2$ the map is the tensor product of the two isomorphisms above. ∎

5.2.4. We can now complete the proof of the Convolution Theorem 5.1. Indeed the assertions (1), (2), (3) of the theorem are just the concatenation of Propositions 5.2.2 and 5.2.3. To prove assertion (4), we may check at a single geometric fibre, say over $t = 1$, where we are calculating the rank of

$$H^1_c(\mathbf{G}_m \otimes \bar{k}, \mathcal{F} \otimes (t \to 1/t)^* \mathcal{G}).$$

Because $\mathcal{F}, \mathcal{G} \in \mathcal{C}$, the sheaf in question is totally wild at both 0 and ∞, so there is no H^0_c or H^2_c, whence the rank of H^1_c is $-\chi_c$, itself given by the Euler–Poincaré formula

$$\begin{aligned} -\chi_c &= \operatorname{Swan}_0(\mathcal{F} \otimes (t \to 1/t)^* \mathcal{G}) + \operatorname{Swan}_\infty(\mathcal{F} \otimes (t \to 1/t)^* \mathcal{G}) \\ &= \operatorname{rank}(\mathcal{F}) \operatorname{Swan}_\infty(\mathcal{G}) + \operatorname{rank}(\mathcal{G}) \operatorname{Swan}_\infty(\mathcal{F}), \end{aligned}$$

the last equalities because $\mathcal{F}$ and $\mathcal{G}$ are both A-free and tame at 0. Because $\mathcal{F} * \mathcal{G}$ lies in $\mathcal{C}$, we have, again by the Euler–Poincaré formula,

$$\operatorname{Swan}_\infty(\mathcal{F} * \mathcal{G}) = \operatorname{rank}\ H^1_c(\mathbf{G}_m \otimes \bar{k}, \mathcal{F} * \mathcal{G}) = \operatorname{rank}\ H^1(\mathbf{G}_m \otimes \bar{k}, \mathcal{F} * \mathcal{G}),$$

and $H^i_c = H^i = 0$ for $i \neq 1$. In particular (5) follows from (6) and (6) has already been established (cf. 5.2.2.1).

Assertion (7) follows from Deligne's main result ([De-5]), which shows that if $\mathcal{F}$ and $\mathcal{G}$ are pure of weights $w(\mathcal{F})$ and $w(\mathcal{G})$ respectively, then $R^1\pi_!(\mathcal{F} \boxtimes \mathcal{G})$ is mixed of weight $\leq 1 + w(\mathcal{F}) + w(\mathcal{G})$. Applying this to $\mathcal{F}^\vee$ and $\mathcal{G}^\vee$, we obtain that $R^1\pi_!(\mathcal{F}^\vee \boxtimes \mathcal{G}^\vee)$ is mixed of weight $\leq 1 - w(\mathcal{F}) - w(\mathcal{G})$. By (3), we know that $R^1\pi_!(\mathcal{F} \boxtimes \mathcal{G})$ and $R^1\pi_!(\mathcal{F}^\vee \boxtimes \mathcal{G}^\vee)$ are lisse and dual with values in $E_\lambda(-1)$, which is pure of weight two, whence $\mathcal{F} * \mathcal{G}$ is necessarily pure of weight $1 + w(\mathcal{F}) + w(\mathcal{G})$.

Assertion (8) is the Lefschetz trace formula, because the $R^1\pi_!(\mathcal{F} \boxtimes \mathcal{G})$ and the $H^1_c(\mathbf{G}_m \otimes \bar{k}, \mathcal{F} \otimes \mathcal{L}_\chi)$ are the only non-vanishing cohomologies.

Assertion (9) has already been verified (cf. 5.2.3.1).

Assertion (10) results from the commutative diagram

$$\begin{array}{ccc} \mathbf{G}_m \times \mathbf{G}_m & \xleftarrow{[N] \times [N]} & \mathbf{G}_m \times \mathbf{G}_m \\ \pi \downarrow & & \downarrow \pi \\ \mathbf{G}_m & \xleftarrow{[N]} & \mathbf{G}_m, \end{array}$$

the fact that $([N] \times [N])_*(\mathcal{F} \boxtimes \mathcal{G}) \simeq ([N]_*\mathcal{F}) \boxtimes ([N]_*\mathcal{G})$, and the fact that and $[N]_*$ maps $\mathcal{C}$ to $\mathcal{C}$ for any $N \geq 1$.

Similarly, (11) results from this commutative diagram, which is also cartesian:

$$\begin{array}{ccc} \mathbf{G}_m \times \mathbf{G}_m & \xleftarrow[\sim]{\mathrm{Trans}_a \times \mathrm{Trans}_b} & \mathbf{G}_m \times \mathbf{G}_m \\ \pi \downarrow & & \downarrow \pi \\ \mathbf{G}_m & \xleftarrow[\mathrm{Trans}_{ab}]{\sim} & \mathbf{G}_m . \end{array}$$

For (12), the diagram figuring in the proof of lemma 5.2.1.1,

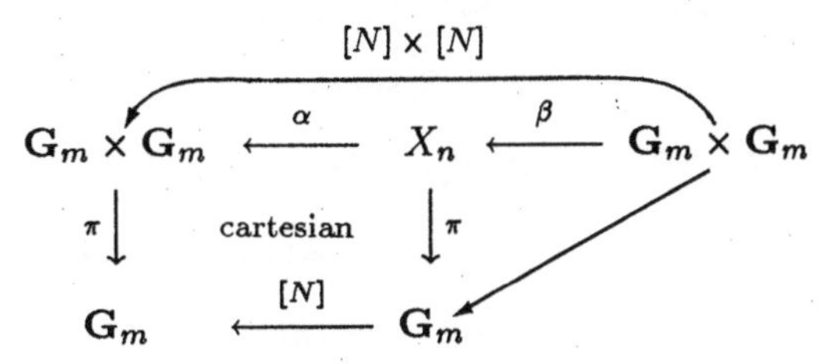

gives us a natural morphism $[N]^*(\mathcal{F} * \mathcal{G}) \to [N]^*(\mathcal{F}) * [N]^*(\mathcal{G})$. Factor N into a power of p and an integer prime to p. For $N = p^v$, all the horizontal arrows in the diagram are finite subjective and radical, so in this case the above morphism is an isomorphism. For N prime to p, Lemma 5.2.1.1 shows that this morphism is injective. If A is not a field, then passing from A to its residue field $\mathbf{F}_\lambda$ and reapplying the same lemma, we see that the map in question remains injective $\otimes \mathbf{F}_\lambda$, whence its cokernel is A-flat. ∎

5.3. Convolution and Duality: Signs. Let $\mathcal{F}$ be a lisse sheaf of free A-modules of finite rank on $\mathbf{G}_m \otimes k$, and (a, ϵ, n) a triple consisting of an element $a \in \boldsymbol{\mu}_2(k)$, an element $\epsilon \in \boldsymbol{\mu}_2(A)$ and an integer n. A "semiduality of type (a, ϵ, n)" on $\mathcal{F}$ is a perfect autoduality of lisse sheaves of free A-modules of finite rank on $\mathbf{G}_m \otimes k$

$$\langle , \rangle : \mathcal{F} \times \mathrm{Trans}_a^*(\mathcal{F}) \to A(-n)$$

such that for any algebraically closed overfield Ω of k, any $t \in \Omega^\times$, and any elements $u \in \mathcal{F}_t$, $v \in \mathcal{F}_{at} = (\mathrm{Trans}_a^*(\mathcal{F}))_t$, the values of the pairings of stalks at the points t and at

$$\langle , \rangle_t : \mathcal{F}_t \times \mathcal{F}_{at} \to A(-n)_t = A$$
$$\langle , \rangle_{at} : \mathcal{F}_{at} \times \mathcal{F}_{a^2t} \to A(-n)_{at} = A$$

are related by $\langle v, u \rangle_{at} = \epsilon \langle u, v \rangle_t$ in A.

5.3.1. Lemma. *Suppose that $\mathcal{F}_1$ and $\mathcal{F}_2$ are lisse sheaves of free A-modules of finite rank, given with semi-dualities of type (a_i, ϵ_i, n_i) for $i = 1, 2$. If*

$a_1 = a_2 = a$, then $\mathcal{F}_1 \otimes \mathcal{F}_2$ carries a semi-duality of type $(a, \epsilon_1\epsilon_2, n_1 + n_2)$ defined by the pairing

$$\begin{array}{c}(\mathcal{F}_1 \otimes \mathcal{F}_2) \times \mathrm{Trans}_a^*(\mathcal{F}_1 \otimes \mathcal{F}_2) \\ \downarrow \wr \\ \mathcal{F}_1 \otimes \mathcal{F}_2 \times (\mathrm{Trans}_a)^*(\mathcal{F}_1) \otimes (\mathrm{Trans}_a)^*(\mathcal{F}_2) \\ \downarrow \text{product} \\ \mathcal{F}_1 \otimes \mathcal{F}_2 \otimes (\mathrm{Trans}_a)^*(\mathcal{F}_1) \otimes (\mathrm{Trans}_a)^*(\mathcal{F}_2) \\ \downarrow \wr\ \text{switch}_{2,3} \\ \mathcal{F}_1 \otimes (\mathrm{Trans}_a)^*(\mathcal{F}_1) \otimes \mathcal{F}_2 \otimes (\mathrm{Trans}_a^*(\mathcal{F}_2) \\ \downarrow \text{contraction} \otimes \text{contraction} \\ A(-n_1) \otimes A(-n_2) \\ \downarrow \wr \\ A(-n_1 - n_2).\end{array}$$

Proof. Obvious. ∎

5.3.2. Lemma. *Given $\mathcal{F}$ as above with a semi-duality of type (a, ϵ, n) and an integer $N \geq 1$, the direct image $[N]_*\mathcal{F}$ carries a semi-duality of type (a^N, ϵ, n) which is given on stalks as follows: for Ω an algebraically closed overfield of k, and any $t \in \Omega^\times$, we have canonical decompositions*

$$([N]_*(\mathcal{F}))_t = \bigoplus_{\substack{x^N = t \\ x \in \Omega^\times}} \mathcal{F}_x, \quad \textit{say } u = \sum_{x^N = t} u_x$$

$$([N]_*(\mathcal{F}))_{a^N t} = \bigoplus_{\substack{x^N = t \\ x \in \Omega^\times}} \mathcal{F}_{ax}, \quad \textit{say } v = \sum_{x^N = t} v_{ax}$$

and the pairing is given by

$$< u, v >_{t,[N]_*\mathcal{F}} = \sum_{x^N = t} < u_x, v_{ax} >_{x,\mathcal{F}} .$$

Proof. Factor N into a power of p and an integer prime to p. For $N = p^v$ there is nothing to prove ($[N]$ being finite radicial and subjective), and for N prime to p this is just an explicit spelling out of duality for a finite etale morphism. ∎

5.3.3. Lemma. *Given objects $\mathcal{F}$ and $\mathcal{G}$ in $\mathcal{C}$ with semi-dualities of type $(a(\mathcal{F}),\ \epsilon(\mathcal{F}),\ n(\mathcal{F}))$ and $(a(\mathcal{G}),\ \epsilon(\mathcal{G}),\ n(\mathcal{G}))$ respectively, the convolution $\mathcal{F}*\mathcal{G}$ carries a semi-duality of type*

$$\begin{aligned} a(\mathcal{F}*\mathcal{G}) &= a(\mathcal{F})a(\mathcal{G}),\\ \epsilon(\mathcal{F}*\mathcal{G}) &= -\epsilon(\mathcal{F})\epsilon(\mathcal{G}),\\ n(\mathcal{F}*\mathcal{G}) &= 1+n(\mathcal{F})+n(\mathcal{G}) \end{aligned}$$

defined by

$$\begin{array}{c} (\mathcal{F}*\mathcal{G})\times(\mathrm{Trans}_{a(\mathcal{F})a(\mathcal{G})})^*(\mathcal{F}*\mathcal{G})\\ \downarrow\wr\\ (\mathcal{F}*\mathcal{G})\times((\mathrm{Trans}_{a(\mathcal{F})})^*(\mathcal{F}))*((\mathrm{Trans}_{a(\mathcal{G})})^*(\mathcal{G}))\\ \Vert\\ R^1\pi_!(\mathcal{F}\boxtimes\mathcal{G})\times R^1\pi_!(\mathrm{Trans}^*_{a(\mathcal{F})}(\mathcal{F})\boxtimes\mathrm{Trans}^*_{a(\mathcal{G})}(\mathcal{G}))\\ \downarrow\text{cup product}\\ R^2\pi_!\big((\mathcal{F}\boxtimes\mathcal{G})\otimes\big(\mathrm{Trans}^*_{a(\mathcal{F})}(\mathcal{F})\boxtimes\mathrm{Trans}^*_{a(\mathcal{G})}(\mathcal{G})\big)\big)\\ \downarrow\text{contraction, via } \langle,\rangle_{\mathcal{F}}\boxtimes\langle,\rangle_{\mathcal{G}}\\ R^2\pi_!(A(-n(\mathcal{F}))\boxtimes A(-n(\mathcal{G})))\\ \wr\downarrow\text{trace}\\ A(-1-n(\mathcal{F})-n(\mathcal{G})). \end{array}$$

Proof. That this pairing is a perfect duality is part (3) of the Convolution Theorem 5.1. That its "sign" is $\epsilon(\mathcal{F}*\mathcal{G})=-\epsilon(\mathcal{F})\epsilon(\mathcal{G})$ results from the skew-symmetry of cup-product on an H^1. ∎

5.4. Multiple Convolution. For $n\geq 3$ and $\mathcal{F}_1,\ldots,\mathcal{F}_n$ objects of $\mathcal{C}$, we define $\mathcal{F}_1*\cdots*\mathcal{F}_n$ to be

$$\mathcal{F}_1*(\mathcal{F}_2*(\cdots*(\mathcal{F}_{n-1}*\mathcal{F}_n))\ldots).$$

More symmetrically, let us introduce the n-fold multiplication map $\pi(n)$: $(\mathbf{G}_m)^n\to\mathbf{G}_m$ taking $(x_1,\ldots,x_n)$ to $x_1\ldots x_n$. Factoring it as successive single multiplications

$$x_1(x_2(\ldots(x_{n-1}x_n))\ldots),$$

and using the smooth base change theorem to control the "not yet multiplied" variables, one shows that for $\mathcal{F}_1, \ldots, \mathcal{F}_n$ in $\mathcal{C}$, we have

$$R^i\pi(n)_!(\mathcal{F}_1 \boxtimes \ldots \boxtimes \mathcal{F}_n) = \begin{cases} 0 & \text{for } i \neq n-1 \\ \mathcal{F}_1 * \mathcal{F}_2 * \cdots * \mathcal{F}_n & \text{for } i = n-1, \end{cases}$$

and the "forget supports" map is an isomorphism

$$R\pi(n)_!(\mathcal{F}_1 \boxtimes \ldots \boxtimes \mathcal{F}_n) \xrightarrow{\sim} R\pi(n)_*(\mathcal{F}_1 \boxtimes \ldots \boxtimes \mathcal{F}_n)$$

whose formation commutes with passage to fibres.

An advantage of this second description is that, as multiplication is associative and commutative, the multiple convolution $\mathcal{F}_1 * \cdots * \mathcal{F}_n$ is canonically associative and commutative. Another advantage is that it exhibits the stalks of $\mathcal{F}_1 * \cdots * \mathcal{F}_n$ as the compact cohomology groups of the varieties $x_1 \ldots x_n = t$ with coefficients in $\mathcal{F}_1 \boxtimes \ldots \boxtimes \mathcal{F}_n$, and shows that

$$H^i_c((x_1 \ldots x_n = t) \underset{k}{\otimes} \Omega, \mathcal{F}_1 \boxtimes \ldots \boxtimes \mathcal{F}_n) \xrightarrow{\sim} H^i(-,-),$$

with both sides vanishing for $i \neq n-1$.

5.5. First Applications to Kloosterman Sheaves. It is now a simple matter to construct the Kloosterman sheaves $\mathrm{Kl}(\psi; \chi_1, \ldots, \chi_n; b_1, \ldots, b_n)$ as the multiple convolutions

$$\mathop{*}_{i=1}^{n} \mathrm{Kl}(\psi; \chi_i; b_i).$$

That so defined they in fact satisfy all the asserted properties of the existence theorem is clear from the case $n = 1$, and the Convolution Theorem 5.1. We can also determine the sign of the autoduality in general (cf. 4.2), as well as analyze the semi-duality which we had already discovered arithmetically.

5.5.1. Proposition. *Suppose that the pairs $(b_i, \overline{\chi}_i)$, $i = 1, \ldots, n$, are just a permutation of the pairs (b_i, χ_i), $i = 1, \ldots, n$, and that each of the characters χ_i of $\mathbf{F}_q^\times$ satisfies $\chi_i(-1) = 1$ (this last condition is always satisfied if we replace $\mathbf{F}_q$ by its quadratic extension). Then $\mathrm{Kl}(\psi; \chi_1, \ldots, \chi_n; b_1, \ldots, b_n)$ is semi-dual of type*

$$((-1)^{\Sigma b_i}, (-1)^{n-1+\delta}, n-1),$$

where δ is the number of i for which $\chi_i^2 = \mathbf{1}$ but $\chi_i \neq \mathbf{1}$.

Proof. Let r be the number of i for which $\chi_i = 1$, and $2s = n - \delta - r$ the number of conjugate pairs (b, χ) and $(b, \overline{\chi})$ with $\chi \neq \overline{\chi}$. Renumbering, we may assume that

$$\begin{cases} (b_{s+i}, \chi_{s+i}) = (b_i, \overline{\chi}_i) \text{ with } \chi_i^2 \neq \mathbf{1} & \text{for } 1 \leq i \leq s, \\ \chi_i = \mathbf{1} & \text{for } 2s+1 \leq i \leq 2s+r, \\ \chi_i^2 = \mathbf{1} \text{ and } \chi_i \neq \mathbf{1} & \text{for } 2s+r+1 \leq i \leq n. \end{cases}$$

The expression of a Kloosterman sheaf as a multiple convolution gives

$$\begin{aligned} \mathrm{Kl}(\psi; \chi_1, \ldots, \chi_n; b_1, \ldots, b_n) &= \mathop{*}_{i=1}^{n} \mathrm{Kl}(\psi; \chi_i; b_i) \\ &\simeq \mathop{*}_{i=1}^{s} \left(\mathrm{Kl}(\psi; \chi_i; b_i) * \mathrm{Kl}(\psi; \overline{\chi}_i, b_i)\right) \mathop{*}_{i=2s+1}^{n} \mathrm{Kl}(\psi; \chi_i; b_i). \end{aligned}$$

The known behavior of semi-dualities under convolution (cf. 5.3) reduces us to proving the theorem in the three cases

$$\begin{cases} \mathrm{Kl}(\psi; \chi; b) * \mathrm{Kl}(\psi; \overline{\chi}; b), & \chi \neq \overline{\chi} \\ \mathrm{Kl}(\psi; \chi; b), & \chi = \mathbf{1} \\ \mathrm{Kl}(\psi; \chi; b), & \chi^2 = \mathbf{1} \text{ and } \chi \neq \mathbf{1}. \end{cases}$$

The expression of $\mathrm{Kl}(\psi; \chi; b)$ as $[b]_*(\mathrm{Kl}(\psi; \chi; 1))$, the formula $[b]_*(\mathcal{F} * \mathcal{G}) \simeq (([b]_*(\mathcal{F}))) * (([b]_*(\mathcal{G})))$, and the known behavior of semi-dualities under direct image further reduces us to the cases

$$\begin{cases} \mathrm{Kl}(\psi; \chi, \overline{\chi}; 1, 1), & \chi \neq \overline{\chi} \\ \mathrm{Kl}(\psi; \mathbf{1}; 1) & \\ \mathrm{Kl}(\psi; \chi; 1) & \chi^2 = \mathbf{1} \text{ and } \chi \neq \mathbf{1}. \end{cases}$$

What must be proven is that each of these sheaves carries a semi-duality, whose type is given by the following table:

sheaf $\mathcal{F}$	$a(\mathcal{F})$	$\epsilon(\mathcal{F})$	$n(\mathcal{F})$
$\mathrm{Kl}(\psi; \chi, \overline{\chi}; 1, 1)$, $\chi \neq \overline{\chi}$	1	-1	1
$\mathrm{Kl}(\psi; \mathbf{1}, 1)$	-1	$+1$	0
$\mathrm{Kl}(\psi; \chi; 1)$, $\chi^2 = \mathbf{1}, \chi \neq \mathbf{1}$	-1	-1	0

By passage to the limit, it will be enough to give canonical constructions of such semi-dualities over finite coefficient rings.

We begin with the case $\mathrm{Kl}(\psi; \mathbf{1}; 1) = \mathcal{L}_\psi | \mathbf{G}_m$. On $\mathbf{G}_a$, we have a canonical isomorphism $\mathrm{sum}^*(\mathcal{L}_\psi) = \mathcal{L}_\psi \boxtimes \mathcal{L}_\psi$ on $\mathbf{G}_a \times \mathbf{G}_a$. Restricting to the subscheme $x + y = 0$ of $\mathbf{G}_a \times \mathbf{G}_a$ we find a symmetric identification

$$\mathcal{L}_\psi \otimes (x \mapsto -x)^* \mathcal{L}_\psi \xrightarrow{\sim} \text{the constant sheaf } (\mathcal{L}_\psi)_0 \simeq A,$$

and this, restricted to $\mathbf{G}_m$, is a semi-duality of type $(-1, 1, 0)$, as required.

The next case, which can only occur for p odd, is

$$\mathrm{Kl}(\psi; \chi; 1) = (\mathcal{L}_\psi | \mathbf{G}_m) \otimes \mathcal{L}_\chi, \quad \chi^2 = \mathbf{1}, \ \chi \neq \mathbf{1}.$$

In view of the tensor product behavior of semi-dualities, it suffices to exhibit on $\mathcal{L}_\chi$ a semi-duality of type $(-1, -1, 0)$. The push-out construction of $\mathcal{L}_\chi$ shows that if we denote by $\mathcal{L}_\chi^\times$ the complement of the zero-section in $\mathcal{L}_\chi$, then $\mathcal{L}_\chi^\times$ as scheme is the quotient

$$A^\times \times \mathbf{G}_m / \text{ the subgroup generated by } (-1, -1),$$

with $A^\times$-torseur structure over $\mathbf{G}_m$ given by the morphism to $\mathbf{G}_m$

$$(a, z) \mapsto z^2.$$

Over $t \in \mathbf{G}_m$, the fibre of $\mathcal{L}_\chi^\times$ is (the sheafication of)

$$\{(a \in A^\times, z \in \mathbf{G}_m \text{ with } z^2 = t\} / (a, z) \sim (-a, -z);$$

Similarly, the fibre over $-t$ is

$$\{(a' \in A^\times, z' \in \mathbf{G}_m \text{ with } (z')^2 = -t\} / (a', z') \sim (a, z).$$

The ratio z/z' is a square root of -1. By hypothesis, $\chi(-1) = 1$, i.e., -1 is a square in $\mathbf{F}_q^\times$. If we fix a choice of $\sqrt{-1} \in \mathbf{F}_q^\times$, then for z, z' both in a connected $\mathbf{F}_q$-algebra, there exists a unique choice of ± 1 in this algebra for which $z/z' = \pm\sqrt{-1}$. Let us denote by

$$\left\langle \frac{z/z'}{\sqrt{-1}} \right\rangle$$

the same ± 1, viewed in $A^\times$. Then we define a pairing

$$\mathcal{L}_\chi^\times \times (t \mapsto -t)^*(\mathcal{L}_\chi^\times) \to \text{ the constant sheaf } A$$

by the explicit formula

$$(a, z) \times (a', z') \mapsto aa' \left\langle \frac{z/z'}{\sqrt{-1}} \right\rangle.$$

This pairing is visibly skew-symmetric, i.e., it is a semi-duality of type $(-1, -1, 0)$.

It remains to treat the sheaf $\mathrm{Kl}(\psi; \chi; \overline{\chi}; 1, 1)$, with $\chi \neq \overline{\chi}$. We must show that it is autodual by a skew-symmetric pairing with values in $\mathcal{O}_\lambda(-1)$, i.e., semi-dual of type $(+1, -1, +1)$. We have already seen (cf. 4.1.11) arithmetically that this sheaf is autodual towards $\mathcal{O}_\lambda(-1)$, and that any

I_∞-equivariant pairing toward $E_\lambda(-1)$, is a multiple of this one. So it suffices to construct an alternating autoduality over $\mathbf{F}_{q^2}$ towards $E_\lambda(-1)$. At a geometric point $t \in \mathbf{G}_m(\Omega)$, Ω an algebraically closed overfield of $\mathbf{F}_q$, the stalk of $\mathrm{Kl}(\psi : \chi, \overline{\chi}; 1, 1)$ is

$$H_c^1\big((xy = t) \otimes \Omega, \mathcal{L}_{\psi(x+y)} \otimes \mathcal{L}_{\chi(x)} \otimes \mathcal{L}_{\overline{\chi}(y)}\big).$$

To make this more explicit, let us introduce the $(\mathbf{F}_q, +) \times (\mathbf{F}_q^\times) \times (\mathbf{F}_q^\times)$-torseur Z over $xy = t$ with equation

$$\begin{aligned} z - z^q &= x + y, \\ u^{1-q} &= x, \\ v^{1-q} &= y. \end{aligned}$$

In E_λ-cohomology, we have

$$H_c^1((xy = t) \otimes \Omega, \mathcal{L}_{\psi(x+y)} \otimes \mathcal{L}_{\chi(x)} \otimes \mathcal{L}_{\overline{\chi}(y)}) \simeq H_c^1(Z \otimes \Omega, E_\lambda)^{(\psi, \chi, \overline{\chi})}.$$

The choice of a $(q-1)$-st root ζ of -1, possible always in $\mathbf{F}_{q^2}$, allows us to define an involution $\mathbf{A} \in \mathrm{Aut}(Z)$ by the explicit formula

$$(x, y, z, u, v) \mapsto (-y, -x, -z, \zeta v, \zeta^{-1} u)$$

which maps

$$H_c^1(Z \otimes \Omega, E_\lambda)^{(\psi, \chi, \overline{\chi})} \xrightarrow{\sim} H_c^1(Z \otimes \Omega, E_\lambda)^{(\overline{\psi}, \overline{\chi}, \chi)}.$$

The cup-product pairing

$$\langle , \rangle : H_c^1(Z \otimes \Omega, E_\lambda) \times H_c^1(Z \otimes \Omega, E_\lambda) \to E_\lambda(-1)$$

establishes, for ψ non-trivial and χ_1 and χ_2 arbitrary, a perfect pairing of isotypical components

$$(H_c^1)^{(\psi, \chi_1, \chi_2)} \times (H_c^1)^{(\overline{\psi}, \overline{\chi}_1, \overline{\chi}_2)} \to E_\lambda(-1).$$

Therefore the pairing

$$(,) : (H_c^1)^{(\psi, \chi, \overline{\chi})} \times (H_c^1)^{(\psi, \chi, \overline{\chi})} \to E_\lambda(-1)$$

defined by

$$(\alpha, \beta) = \langle \alpha, \mathbf{A}^*(\beta) \rangle$$

is an autoduality. It remains only to verify that it is alternating. We readily compute

$$\begin{aligned} (\beta, \alpha) &= \langle \beta, \mathbf{A}^*(\alpha) \rangle = \langle \mathbf{A}^*(\beta), \mathbf{A}^*\mathbf{A}^*(\alpha) \rangle \\ &= \langle \mathbf{A}^*(\beta), \alpha \rangle = -\langle \alpha, \mathbf{A}^*(\beta) \rangle = -(\alpha, \beta). \ \blacksquare \end{aligned}$$

5.5.2. Remark. Here as an alternate proof of 5.5.1 for $\mathrm{Kl}(\psi;\chi,\overline{\chi};1,1)$. Because this sheaf is of rank two, 5.5.1 is the assertion that its *determinant* is isomorphic to $E_\lambda(-1)$. We see later (but without any circularity) that this is in fact the case (cf. 7.2, 7.4.1).

5.6. Direct images of Kloosterman Sheaves, via Hasse–Davenport. In this section, we will give two different computations of the direct image of a Kloosterman sheaf under the N-th power map

$$[N] : \mathbf{G}_m \otimes \mathbf{F}_q \to \mathbf{G}_m \otimes \mathbf{F}_q$$
$$x \mapsto x^N$$

for integers $N \geq 1$. The first is entirely elementary.

5.6.1. Proposition. *For any integer $N \geq 1$, we have a canonical isomorphism of lisse $\mathcal{O}_\lambda$-sheaves on $\mathbf{G}_m \otimes \mathbf{F}_q$*

$$[N]_*(\mathrm{Kl}(\psi;\chi_1,\ldots,\chi_n;b_1\cdots{}_n) \simeq \mathrm{Kl}(\psi;\chi_1,\ldots,\chi_n;Nb_1,\ldots,Nb_n).$$

Proof. By construction, we have

$$\mathrm{Kl}(\psi;\chi_1,\ldots,\chi_n;b_1,\ldots,b_n) = \mathop{*}_{i=1}^{n} \mathrm{Kl}(\psi;\chi_i,b_i) = \mathop{*}_{i=1}^{n}([b_i]_*(\mathcal{L}_\psi \otimes \mathcal{L}_{\chi_i})).$$

By property (10) of the convolution theorem (5.1), we have

$$[N]_*(\mathcal{F} * \mathcal{G}) \simeq ([N]_*(\mathcal{F})) * ([N]_*(\mathcal{G})).$$

Applying this to the multiple convolution expression above, we find

$$\begin{aligned}[N]_*(\mathrm{Kl}(\psi;\chi_1,\ldots,\chi_n,b_1,\ldots,b_n)) &= \mathop{*}_{i=1}^{n}[Nb_i]_*(\mathcal{L}_\psi \otimes \mathcal{L}_{\chi_i}) \\ &\simeq \mathrm{Kl}(\psi;\chi_n,\ldots,\chi_i;Nb_1,\ldots,Nb_n). \ \blacksquare\end{aligned}$$

The second computation we shall give is based on one of the Hasse–Davenport identities. Let $\mathbf{F}_q$ be a finite field, N a divisor of $q-1$, ψ a non-trivial additive character of $\mathbf{F}_q$, and ψ_N the additive character

$$\psi_N(x) = \psi(Nx).$$

Let us denote by $\rho_1,\ldots,\rho_N : \mathbf{F}_q^\times \to \boldsymbol{\mu}_N$ the N characters of $\mathbf{F}_q^\times$ of order dividing N. Then for any multiplicative character χ of $\mathbf{F}_q^\times$, the Hasse–Davenport identity ([H-D], 0.9_1) asserts that

$$(-g(\psi_N,\chi^N))\prod_{i=1}^{N}(-g(\psi,\rho_i)) = \prod_{i=1}^{N}(-g(\psi,\chi\rho_i)).$$

We now "translate" this identity into a statement about direct images of Kloosterman sheaves. To formulate the result, it is convenient to adopt the following notation: for any unit ϵ in $\mathcal{O}_\lambda^*$, we denote by $[\epsilon]^{\deg}$ the lisse geometrically constant $\mathcal{O}_\lambda$-sheaf of rank one on $\mathbf{G}_m \otimes \mathbf{F}_q$ which has Frobeniuses acting by $\operatorname{trace}(F_x|([\epsilon]^{\deg})_{\bar{x}}) = \epsilon^{\deg(x)}$, for every closed point x.

5.6.2. Proposition. *If N divides $q-1$, we have, for any integer $n \geq 1$ and any n multiplicative characters $\chi_1, \ldots, \chi_n$ of $\mathbf{F}_q^\times$, an isomorphism of lisse $\mathcal{O}_\lambda$-sheaves on $\mathbf{G}_m \otimes \mathbf{F}_q$*

$$[N]_*(\mathrm{Kl}(\psi_N; \chi_1^N, \ldots, \chi_n^N; 1, \ldots, 1)) \otimes_{\mathcal{O}_\lambda} \left[\prod_{i=1}^{N}(-g(\psi, \rho_i))^n\right]^{\deg}$$
$$\simeq \mathrm{Kl}(\psi;\ \textit{all } \chi_i\rho_j \textit{ for } i = 1, \ldots, n,\ j = 1, \ldots, N;\ 1, \ldots, 1),$$

where $\rho_1, \ldots, \rho_N$ denote the N characters of $\mathbf{F}_q^\times$ of order dividing N.

Proof. Expressing both sides as convolutions, we reduce to the case $n = 1$:

$$[N]_*(\mathcal{L}_{\psi_N} \otimes \mathcal{L}_{\chi^N}) \otimes \left[\prod_{i=1}^{N}(-g(\psi_1, \rho_i))\right]^{\deg} \simeq \mathrm{Kl}(\psi; \chi\rho_1, \ldots, \chi\rho_N; 1, \ldots, 1).$$

Tensoring both sides by $\mathcal{L}_{\overline{\chi}}$ and applying the projection formula, this reduces to the special case $\chi = \mathbf{1}$:

$$[N]_*(\mathcal{L}_{\psi_N}) \otimes \left[\prod_{i=1}^{N}(-g(\psi, \rho_i)\right]^{\deg} \simeq \mathrm{Kl}(\psi; \rho_1, \ldots, \rho_N; 1, \ldots, 1).$$

It suffices, in view of 4.1.2, to show that both sides have the same local traces. By multiplicative Fourier transform, it suffices to show that over any finite overfield k of $\mathbf{F}_q$, for any multiplicative character χ of $k^\times$, both sides when tensored with $\mathcal{L}_\chi$ on $\mathbf{G}_m \otimes k$ have equal traces of Frobenius on $H_c^1(\mathbf{G}_m \otimes k^{\mathrm{sep}}, \text{—})$. Replacing ψ by $\psi \circ \operatorname{trace}_{k/\mathbf{F}_q}$, and the ρ_i by $\rho_i \circ \mathbf{N}_{k/\mathbf{F}_q}$, we are reduced to universally treating the case $k = \mathbf{F}_q$. We must show that

$$\operatorname{trace}\big(F\big|H_c^1(\mathbf{G}_m \otimes \overline{\mathbf{F}}_q, [N]_*(\mathcal{L}_{\psi_N} \otimes \mathcal{L}_{\chi^N})\big) \prod_{i=1}^{N}(-g(\psi, \rho_i))$$
$$= \operatorname{trace}(F|H_c^1(\mathbf{G}_m \otimes \overline{\mathbf{F}}_q, \mathrm{Kl}(\psi; \chi\rho_1, \ldots, \chi\rho_N; 1, \ldots, 1))),$$

and this is precisely the Hasse–Davenport identity. ∎

5.6.3. Remarks.

(1) It is also possible to give a "direct," albeit somewhat convoluted, proof of this last proposition, and so a "new" proof of the Hasse–Davenport identity.

(2) The fact that $\mathrm{Kl}(\psi; \rho_1, \ldots, \rho_n; 1, \ldots, 1)$ as representation of $\pi_1^{\mathrm{geom}} = \pi_1(\mathbf{G}_m \otimes \overline{\mathbf{F}}_q, \bar{\eta})$ is *induced* from a character of finite order, namely $\mathcal{L}_{\psi_N}$, of a normal subgroup of finite index (N), show that the sheaf $\mathrm{Kl}(\psi; \rho_1, \ldots, \rho_N; 1, \ldots, 1)$ has *finite* geometric monodromy. We will see in Chapters 11 and 12 that the situation is quite different for $\mathrm{Kl}(\psi; 1, \ldots, 1; 1, \ldots, 1)$.

CHAPTER 6

Local Convolution

6.0. Let k be an algebraically closed field of characteristic $p > 0$, l a prime number $l \neq p$, and A an l-adic coefficient ring. Following a suggestion of O. Gabber and G. Laumon, we will show that for $\mathcal{F}, \mathcal{G}$ in $\mathcal{C}$, the convolution $\mathcal{F} * \mathcal{G}$ *as* I_∞-representation depends only on $\mathcal{F}$ and $\mathcal{G}$ *as* I_∞-representations. In fact, denoting by

$$\mathcal{C}_{\text{loc}} = \begin{cases} \text{the category of totally wild representations} \\ \text{of } I_\infty \text{ on finite free } A\text{-modules,} \end{cases}$$

we will construct a bi-exact local convolution

$$\begin{aligned} \mathcal{C}_{\text{loc}} \times \mathcal{C}_{\text{loc}} &\longrightarrow \mathcal{C}_{\text{loc}} \\ M \times N &\longmapsto M *_{\text{loc}} N \end{aligned}$$

such that, denoting by

$$\begin{aligned} \mathcal{C} &\longrightarrow \mathcal{C}_{\text{loc}} \\ \mathcal{F} &\longmapsto \mathcal{F}_{\text{loc}} \end{aligned}$$

the functor "restriction to I_∞", we have a commutative diagram

$$\begin{array}{ccc} \mathcal{C} \times \mathcal{C} & \xrightarrow{\ *\ } & \mathcal{C} \\ \big\downarrow \quad \big\downarrow & & \big\downarrow \\ \mathcal{C}_{\text{loc}} \times \mathcal{C}_{\text{loc}} & \xrightarrow{\ *_{\text{loc}}\ } & \mathcal{C}_{\text{loc}} \end{array}$$

i.e., for $\mathcal{F}, \mathcal{G}$ in $\mathcal{C}$ we have a canonical isomorphism in $\mathcal{C}_{\text{loc}}$

$$(\mathcal{F} * \mathcal{G})_{\text{loc}} \xrightarrow{\sim} (\mathcal{F}_{\text{loc}}) *_{\text{loc}} (\mathcal{G}_{\text{loc}}).$$

6.1. Remark. According to [Ka-2], the functor "restriction to I_∞"

$$\mathcal{C} \to \mathcal{C}_{\text{loc}}$$

is essentially surjective. Therefore if a "local convolution" is to exist as a bi-additive functor sitting in a commutative diagram as above, our knowledge of global convolution forces the following numerical formulas: for $M, N \in \mathcal{C}_{\text{loc}}$, we have

$$\begin{aligned}
\text{Swan}_\infty(M *_{\text{loc}} N) &= \text{Swan}_\infty(M)\,\text{Swan}_\infty(N) \\
\text{rank}(M *_{\text{loc}} N) &= \text{rank}(M)\,\text{Swan}_\infty(N) + \text{rank}(N)\,\text{Swan}_\infty(M).
\end{aligned}$$

6.2. For convenience, we will perform a multiplicative inversion ($t \mapsto 1/t$) in order to center our calculations at zero rather than at ∞. Because inversion is a group homomorphism, we have a commutative diagram

$$\begin{array}{ccc}
\mathbf{G}_m \times \mathbf{G}_m & \xleftarrow[\sim]{\text{inv} \times \text{inv}} & \mathbf{G}_m \times \mathbf{G}_m \\
\downarrow{\scriptstyle \pi = xy} & & \downarrow{\scriptstyle \pi = xy} \\
\mathbf{G}_m & \xleftarrow[\sim]{\text{inv}} & \mathbf{G}_m .
\end{array}$$

Therefore, for any two sheaves $\mathcal{F}, \mathcal{G}$ on $\mathbf{G}_m \otimes k$, we have

$$\begin{aligned}
\text{inv}_!(R\pi_!(\mathcal{F} \boxtimes \mathcal{G})) &= R\pi_!(\text{inv}_!\,\mathcal{F} \boxtimes \text{inv}_!\,\mathcal{F}) \\
\text{inv}_*(R\pi_*(\mathcal{F} \boxtimes \mathcal{G})) &= R\pi_*(\text{inv}_*\,\mathcal{F} \boxtimes \text{inv}_*\,\mathcal{G}).
\end{aligned}$$

Because inv is an involution, we have $\text{inv}_* = inv^*$; because it is proper we have $\text{inv}_! = inv_*$.

In particular, we have, for $\mathcal{F}, \mathcal{G}$ in $\mathcal{C}$,

$$\begin{aligned}
inv_!(\mathcal{F} * \mathcal{G}) &= \text{inv}_!(R^1\pi_!(\mathcal{F} \boxtimes \mathcal{G})) \\
&= R^1\pi_!(\text{inv}_!\,\mathcal{F} \boxtimes \text{inv}_!\,\mathcal{G}).
\end{aligned}$$

Let us therefore denote temporarily $\mathcal{MC}$ ($\mathcal{M}$ for "mirror") the category of lisse sheaves of free finitely generated A-modules on $\mathbf{G}_m \otimes k$ which are tame at ∞ and totally wild at zero, and by $\mathcal{MC}_{\text{loc}}$ the category of totally wild representations of I_0 on free finitely generated A-modules.

6.3. For $\mathcal{F}, \mathcal{G}$ in $\mathcal{MC}$, we denote by $\mathcal{F}, \mathcal{G}$ in $\mathcal{MC}$ their convolution, defined by

$$\mathcal{F} * \mathcal{G} = R^1\pi_!(\mathcal{F} \boxtimes \mathcal{G});$$

it is simply the mirror image of usual convolution by the equivalence of categories

$$\mathcal{C} \quad \underset{\text{inv}_! = \text{inv}_* = \text{inv}^*}{\overset{\sim}{\rightleftarrows}} \quad \mathcal{MC}.$$

6.4. In order to study the multiplication morphism

$$\begin{array}{ccc} \mathbf{G}_m \times \mathbf{G}_m & & (x,y) \\ \downarrow \pi & & \downarrow \\ \mathbf{G}_m & & t = xy \end{array}$$

near $t = 0$, we will employ the method of vanishing cycles. The method is similar to that used in 5.2.1.3 to prove tameness at zero, but here we use a more symmetrical compactification.

Consider the subvariety V of $\mathbf{P}^2 \times \mathbf{A}^1$ defined in the (projective) coordinates (X, Y, Z) of $\mathbf{P}^2$ and the coordinate t of $\mathbf{A}^1$ by the single equation

$$XY = tZ^2.$$

By the second projection, we have a proper map "t"

$$\begin{array}{c} V \\ \downarrow t \\ \mathbf{A}^1. \end{array}$$

Over the open set $V[1/Z]$ of V where Z is invertible, we have

$$\begin{array}{ccc} V[1/Z] & \xrightarrow[\sim]{(X/Z,\,Y/Z)} & \mathbf{A}^1 \times \mathbf{A}^1 \\ \downarrow t & & \downarrow xy \\ \mathbf{A}^1 & = & \mathbf{A}^1. \end{array}$$

The fibre over $t = 0$ of $V \to \mathbf{A}^1$ is the subvariety V_0 of V defined by

$$V_0 : XY = 0 \text{ in } \mathbf{P}^2.$$

Thus V_0 is the union of two lines in $\mathbf{P}^2$ which cross at the point $(0, 0, 1)$. On V_0, there are three points which will require our special attention:

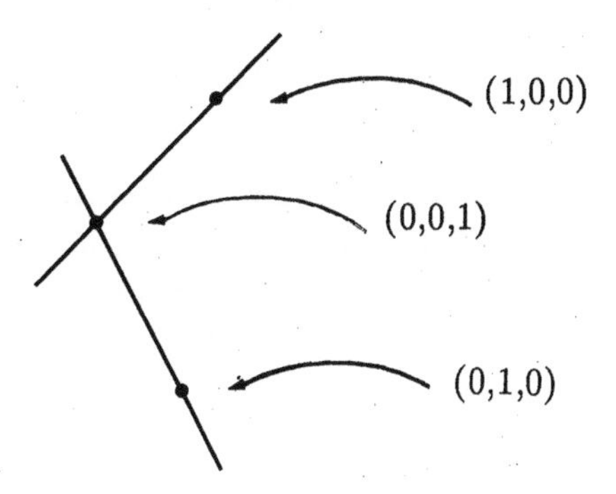

The "picture" of $V \to \mathbf{A}^1$ is

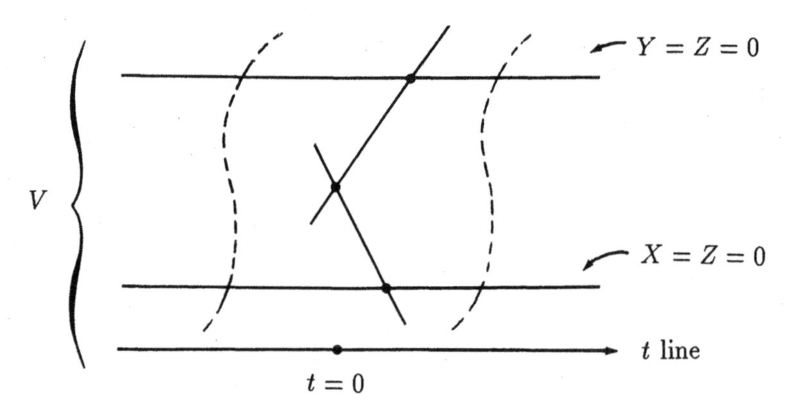

The sheaf on V with which we are dealing is the extension by *zero* of the lisse sheaf on the open region

$$V[1/Z] \cap (t \text{ invertible}) \xrightarrow[\sim]{(X/Z,Y/Z)} \mathbf{G}_m \times \mathbf{G}_m.$$

given by $\mathcal{F} \boxtimes \mathcal{G}$, where $\mathcal{F}, \mathcal{G}$ both lie in $\mathcal{MC}$.

6.5. Theorem. *In the above situation, the vanishing cycle sheaves on V_0 for $V \xrightarrow{t} \mathbf{A}^1$ over $t = 0$ and the sheaf "$\mathcal{F} \boxtimes \mathcal{G}$ extended by zero" are given as follows, for $\mathcal{F}, \mathcal{G}$ in $\mathcal{MC}$:*

$$\begin{cases} R^i\phi = 0 \quad \text{for } i \neq 1 \\ \\ R^1\phi = \begin{array}{l} \text{a punctual sheaf on } V_0, \text{ supported at } (0,0,1), \\ \text{which as } I_0\text{-representation is totally wild.} \end{array} \end{cases}$$

Moreover, we have an isomorphism of I_0-representations

$$\mathcal{F} * \mathcal{G} \xrightarrow{\sim} (R^1\phi)_{(0,0,1)}.$$

We will prove this by a series of lemmas.

6.5.1. Lemma. *For $\mathcal{F}, \mathcal{G}$ lisse on $\mathbf{G}_m$ and totally wild at zero, the $R^i\phi$ all vanish on $V_0 - \{(0,0,1),(1,0,0),(0,1,0)\}$.*

Proof. At a point of V_0 outside the three named exceptions, Z is invertible and one of X or Y is invertible. By the symmetry in X, Y of the situation, we may suppose that ZX is invertible. In the coordinates $x = X/Z$, $y = Y/Z$ on $V[1/Z]$, a Zariski open neighborhood of $V_0 \cap (ZX \text{ invertible})$ in

V is given by $V[1/ZX]$ = the open set $\mathbf{G}_m \times \mathbf{A}^1$ of $\mathbf{A}^1 \times \mathbf{A}^1$ where x is invertible. On this open set, our map is

$$\begin{array}{ccccc} V_0[1/ZX] = \mathbf{G}_m & \hookrightarrow & \mathbf{G}_m \times \mathbf{A}^1 & \xrightarrow[\sim]{(x,y)\mapsto(x,xy)} & \mathbf{G}_m \times \mathbf{A}^1;\ \text{coordinates } (x,t) \\ \downarrow & & \downarrow {\scriptstyle (x,y) \mapsto xy} & & \downarrow t \\ t=0 & \hookrightarrow & \mathbf{A}^1 & = & \mathbf{A}^1 \end{array}$$

and the sheaf on $\mathbf{G}_m \times \mathbf{A}^1$ with coordinates (x,t) is $\mathcal{F}_x \otimes \mathcal{G}_{t/x}$ on $\mathbf{G}_m \times \mathbf{G}_m$, extended by zero. Because $\mathcal{F}$ is *lisse* on $\mathbf{G}_m$, we have

$$\begin{aligned} R\phi(\mathcal{F} \boxtimes \mathcal{G}) \mid V_0[1/ZX] &= R\phi(\mathcal{F}_x \otimes \mathcal{G}_{t/x}) \mid \mathbf{G}_m \text{ of } x\text{'s} \\ &= \mathcal{F} \otimes (R\phi(\mathcal{G}_{t/x}) \mid V_0[1/Z]). \end{aligned}$$

Thus to prove Lemma 6.5.1 we are reduced to showing

6.5.2. Lemma. *For $\mathcal{G} \in \mathcal{MC}$, the morphism*

$$\begin{array}{ccc} \mathbf{G}_m \times \mathbf{A}^1 & & (x,t) \\ \downarrow & & \downarrow \\ \mathbf{A}^1 & & t \end{array}$$

has no vanishing cycles over $t=0$ for $\mathcal{G}_{t/x}$ on $\mathbf{G}_m \times \mathbf{G}_m$ extended by zero.

Proof. To prove 6.5.2, we first need

6.5.3. Lemma. *For any lisse $\mathcal{G}$ on $\mathbf{G}_m$, the vanishing cycles for*

$$\begin{array}{ccc} \mathbf{G}_m \times \mathbf{A}^1 & & (x,t) \\ \downarrow & & \downarrow \\ \mathbf{A}^1 & & t \end{array}$$

and the sheaf $\mathcal{G}_{t/x}$ on $\mathbf{G}_m \times \mathbf{G}_m$ extended by zero satisfy:

each $R^i\phi$ is lisse on $\mathbf{G}_m$, and I_0 acts tamely on it.

Proof of 6.5.3. For any $a \in k^\times = \mathbf{G}_m(k)$, we have a commutative diagram

$$\begin{array}{ccc} \mathbf{G}_m \times \mathbf{A}^1 & \xrightarrow[\sim]{(x,t)\mapsto(ax,at)} & \mathbf{G}_m \times \mathbf{A}^1 \\ \downarrow & & \downarrow \\ \mathbf{A}^1 & \xrightarrow[\sim]{t\mapsto at} & \mathbf{A}^1. \end{array}$$

Therefore each $R^i\phi$ is endowed with an action of $\mathbf{G}_m(k)$ covering the translation action of $\mathbf{G}_m(k)$ on $\mathbf{G}_m \otimes k$. Therefore $R^i\phi$, being in any case a constructible sheaf on $\mathbf{G}_m \otimes k$, is certainly lisse on $\mathbf{G}_m \otimes k$, just because $\mathbf{G}_m(k)$ is infinite (k is algebraically closed). The same diagram shows that the action of I_0 on $(R^i\phi)_a$ is isomorphic to the pull-back by $t \mapsto at$ of the action of I_0 on $(R^i\phi)_1$. Because $R^i\phi$ is a lisse sheaf on $\mathbf{G}_m \otimes k$, the action of I_0 on it gives rise to a representation ρ_a of I_0 on $(R^i\phi)_a$ whose isomorphism class is independent of $a \in \mathbf{G}_m(k)$. Therefore the representation ρ_1 is isomorphic to its pull-back by $t \mapsto at$ for all $a \in k^\times$. By Verdier's result (4.1.6(1)), ρ_1 is tame. ∎

We now conclude the proof of Lemma 6.5.2 (and so of Lemma 6.5.1). Because the $R^i\phi$ are *tame*, they are equal to the "tame vanishing cycles" (cf. SGA 7, I, Intro.). As the sheaf "$\mathcal{G}_{t/x}$ extended by 0" vanishes on the special fibre $t = 0$, we have $R\phi = R\psi$, whence

$$R\phi = \varinjlim_{\substack{N \text{ prime to } p \\ N \geq 1}} i * R(j_N)_*((j_N)^*(\mathcal{G}_{t/x}))$$

where i is the inclusion of the special fibre, and where j_N is the map

$$\begin{aligned} \mathbf{G}_m \times \mathbf{G}_m &\longrightarrow \mathbf{G}_m \times \mathbf{A}^1 \\ (x,t) &\longmapsto (x, t^N). \end{aligned}$$

We will show that we have the more precise vanishing

$$i^* R(j_N)_*(j_N^*(\mathcal{G}_{t/x})) = 0 \qquad \text{for } N \geq 1 \text{ prime to } p.$$

Equivalently, we must show that, denoting by

$$\begin{aligned} j_1 = j : \mathbf{G}_m \times \mathbf{G}_m &\hookrightarrow \mathbf{G}_m \times \mathbf{A}^1 \\ (x,t) &\mapsto (x,t), \end{aligned}$$

we have

$$j_!(\mathcal{G}_{t^N/x}) \xrightarrow{\sim} Rj_*(\mathcal{G}_{t^N/x}).$$

Making the finite etale base change $(x,t) \mapsto (x^N, t)$, it suffices to prove that

$$j_!(\mathcal{G}_{(t/x)^N}) \xrightarrow{\sim} Rj_*(\mathcal{G}_{(t/x)^N}).$$

Making the shearing automorphism $(x,t) \mapsto (x, tx)$, it suffices to prove

$$j_!(\mathcal{G}_{t^N}) \xrightarrow{\sim} Rj_*(\mathcal{G}_{t^N}).$$

By the smooth base change theorem, applied to

$$\begin{array}{ccccc} \mathbf{G}_m \times \mathbf{G}_m & \longrightarrow & \mathbf{G}_m & & t \\ \int j & & \int j_0 & & \downarrow \\ \mathbf{G}_m \times \mathbf{A}^1 & \longrightarrow & \mathbf{A}^1 & & t \\ (x,t) & \longmapsto & t, & & \end{array}$$

it suffices to prove

$$(j_0)_!(\mathcal{G}_{t^N}) \xrightarrow{\sim} R(j_0)_*(\mathcal{G}_{t^N}),$$

and this holds because $\mathcal{G}_{t^N}$ is totally wild at zero by hypothesis. ∎

6.5.4. Lemma. *Hypotheses as in the theorem, the stalks of the $R^i\phi$ at the points (1,0,0) and (0,1,0), i.e., at the points of $V_0 \cap (Z = 0)$, are tame representations of I_0.*

Proof. Again by X, Y symmetry, it suffices to prove this at (1,0,0). In the neighborhood $V[1/X]$ of $(1,0,0) \in V_0$ in the ambient space V, we have functions Y/X, Z/X, t which define $V[1/X]$ in $\mathbf{A}^3$ as the hypersurface

$$Y/X = t(Z/X)^2.$$

Putting $w = Y/X$, $z = Z/X$, the sheaf "$\mathcal{F} \boxtimes \mathcal{G}$ extended by zero" is

$$\mathcal{F}_{X/Z} \otimes \mathcal{G}_{Y/Z} = \mathcal{F}_{1/z} \otimes \mathcal{G}_{tz}$$

on $\mathbf{G}_m \times \mathbf{G}_m$ with coordinates (z,t), extended by zero to $\mathbf{A}^1 \times \mathbf{A}^1$ with coordinates (z,t). We must show that its $R^i\phi$ at $(z = 0, t = 0)$ for the map $(z,t) \mapsto t$ are *tame*.

For this, we may reduce the case of finite coefficients $\mathbf{F}_\lambda$. Because $\mathcal{F} \in \mathcal{MC}$ is tame at ∞, $\mathcal{F}_{1/z}$ is tame at $z = 0$. Passing to a finite extension field of $\mathbf{F}_\lambda$, we may write $\mathcal{F}_{1/z}$ for z near zero as a successive extension of sheaves $\mathcal{L}_{\chi(z)}$, for characters χ of I_0 of finite order prime to p.

Thus we are reduced to proving the tameness of the vanishing cycles at $z = t = 0$ of the sheaf $\mathcal{L}_{\chi(z)} \otimes \mathcal{G}_{tz}$ on $\mathbf{G}_m \times \mathbf{G}_m$, extended by zero to $\mathbf{A}^1 \times \mathbf{A}^1$, for the map $(z,t) \mapsto t$.

Let $N \geq 1$ annihilate χ. Pass to the finite $\mathbf{A}^1 \times \mathbf{A}^1$-scheme

$$\begin{array}{ccc} \mathbf{A}^1 \times \mathbf{A}^1 & & (z,t) \\ f \downarrow & & \downarrow \\ \mathbf{A}^1 \times \mathbf{A}^1 & & (z^N, t). \end{array}$$

We have

$$f_*f^*(\mathcal{G}_{tz}\text{ ext. by }0) = \bigoplus_{\chi^N=1}(\mathcal{L}_{\chi(z)}\otimes\mathcal{G}_{tz}\text{ ext. by }0).$$

By general properties of vanishing cycles, we have for any finite f and any sheaf $\mathcal{H}$ "upstairs,"

$$f_*(R\phi(\mathcal{H}, t\circ f)) = R\phi(f_*\mathcal{H}, t).$$

Applying this with $\mathcal{H} = f^*(\mathcal{G}_{tz}\text{ ext. by }0)$, we get

$$f_*(R\phi(f^*\mathcal{G}_{tz}, t\circ f)) = \bigoplus_{\chi^N=1} R\phi(\mathcal{L}_{\chi(z)}\otimes\mathcal{G}_{tz}, t).$$

So it suffices to show that

$$R\phi(f^*\mathcal{G}_{tz}, t\circ f) = R\phi(\mathcal{G}_{tz^N}, t)$$

is tame as I_0-representation.

For this, we consider the diagram, for each $a\in k^\times$

$$\begin{array}{ccc} \mathbf{A}^1\times\mathbf{A}^1 & \xrightarrow{(z,t)\mapsto(az,t/a^N)} & \mathbf{A}^1\times\mathbf{A}^1 \\ \downarrow t & & \downarrow t \\ \mathbf{A}^1 & \xrightarrow{t\mapsto t/a^N} & \mathbf{A}^1, \end{array}$$

which shows that the representation ρ of I_0 on $R\phi(\mathcal{G}_{tz^N}, t)_{(0,0)}$ is isomorphic to its pull-back by $t\mapsto t/a^N$, for any $a\in k^\times$. Because $k^\times$ is infinite, it follows from Verdier (4.1.6(1)) once again that this representation is tame. ∎

Combining Lemmas 6.5.1 and 6.5.4, we see that the $R\phi$ are punctual sheaves on V_0, supported at the three points (0,0,1), (1,0,0), (0,1,0), and that at the last two of these points, I_0 acts tamely. Because the sheaf in question on V vanishes on V_0, the vanishing cycle exact sequence gives an isomorphism of I_0-representations

$$R^i\pi_!(\mathcal{F}\boxtimes\mathcal{G}) \simeq (R^i\phi)_{(0,0,1)}\bigoplus(R^i\phi)_{(1,0,0)}\bigoplus(R^i\phi)_{(0,1,0)},$$

for $\mathcal{F},\mathcal{G}$ in $\mathcal{MC}$.

But we know that for $\mathcal{F},\mathcal{G}$ in $\mathcal{MC}$, we have

$$R^i\pi_!(\mathcal{F}\boxtimes\mathcal{G}) = \begin{cases} 0 & \text{for } i\neq 1 \\ \text{totally wild as } I_0-\text{rep.} & \text{for } i=1. \end{cases}$$

Therefore we have, by punctuality of the $R\phi$,

$$R^i\phi = 0 \qquad \text{for } i \neq 1.$$

By the tameness of $R^i\phi$ at (1,0,0) and (0,1,0), we conclude that

$R^1\phi$ is punctual, supported at (0,0,1),
and as I_0-representation is totally wild,

and that we have an isomorphism of I_0-representations

$$\mathcal{F} * \mathcal{G} \xrightarrow{\sim} (R^1\phi)_{(0,0,1)}.$$

This concludes the proof of the theorem as stated. ∎

6.6. Construction of Local Convolution. Concretely, for $\mathcal{F}, \mathcal{G}$ in $\mathcal{MC}$, we have an I_0-isomorphism

$$\mathcal{F} * \mathcal{G} \simeq H^1\left((\mathbf{A}^2)^{\text{h.s.}}_{(0,0)} \underset{\xrightarrow[xy]{} (\mathbf{A}^1)^{\text{h.s.}}_{(0,0)}}{\times} \operatorname{Spec}((K_{\{0\}})^{\text{sep}}); \mathcal{F} \boxtimes \mathcal{G} \right),$$

$K_{\{0\}}$ denoting the fraction field of the strict henselization of $\mathbf{A}^1 \otimes k$ at $t = 0$. This last expression visibly depends only on $\mathcal{F}$ and $\mathcal{G}$ themselves *as* I_0-representations.

Given any object N in $\mathcal{MC}_{\text{loc}}$, let us denote by $\tilde{N}$ the constructible sheaf on $(\mathbf{A}^1)^{\text{h.s.}}_{(0)}$ which is "N on $\operatorname{Spec}(K_{\{0\}})$, extended by zero." Then for M, N in $\mathcal{MC}_{\text{loc}}$, we define their local convolution to be

$$M *_{\text{loc}} N \overset{\text{dfn}}{=} H^1\left((\mathbf{A}^2)^{\text{h.s.}}_{(0,0)} \underset{\xrightarrow[xy]{} (\mathbf{A}^1)^{\text{h.s.}}_{(0)}}{\times} \operatorname{Spec}((K_{\{0\}})^{\text{sep}}); \tilde{M} \boxtimes \tilde{N} \right),$$

viewed as I_0-representation. Because $\mathcal{MC} \to \mathcal{MC}_{\text{loc}}$ is essentially surjective by [Ka-2], we see that $M *_{\text{loc}} N$ as defined above is in fact an object of $\mathcal{MC}_{\text{loc}}$, that all the other H^i vanish (this gives the bi-exactness), and that, so defined, we have canonical isomorphisms

$$(\mathcal{F} * \mathcal{G})_{\text{loc}} \xrightarrow{\sim} (\mathcal{F}_{\text{loc}}) *_{\text{loc}} (\mathcal{F}_{\text{loc}})$$

for any $\mathcal{F}, \mathcal{G}$ in $\mathcal{MC}$.

CHAPTER 7

Local Monodromy at Zero of a Convolution: Detailed Study

7.0. General Review

7.0.1. Let C be a smooth geometrically connected curve over $\mathbf{F}_q$, $x \in C(\mathbf{F}_q)$ a rational point, E_λ a finite extension of $\mathbf{Q}_l$, $l \neq p = \text{char}(\mathbf{F}_q)$, and $\mathcal{F}$ a lisse E_λ-sheaf on $C - \{x\}$. In this section we will recall the known structure of $\mathcal{F}$ as a representation of the groups

$$D_x \supset I_x \supset P_x.$$

7.0.2. Shrinking C if necessary, we may assume that C is affine and that there exists a function $t : C \to \mathbf{A}^1$ on C which is invertible on $C - \{x\}$ and which has a simple zero at x. For $\chi : \mathbf{F}_q^\times \to E_\lambda^\times$ a multiplicative character, we have the lisse rank-one E_λ-sheaf $\mathcal{L}_\chi$ on $\mathbf{G}_m$ over $\mathbf{F}_q$. Its inverse image $t^*(\mathcal{L}_\chi)$ on $C - \{x\}$ will be denoted simply $\mathcal{L}_\chi$ when no confusion is possible (e.g. in an expression like $\mathcal{F} \otimes \mathcal{L}_\chi$). These $\mathcal{L}_\chi$'s, when restricted (as representations) to I_x, are precisely the E_λ-valued characters of I_x of order dividing $q-1$. In particular, every tame character of I_x of finite order with values in any extension of E_λ is given by an $\mathcal{L}_\chi$, if we allow ourselves to replace both $\mathbf{F}_q$ and E_λ by arbitrary finite extensions of themselves in the preceding constructions.

7.0.3. For $\mathcal{F}$ a lisse E_λ-sheaf on $C - \{x\}$, $\mathcal{F}$ as P_x-representation has a break-decomposition,

$$\mathcal{F} = \mathcal{F}^{P_x} \oplus \bigoplus_{y>0} \mathcal{F}(\text{ break } y),$$

each summand of which is D_x-stable (cf. 1.8). We define

$$\begin{cases} \mathcal{F}^{\text{tame}} = \mathcal{F}^{P_x} \\ \\ \mathcal{F}^{\text{wild}} = \bigoplus_{y>0} \mathcal{F}(\text{ break } y), \end{cases}$$

whence we obtain a decomposition

$$\mathcal{F} = \mathcal{F}^{\text{tame}} \bigoplus \mathcal{F}^{\text{wild}}, \qquad \text{as } D_x\text{-module.}$$

For any $\mathcal{L}_\chi$ (indeed for any lisse E_λ-sheaf on $C - \{x\}$ which is tame at x), we have

$$(\mathcal{F} \otimes \mathcal{L}_\chi)^{\text{tame}} = \mathcal{F}^{\text{tame}} \otimes \mathcal{L}_\chi$$
$$(\mathcal{F} \otimes \mathcal{L}_\chi)^{\text{wild}} = \mathcal{F}^{\text{wild}} \otimes \mathcal{L}_\chi.$$

7.0.4. The tame part $\mathcal{F}^{\text{tame}}$ of $\mathcal{F}$ may be further decomposed according to the Jordan decomposition of the action of a topological generator γ^{tame} of I_x^{tame} acting on $\mathcal{F}^{\text{tame}}$

$$\mathcal{F}^{\text{tame}} = \bigoplus_{\substack{\text{chars } \chi \text{ of } I_x \\ \text{of finite order} \\ \text{prime to } p}} \mathcal{F}^{\chi-\text{unipotent}}, \quad \text{an } I_x\text{-decomposition}$$

after extending scalars from E_λ to a finite extension containing all the eigenvalues of $\gamma^{\text{tame}}|\mathcal{F}^{\text{tame}}$. Intrinsically, $\mathcal{F}^{\chi\text{-unipotent}}$ is the largest I_x-subrepresentation of $\mathcal{F}$ of the form

$$\mathcal{F}^{\chi\text{-unipotent}} = (\text{a unipotent representation of } I_x) \otimes \mathcal{L}_\chi.$$

Thus we may rewrite the above decomposition as

$$\mathcal{F}^{\text{tame}} = \bigoplus_\chi (\mathcal{F} \otimes \mathcal{L}_{\overline{\chi}})^{I_x-\text{unip}} \otimes \mathcal{L}_\chi, \quad \text{an } I_x\text{-decomposition}$$

where $(\mathcal{F} \otimes \mathcal{L}_{\overline{\chi}})^{I_x-\text{unip}}$ is the largest unipotent I_x-subrepresentation of $\mathcal{F} \otimes \mathcal{L}_{\overline{\chi}}$. In general the χ occurring above will not all have order dividing $q-1$, so the above decomposition is not in general D_x-stable; D_x/I_x may permute the χ's which occur. Over a finite extension $\mathbf{F}_{q^d}$ of $\mathbf{F}_q$ (d chosen so that all χ occurring have order dividing $q^d - 1$), this decomposition becomes D_x-stable.

By the local monodromy theorem, $\mathcal{F}$ is quasi-unipotent as a representation of I_x. Denoting by ρ the corresponding representation, we have

7.0.5. Lemma. *The subset of I_x defined by*

$$\{\gamma \in I_x \textit{ such that } \rho(\gamma) \textit{ is unipotent}\}$$

is an open subgroup of I_x, normal in D_x, and if γ_1 and γ_2 are any two elements of this subgroup, then $\log(\rho(\gamma_1))$ *and* $\log(\rho(\gamma_2))$ *are* $\mathbf{Q}_l$*-proportional.*

Proof. Let us denote by

$$t_l : I_x \twoheadrightarrow I_x^{\text{tame}} \simeq \prod_{l \neq p} \mathbf{Z}_l(1) \twoheadrightarrow \mathbf{Z}_l(1)$$

the canonical projection of $I_x = I$ onto $\mathbf{Z}_l(1)$. By the local monodromy theorem, there exists a nilpotent endomorphism

$$N \in \operatorname{End}(\mathcal{F}) \bigotimes_{\mathbf{Z}_l} \mathbf{Z}_l(-1)$$

and an open subgroup $I' \subset I$ such that

$$\rho(\gamma) = \exp(t_l(\gamma) \cdot N) \quad \text{for } \gamma \in I'.$$

But if $\gamma_1 \in I$ has $\rho(\gamma_1)$ unipotent, then for any integer $n \geq 1$, $\rho((\gamma_1)^n) = \rho(\gamma_1)^n$ is unipotent, and

$$\rho(\gamma_1) = \exp\left(\frac{1}{n} \log(\rho((\gamma_1)^n))\right).$$

But for suitable $n \geq 1$, $(\gamma_1)^n$ lies in I', whence

$$\log(\rho((\gamma_1)^n)) = t_l((\gamma_1)^n) \cdot N = n \cdot t_l(\gamma_1) N,$$

so that

$$\rho(\gamma_1) = \exp(t_l(\gamma_1) \cdot N)$$

for all $\gamma_1 \in I$ for which $\rho(\gamma_1)$ is unipotent. Thus for $\gamma \in I$,

$$\rho(\gamma) \text{ is unipotent} \iff \rho(\gamma) = \exp(t_l(\gamma) \cdot N).$$

The set of $\gamma \in I$ where $\rho(\gamma)$ is unipotent is clearly stable by D_x-conjugation, while the above equivalence shows that the subset of such γ is a *subgroup* of I (open because it contains the open subgroup I' of I) and that the logarithms of all such γ are $\mathbf{Q}_l$-proportional. ∎

7.0.6. The monodromy filtration of $\mathcal{F}$ as D_x-representation is an increasing filtration

$$\cdots W_i \subset W_{i+1} \subset \cdots$$

which is characterized by the following properties (cf. [De-5], 1.7.2.2):

(1) If $\rho(\gamma) = 1$ on an open subgroup of I_x, then

$$W_{-1} = 0 \text{ and } W_0 = \mathcal{F}.$$

(2) If ρ is non-trivial on arbitrarily small open subgroups of I, then for any $\gamma \in I$ with $\rho(\gamma)$ unipotent but $\rho(\gamma) \neq 1$, we have

$$(2)_\gamma \begin{cases} \log \rho(\gamma) \text{ maps } W_i \text{ to } W_{i-2} \\ \\ \forall\, i \geq 0, (\log(\rho(\gamma)))^i \text{ maps } gr_i^W \xrightarrow{\sim} gr_{-i}^W. \end{cases}$$

In case (2), one sees easily that the monodromy filtration is already *uniquely characterized* by condition $(2)_\gamma$ for any single $\gamma \in I$ with $\rho(\gamma)$ unipotent and non-trivial. By 7.0.5, we see that the monodromy filtration is D_x-stable, and that it respects the decompositions

$$\mathcal{F} = \mathcal{F}^{\text{tame}} \bigoplus \mathcal{F}^{\text{wild}}$$

$$\mathcal{F}^{\text{tame}} = \bigoplus_{\chi} (\mathcal{F} \otimes \mathcal{L}_{\overline{\chi}})^{I_x-\text{unip}} \otimes \mathcal{L}_{\chi}.$$

(Indeed the monodromy filtration respects *any* decomposition which is $\rho(\gamma)$-stable for any single γ with $\rho(\gamma)$ unipotent and non-trivial.)

According to one of the basic results of Deligne in Weil II, we have

7.0.7. Theorem. *Suppose $\mathcal{F}$ is pure of some weight w. Then*

(1) *$gr_i^W(\mathcal{F})$ is pure of weight $w + i$.*

(2) *If χ has order dividing $q - 1$, and if $(\mathcal{F} \otimes \mathcal{L}_{\overline{\chi}})^{I_x}$ is of dimension $k \geq 1$, with F_x-eigenvalues $\alpha_1, \ldots, \alpha_k$ of weights $w - i_1, \ldots, w - i_k$, then the weight drops $i_1, \ldots, i_k$ are non-negative integers, we have*

$$(\mathcal{F} \otimes \mathcal{L}_{\overline{\chi}})^{I_x-\text{unip}} = \bigoplus_{v=1,\ldots,k} \left(\begin{array}{l} \text{a Jordan block} \\ \text{of dimension } \; 1 + i_v \end{array} \right),$$

as I_x-module,

and the eigenvalues of any element $F_x \in D_x$ of degree 1 on $(\mathcal{F} \otimes \mathcal{L}_{\overline{\chi}})^{I_x-\text{unip}}$ are

$$\left\{ \underbrace{\alpha_1, q\alpha_1, \ldots, q^{i_1}\alpha_1}_{1+i_1}, \ldots, \underbrace{\alpha_k, q\alpha_k, \ldots, q^{i_k}\alpha_k}_{1+i_k} \right\}.$$

Proof. This is just ([De-5], 1.8.4 and 1.67.14.2-3) spelled out. ∎

For applications, we will use these results in the following form.

7.0.8. Corollary. *Suppose that $\mathcal{F}$ is pure of weight w, and lisse, on $C-\{x\}$. Fix a multiplicative character $\chi : \mathbf{F}_q^\times \to E_\lambda^\times$, with $(\mathcal{F}\otimes\mathcal{L}_{\overline{\chi}})^{I_x} \neq 0$. For each eigenvalue α of F_x on $(\mathcal{F}\otimes\mathcal{L}_{\overline{\chi}})^{I_x}$, denote by $w(\alpha)$ its weight. Then*

(1)
$$\dim(\mathcal{F}\otimes\mathcal{L}_{\overline{\chi}})^{I_x-\mathrm{unip}} = \sum(1+w-w(\alpha)),$$
the sum over the eigenvalues α of F_x on $(\mathcal{F}\otimes\mathcal{L}_{\overline{\chi}})^{I_x}$, and the individual terms $1+w-w(\alpha)$ are the strictly positive dimensions of the Jordan blocks of $(\mathcal{F}\otimes\mathcal{L}_{\overline{\chi}})^{I_x-\mathrm{unip}}$ as I_x-representation.

(2) *For any finite set S of characters χ as above, we have an inequality*
$$\mathrm{rank}(\mathcal{F}) \geq \sum_{\chi\in S} \dim(\mathcal{F}\otimes\mathcal{L}_{\overline{\chi}})^{I_x-\mathrm{unip}},$$
with equality if and only if both of the following conditions are satisfied:

(2a) $\mathcal{F}$ is tame at x, i.e., $\mathcal{F}$ is tame as representation of I_x.

(2b) All the characters of I_x^{tame} occurring in $\mathcal{F}^{\mathrm{tame}} = \mathcal{F}$ are of order dividing $q-1$, E_λ-valued, and lie in S.

(3) *For χ as above, $(\mathcal{F}\otimes\mathcal{L}_{\overline{\chi}})^{I_x-\mathrm{unip}}$ as representation of D_x is unipotent on I_x, so its determinant is an unramified representation of D_x/I_x, given on F_x by the formula*
$$\det(F_x|(\mathcal{F}\otimes\mathcal{L}_{\overline{\chi}})^{I_x-\mathrm{unip}}) = \prod\left(q^{\frac{w-w(\alpha)}{2}}\alpha\right)^{1+w-w(\alpha)},$$
the product over the eigenvalues α of F_x on $(\mathcal{F}\otimes\mathcal{L}_{\overline{\chi}})^{I_x}$.

(4) *If the conditions (2a) and (2b) of (2) above are both fulfilled, then as characters of D_x, we have*
$$\det(\mathcal{F})\Big/\prod_{\chi\in S}(\mathcal{L}_\chi)^{\otimes(\text{rank of }(\mathcal{F}\otimes\mathcal{L}_{\overline{\chi}})^{I_x-\mathrm{unip}})}$$
$$= \prod_{\chi\in S}\det((\mathcal{F}\otimes\mathcal{L}_{\overline{\chi}})^{I_x-\mathrm{unip}}).$$

7.1. Application to a "Product Formula" for a Convolution of Pure Sheaves

7.1.1. We now return to $\mathbf{G}_m$ over $\mathbf{F}_q$, and consider a lisse E_λ-sheaf $\mathcal{F}$ on $\mathbf{G}_m\otimes\mathbf{F}_q$, which is both
$$\begin{cases}\text{pure of weight } w \text{ on } \mathbf{G}_m\otimes\mathbf{F}_q\\ \text{totally wild at } \infty.\end{cases}$$

For such an $\mathcal{F}$, if we denote by

$$j : \mathbf{G}_m \hookrightarrow \mathbf{P}^1$$

the inclusion, we have a short exact sequence of sheaves on $\mathbf{P}^1 \otimes \mathbf{F}_q$

$$0 \to j_!\mathcal{F} \to j_*\mathcal{F} \to \mathcal{F}^{I_0} \to 0$$

where $\mathcal{F}^{I_0}$ is viewed as a punctual sheaf at zero, extended by zero to $\mathbf{P}^1$ (the total wildness of $\mathcal{F}$ guarantees $\mathcal{F}^{I_\infty} = 0$). In the long exact cohomology sequence on $\mathbf{P}^1 \otimes \overline{\mathbf{F}}_q$, we have $H^0(\mathbf{P}^1 \otimes \overline{\mathbf{F}}_q, j_*\mathcal{F}) = 0$ by the total wildness at ∞, whence a short exact sequence

$$\begin{array}{c} 0 \to \mathcal{F}^{I_0} \to H^1(\mathbf{P}^1 \otimes \overline{\mathbf{F}}_q, j_!\mathcal{F}) \to H^1(\mathbf{P}^1 \otimes \overline{\mathbf{F}}_q, j_*\mathcal{F}) \to 0 \\ \| \\ H^1_c(\mathbf{G}_m \otimes \overline{\mathbf{F}}_q, \mathcal{F}). \end{array}$$

By Deligne once again, we know that for $\mathcal{F}$ pure of weight w and lisse on $\mathbf{G}_m \otimes \overline{\mathbf{F}}_q$, we have

$$\begin{cases} \mathcal{F}^{I_0} \text{ is mixed of weight } \leq w \\ H^1_c(\mathbf{G}_m \otimes \overline{\mathbf{F}}_q, \mathcal{F}) \text{ is mixed of weight } \leq 1 + w \\ H^1(\mathbf{P}^1 \otimes \overline{\mathbf{F}}_q, j_*\mathcal{F}) \text{ is pure of weight } 1 + w. \end{cases}$$

Thus we obtain the following

7.1.2. Proposition. *For $\mathcal{F}$ a lisse E_λ-sheaf on $\mathbf{G}_m \otimes \mathbf{F}_q$ which is pure of weight w and totally wild at ∞, we have a canonical isomorphism of $D_0/I_0 \simeq$ gal $(\overline{\mathbf{F}}_q/\mathbf{F}_q)$-modules*

$$\mathcal{F}^{I_0} \xrightarrow{\sim} \textit{the part of } H^1_c(\mathbf{G}_m \otimes \overline{\mathbf{F}}_q, \mathcal{F}) \textit{ of weight } \leq w.$$

7.1.3. For $\mathcal{F}$ as in the above proposition, we define an associated polynomial

$$f(\mathcal{F}, T) \in \mathbf{Z}[T]$$

by defining

$$f(\mathcal{F}, T) = \sum T^{(1+w-w(\alpha))},$$

the sum over the eigenvalues of F on $H^1_c(\mathbf{G}_m \otimes \overline{\mathbf{F}}_q, \mathcal{F})$.

In terms of the Jordan blocks of $\mathcal{F}^{I_0-\text{unip}}$, the number of which is the dimension of $\mathcal{F}^{I_0}$, we are attributing $h^1_c(\mathcal{F}) - \dim(\mathcal{F}^{I_0})$ "extra" Jordan blocks, *each of size zero*, to arrive at a "total" number $h^1_c(\mathcal{F})$ of Jordan blocks.

In terms of "all" the $h_c^1(\mathcal{F})$ Jordan blocks of $\mathcal{F}^{I_0-\text{unip}}$, we have

$$f(\mathcal{F},T) = \sum_{\substack{\text{"all" the } h_c^1(\mathcal{F}) \\ \text{Jordan blocks of} \\ \mathcal{F}^{I_0-\text{unip}}}} T^{(\text{size of Jordan block})}.$$

Thus we have, for $\mathcal{F}$ as above (lisse and pure on $\mathbf{G}_m$, totally wild at ∞),

$$\begin{cases} f(\mathcal{F},1) = h_c^1(\mathcal{F}) \\ f'(\mathcal{F},1) = \dim(\mathcal{F}^{I_0-\text{unip}}). \end{cases}$$

7.1.4. Theorem (Product Formula). *Suppose that $\mathcal{F}$ and $\mathcal{G}$ are lisse E_λ-sheaves on $\mathbf{G}_m \otimes \mathbf{F}_q$, pure of weights $w(\mathcal{F})$ and $w(\mathcal{G})$ respectively, and that both $\mathcal{F}$ and $\mathcal{G}$ lie in $\mathcal{C}$, i.e., they are each tame at zero and totally wild at ∞. Then we have*

$$f(\mathcal{F} * \mathcal{G},T) = f(\mathcal{F},T)f(\mathcal{G},T).$$

Proof. By the convolution theorem, $\mathcal{F} * \mathcal{G}$ is lisse, lies in $\mathcal{C}$, and is pure of weight $1 + w(\mathcal{F}) + w(\mathcal{G})$. Therefore $f(\mathcal{F} * \mathcal{G},T)$ may be defined in terms of the absolute values of the eigenvalues of F on $H_c^1(\mathbf{G}_m \otimes \overline{\mathbf{F}}_q, \mathcal{F} * \mathcal{G})$. The formula to be proven,

$$f(\mathcal{F} * \mathcal{G},T) = f(\mathcal{F},T)f(\mathcal{G},T),$$

results immediately from the canonical isomorphism (cf. 5.1 (6))

$$H_c^1(\mathbf{G}_m \otimes \overline{\mathbf{F}}_q, \mathcal{F} * \mathcal{G}) \simeq H_c^1(\mathbf{G}_m \otimes \overline{\mathbf{F}}_q, \mathcal{F}) \underset{E_\lambda}{\otimes} H_c^1(\mathbf{G}_m \otimes \overline{\mathbf{F}}_q, \mathcal{G}). \quad \blacksquare$$

7.1.5. Corollary. *For $\mathcal{F},\mathcal{G}$ as in the theorem above, suppose that all the $\mathcal{L}_\chi$ occurring in both $\mathcal{F}$ and $\mathcal{G}$ as I_0-representations lie in a given set S of E_λ-valued characters of $\mathbf{F}_q^\times$. Then the same is true of $\mathcal{F} * \mathcal{G}$, and for each $\chi \in S$ we have*

$$f((\mathcal{F} * \mathcal{G}) \otimes \mathcal{L}_{\overline{\chi}},T) = f(\mathcal{F} \otimes \mathcal{L}_{\overline{\chi}},T)f(\mathcal{G} \otimes \mathcal{L}_{\overline{\chi}},T).$$

Proof. Replacing both $\mathbf{F}_q$ and E_λ by finite extensions of themselves, we may suppose that all the $\mathcal{L}_\chi$ occurring in $\mathcal{F},\mathcal{G}$, or $\mathcal{F}*\mathcal{G}$ as I_0-representations have χ an E_λ-valued character of $\mathbf{F}_q^\times$. For any χ, the canonical isomorphism (cf. 5.1 (9))

$$(\mathcal{F} * \mathcal{G}) \otimes \mathcal{L}_{\overline{\chi}} \simeq (\mathcal{F} \otimes \mathcal{L}_{\overline{\chi}}) * (\mathcal{G} \otimes \mathcal{L}_{\overline{\chi}}),$$

together with the theorem, shows that

$$f((\mathcal{F} * \mathcal{G}) \otimes \mathcal{L}_{\overline{\chi}},T) = f(\mathcal{F} \otimes \mathcal{L}_{\overline{\chi}},T)f(\mathcal{G} \otimes \mathcal{L}_{\overline{\chi}},T).$$

Now $\mathcal{L}_\chi$ actually occurs in $\mathcal{F}$ as I_0-representation if and only if $(\mathcal{F}*\mathcal{L}_{\overline{\chi}})^{I_0}$ is non-zero, i.e., if and only if $f(\mathcal{F}\otimes\mathcal{L}_{\overline{\chi}},T)$ is a *non-constant* polynomial. Applying this to $\mathcal{F},\mathcal{G}$ and to $\mathcal{F}*\mathcal{G}$, we see by the above product formula that $\mathcal{L}_\chi$ occurs in $\mathcal{F}*\mathcal{G}$ if and only if it occurs in at least one of $\mathcal{F}$ or $\mathcal{G}$. ∎

7.1.6. Numerical Check. We have

$$\begin{aligned}\operatorname{rank}(\mathcal{F}) &= \sum_\chi \operatorname{rank}\big((\mathcal{F}\otimes\mathcal{L}_{\overline{\chi}})^{I_0-\mathrm{unip}}\big)\\ &= \sum_\chi f'(\mathcal{F}\otimes\mathcal{L}_{\overline{\chi}},1)\end{aligned}$$

$$\begin{aligned}\operatorname{Swan}_\infty(\mathcal{F}) = \operatorname{Swan}_\infty(\mathcal{F}\otimes\mathcal{L}_{\overline{\chi}}) &= h^1_c(\mathcal{F}\otimes\mathcal{L}_{\overline{\chi}})\\ &= f(\mathcal{F}\otimes\mathcal{L}_{\overline{\chi}},1),\end{aligned}$$

and similarly for $\mathcal{G}$, and for $\mathcal{F}*\mathcal{G}$.

Let us check that these formulas and the product formula

$$f((\mathcal{F}*\mathcal{G})\otimes\mathcal{L}_{\overline{\chi}},T) = f(\mathcal{F}\otimes\mathcal{L}_{\overline{\chi}},T)f(\mathcal{G}\otimes\mathcal{L}_{\overline{\chi}},T)$$

imply the earlier-established formulas (cf. 5.1, (4) and (5))

$$\begin{cases}\operatorname{Swan}_\infty(\mathcal{F}*\mathcal{G}) = \operatorname{Swan}_\infty(\mathcal{F})\operatorname{Swan}_\infty(\mathcal{G})\\ \operatorname{rank}(\mathcal{F}*\mathcal{G}) = \operatorname{rank}(\mathcal{F})\operatorname{Swan}_\infty(\mathcal{G}) + \operatorname{rank}(\mathcal{G})\operatorname{Swan}_\infty(\mathcal{F}).\end{cases}$$

For the first, we simply *evaluate* the product formula at $T=1$. For the second, we write

$$\operatorname{rank}(\mathcal{F}*\mathcal{G}) = \sum_\chi f'((\mathcal{F}*\mathcal{G})\otimes\mathcal{L}_{\overline{\chi}},1).$$

Applying the product rule of differentiation to

$$f((\mathcal{F}*\mathcal{G})\otimes\mathcal{L}_{\overline{\chi}},T) = f(\mathcal{F}\otimes\mathcal{L}_{\overline{\chi}},T)f(\mathcal{G}\otimes\mathcal{L}_{\overline{\chi}},T)$$

we find

$$\begin{aligned}f'((\mathcal{F}*\mathcal{G})\otimes\mathcal{L}_{\overline{\chi}},1) &= f'(\mathcal{F}\otimes\mathcal{L}_{\overline{\chi}},1)f(\mathcal{G}\otimes\mathcal{L}_{\overline{\chi}},1)\\ &\quad + f(\mathcal{F}\otimes\mathcal{L}_{\overline{\chi}},1)f'(\mathcal{G}\otimes\mathcal{L}_{\overline{\chi}},1)\\ &= f'(\mathcal{F}\otimes\mathcal{L}_{\overline{\chi}},1)\operatorname{Swan}_\infty(\mathcal{G})\\ &\quad + f'(\mathcal{G}\otimes\mathcal{L}_{\overline{\chi}},1)\operatorname{Swan}_\infty(\mathcal{F}).\end{aligned}$$

Summing over χ now yields

$$\operatorname{rank}(\mathcal{F}*\mathcal{G}) = \operatorname{rank}(\mathcal{F})\operatorname{Swan}_\infty(\mathcal{G}) + \operatorname{rank}(\mathcal{G})\operatorname{Swan}_\infty(\mathcal{F}),$$

as required.

7.2. Application to Sheaves with $\mathrm{Swan}_\infty = 1$

7.2.1. Let us now consider a lisse E_λ-sheaf $\mathcal{F}$ on $\mathbf{G}_m \otimes \mathbf{F}_q$ which satisfies

$$\begin{cases} \mathcal{F} \text{ is pure of weight } w(\mathcal{F}). \\ \mathcal{F} \text{ is tame at 0 and totally wild at } \infty. \\ \mathrm{Swan}_\infty(\mathcal{F}) = 1. \end{cases}$$

Then for every χ, we have

$$\dim H^1_c(\mathbf{G}_m \otimes \overline{\mathbf{F}}_q, \mathcal{F} \otimes \mathcal{L}_{\overline{\chi}}) = 1.$$

Passing to a larger finite field if necessary, we may speak of the weight $w(H^1_c(\mathcal{F} \otimes \mathcal{L}_{\overline{\chi}}))$ of the action of Frobenius on this one-dimensional space. Applying the previous theory, we find that as I_0-representation, we have a canonical decomposition

$$\mathcal{F} \simeq \bigoplus_{\substack{\chi \text{ such that} \\ w(H^1_c(\mathcal{F}\otimes\mathcal{L}_{\overline{\chi}}))\leq w}} \mathcal{L}_\chi \otimes \begin{pmatrix} \text{a single unipotent} \\ \text{Jordan block of} \\ \text{dimension} \\ 1+w-w(H^1_c(\mathcal{F}\otimes\mathcal{L}_{\overline{\chi}})) \end{pmatrix}.$$

Furthermore, if all the χ which occur above have order dividing $q-1$, then the above decomposition is D_0-stable, and if for each such χ we denote by

$$\begin{cases} \alpha(\chi) \quad = F|H^1_c(\mathbf{G}_m \otimes \overline{\mathbf{F}}_q, \mathcal{F} \otimes \mathcal{L}_{\overline{\chi}}) \\ \\ \mathrm{drop}(\chi) = 1 + w - w(\alpha(\chi)) \end{cases}$$

then the eigenvalues of F_0 on $(\mathcal{F} \otimes \mathcal{L}_{\overline{\chi}})^{I_0-\mathrm{unip}}$ are the drop(χ)-quantities

$$\{\alpha(\chi), q \cdot \alpha(\chi), \ldots, q^{\mathrm{drop}(\chi)-1}\alpha(\chi)\}.$$

7.2.2. Corollary. *Suppose that $\mathcal{F}$ as above lisse and pure on $\mathbf{G}_m \otimes \mathbf{F}_q$, tame at 0, totally wild at ∞, $\mathrm{Swan}_\infty(\mathcal{F}) = 1$ has all characters χ occurring in its local monodromy at zero of order dividing $q-1$. Then if $\mathrm{rk}(\mathcal{F}) \geq 2$, the lisse rank-one sheaf on $\mathbf{G}_m \otimes \mathbf{F}_q$*

$$\frac{\det(\mathcal{F})}{\prod_\chi \mathcal{L}_\chi^{\otimes \mathrm{drop}(\chi)}}$$

is geometrically constant. For any closed point X of $\mathbf{G}_m \otimes \mathbf{F}_q$, the Frobenius F_x operates by the scalar $A^{\deg(x)}$, with

$$A = \prod_\chi \left(q^{\frac{\mathrm{drop}(\chi)-1}{2}} \alpha(\chi)\right)^{\mathrm{drop}(\chi)}.$$

Proof. For $\operatorname{rank}(\mathcal{F}) \geq 2$, the fact that $\mathcal{F}$ is totally wild at ∞ with $\operatorname{Swan}_\infty(\mathcal{F}) = 1$ implies that all the breaks of $\mathcal{F}$ at ∞ are equal to $1/\operatorname{rank}(\mathcal{F})$, so $\leq 1/2$. Therefore $\det(\mathcal{F})$ has its break at ∞ also $\leq 1/2$, so by the Hasse–Arf theorem $\det(\mathcal{F})$ is *tame* at ∞. Thus $\det(\mathcal{F})$ is tame at both zero and ∞. In view of the structure of the local monodromy at zero of $\mathcal{F}$ (cf. 7.0.8 (4)), we see that the ratio

$$\frac{\det(\mathcal{F})}{\prod_{\chi} \mathcal{L}_\chi^{\otimes \operatorname{drop}(\chi)}}$$

is a lisse rank-one sheaf on $\mathbf{G}_m \otimes \mathbf{F}_q$, which is both tame at ∞ and unramified at zero. It is therefore geometrically constant ($\mathbf{P}^1 - \{\infty\} = \mathbf{A}^1$ has no non-trivial tame at ∞ finite etale coverings over a separably closed field), so it extends to a lisse rank-one sheaf on $\mathbf{P}^1 \otimes \mathbf{F}_q$ on which F_x, x any closed point of $\mathbf{P}^1$, operates by $A^{\deg(x)}$ for some scalar A. Taking for x the origin, we may compute A as the action of any element $F_0 \in D_0$ of degree one on

$$\frac{\det(\mathcal{F})}{\prod (\mathcal{L}_\chi)^{\otimes \operatorname{drop}(\chi)}}.$$

But as D_0-representation, this is the character

$$\prod_{\chi} \det\left((\mathcal{F} \otimes \mathcal{L}_{\overline{\chi}})^{I_0\text{-unip}}\right),$$

on which F_0 does indeed act as the asserted scalar A. ∎

7.3. Application to Kloosterman Sheaves. Let us now consider in detail the case of a Kloosterman sheaf

$$\mathrm{Kl}(\psi; \chi_1, \ldots, \chi_n; b_1, \ldots, b_n)$$

on $\mathbf{G}_m \otimes \mathbf{F}_q$, where the χ_i are all multiplicative characters of $\mathbf{F}_q^\times$.

7.3.1. Lemma. *For any multiplicative character χ of $\mathbf{F}_q^\times$, the action of F on the one-dimensional space*

$$H_c^1(\mathbf{G}_m \otimes \overline{\mathbf{F}}_q, \mathrm{Kl}(\psi; \chi_1, \ldots, \chi_n; b_1, \ldots, b_n) \otimes \mathcal{L}_{\overline{\chi}})$$

is by the scalar

$$\alpha(\chi) = \prod_{i=1}^{n} (-g(\psi, \chi_i/\chi^{b_i})),$$

and the drop of weight $\operatorname{drop}(\chi)$ *is given by*

$$\operatorname{drop}(\chi) = \textit{ the number of indices } i \textit{ for which } \chi^{b_i} = \chi_i$$

Proof. Because Kl is totally wild at ∞, the $H_c^2(\mathrm{Kl}\otimes\mathcal{L}_{\overline{\chi}})$ vanishes, so the Lefschetz trace formula gives

$$\alpha(\chi) = \mathrm{trace}(F|H_c^1(\mathrm{Kl}\otimes\mathcal{L}_{\overline{\chi}})) = -\sum_{a\in\mathbf{F}_q^\times} \overline{\chi}(a)\,\mathrm{trace}(F_a|(\mathrm{Kl})_{\overline{a}})$$

$$= -\sum_{a\in\mathbf{F}_q^\times} \overline{\chi}(a)(-1)^{n-1}\,\mathrm{Kl}(\psi;\chi_1,\ldots,\chi_n;b_1,\ldots,b_n)(\mathbf{F}_q,a)$$

$$= (-1)^n \times \prod_{i=1}^{n} g(\psi,\chi_i\overline{\chi}^{b_i}),$$

the last equality being the expression of monomials in gauss sums as multiplicative Fourier transforms of Kloosterman sums (cf. 4.0).

The Kloosterman sheaf in question being pure of weight $n-1$, we see that

$$\mathrm{drop}(\chi) = n - w(\alpha(\chi))$$

$$= \sum_{i=1}^{n}(1 - w(g(\psi,\chi_i/\chi^{b_i}))).$$

As the weight of $g(\psi,\chi)$ for ψ non-trivial is *one* unless χ is trivial, which case the weight is *zero*, we obtain the assorted formula for the drop. ∎

Applying the general results of the previous section, we find

7.3.2. Theorem. *Suppose that for each i, the prime-to-p part b_i' of b_i divides $q-1$, and that χ_i has order dividing $(q-1)/b_i'$. Then*

(1) *the characters χ occurring in the local monodromy at zero of*

$$\mathrm{Kl}(\psi;\chi_1,\ldots,\chi_n;b_1,\ldots,b_n)$$

are precisely those characters χ of $\mathbf{F}_q^\times$ such that for some value of $i=1,\ldots,n$, we have

$$\chi^{b_i} = \chi_i.$$

(2) *Each character χ that occurs in (1) above occurs with a single Jordan block, of dimension equal to*

$$\mathrm{drop}(\chi) = \textit{ the number of indices } i \textit{ for which } \chi^{b_i} = \chi_i.$$

(3) *As D_0-representation, we have a decomposition*

$$\mathrm{Kl} \simeq \bigoplus_{\substack{\chi \text{ such that} \\ \chi^{b_i}=\chi_i \text{ for at} \\ \text{least one } i}} \mathcal{L}_\chi \otimes (\mathrm{Kl} \otimes \mathcal{L}_{\overline{\chi}})^{I_0-\mathrm{unip}},$$

where $\mathrm{Kl} \otimes \mathcal{L}_{\overline{\chi}})^{I_0-\mathrm{unip}}$ *is a D_0-representation of dimension equal to* $\mathrm{drop}(\chi)$*, which under I_0 is unipotent and has a single Jordan block, and on which any element $F_0 \in D_0$ of degree one acts with eigenvalues*

$$\alpha(\chi), q \cdot \alpha(\chi), \ldots, q^{\mathrm{drop}(\chi)-1}\alpha(\chi),$$

with

$$\alpha(\chi) = \prod_{i=1}^{n} (-g(\psi, \chi_i/\chi^{b_i})).$$

(4) *If in addition* $\mathrm{rk}(\mathrm{Kl}) = \Sigma b_i'$ *is* ≥ 2*, then the ratio*

$$\frac{\det(\mathrm{Kl})}{\prod_\chi \mathcal{L}_\chi^{\otimes \mathrm{drop}(\chi)}}$$

is a lisse rank-one sheaf on $\mathbf{G}_m \otimes \mathbf{F}_q$ *which is geometrically constant, on which F_x, for x a closed point of* $\mathbf{G}_m \otimes \mathbf{F}_q$*, acts by* $A^{\det(x)}$*, for*

$$A = \prod_\chi \left(q^{\frac{\mathrm{drop}(\chi)-1}{2}} \alpha(\chi)\right)^{\mathrm{drop}(\chi)}.$$

Proof. The hypothesis concerning the χ_i and the b_i guarantees that any character χ of any finite extension of $\mathbf{F}_q$ such that $\chi^{b_i} = \chi_i \circ \mathrm{Norm}$ is already (the composition with the norm of) a character of $\mathbf{F}_q^\times$. ∎.

7.4. Some Special Cases

7.4.1. Let us consider the case when all the $b_i = 1$, i.e., we consider

$$\mathrm{Kl}(\psi; \chi_1, \ldots, \chi_n) \stackrel{\mathrm{dfn}}{=} \mathrm{Kl}(\psi; \chi_1, \ldots, \chi_n; 1, \ldots, 1).$$

For this Kloosterman sheaf, the χ occurring in the local monodromy at zero are exactly the χ_i *themselves*, and so we have

$$\mathrm{Kl}(\psi; \chi_1, \ldots, \chi_n) \simeq \bigoplus_{\substack{\chi \text{ among} \\ \{\chi_1, \ldots, \chi_n\}}} \mathcal{L}_\chi \otimes \begin{pmatrix} \text{a single unipotent Jordan block} \\ \text{of size =(the number of } i \text{ for} \\ \text{which } \chi = \chi_i) = \mathrm{drop}(\chi) \end{pmatrix}$$

as I_0-representation. If $n = \text{rank}(\mathrm{Kl})$ is ≥ 2, then, by 7.3.2 (4),

$$\frac{\det(\mathrm{Kl}(\psi;\chi_1,\ldots,\chi_n))}{\prod_{i=1}^{n} \mathcal{L}_{\chi_i}}$$

is geometrically constant on $\mathbf{G}_m \otimes \mathbf{F}_q$, with F_x acting by $A^{\deg(x)}$, and

$$A = \prod_{\substack{\chi \text{ among} \\ \{\chi_1,\ldots,\chi_n\}}} \left(q^{\frac{\text{drop}(\chi)-1}{2}} \prod_{i=1}^{n} (-g(\psi,\chi_i/\chi)) \right)^{\text{drop}(\chi)}$$

(7.4.1.1)

$$= (q)^{\sum_{\chi} \frac{1}{2}\text{drop}(\chi)(\text{drop}(\chi)-1)} \prod_{1\leq i,j\leq n} (-g(\psi,\chi_i/\chi_j)).$$

7.4.1.2. Lemma. *Hypotheses and notations as in 7.4.1 above, if* $n \geq 2$ *we have the formula*

$$A = (q)^{\frac{n(n-1)}{2}} \prod_{1\leq i<j\leq n} (\chi_i(-1)/\chi_j(-1)).$$

Proof. That this formula is correct up to a multiplicative factor which is an integral power of q is clear from rewriting the product

$$\prod_{1\leq i,j\leq n} (-g(\psi,\chi_i/\chi_j))$$

in the last line of 7.4.1.1 as

$$\prod_{1\leq i<j\leq n} [(-g(\psi,\chi_i/\chi_j))(-g(\psi,\chi_j/\chi_i))]$$

(this is legitimate because $-g(\psi,\mathbf{1}) = -1$) and remembering that for any χ ($\chi = \chi_i/\chi_j$ in our case), we have

$$g(\psi,\chi)\cdot g(\psi,\overline{\chi}) = \begin{cases} 1 & \text{if } \chi = \mathbf{1} \\ \chi(-1)\cdot q & \text{if } \chi \neq \mathbf{1}. \end{cases}$$

To see that we have the *correct* power of q, it suffices to remark that A is pure of weight equal to the weight of $\det(\mathrm{Kl}(\psi;\chi_1,\ldots,\chi_n))$, i.e.,

$$|A| = (q)^{\frac{n(n-1)}{2}}. \quad \blacksquare$$

7.4.1.3. Corollary. *Hypotheses and notations as in 7.4.1 above, if $n \geq 2$ then for any closed point x of $\mathbf{G}_m \otimes \mathbf{F}_q$, we have the formula*

$$\det(F_x | \mathrm{Kl}(\psi; \chi_1, \ldots, \chi_n)_{\bar{x}}) = A^{\deg(x)} \left(\prod_{i=1}^{n} \chi_i \right) (\mathbf{N}_{\mathbf{F}_q(x)/\mathbf{F}_q}(x)),$$

where

$$A = (q)^{\frac{n(n-1)}{2}} \prod_{1 \leq i < j \leq n} \chi_i(-1)/\chi_j(-1)).$$

7.4.1.4. Remark. The formulas 7.4.1.2 and 7.4.1.3 are both simplified by the identity

$$\prod_{1 \leq i < j \leq n} (\chi_i(-1)/\chi_j(-1)) = \left(\left(\prod_{i=1}^{n} \chi_i \right) (-1) \right)^{n-1},$$

whose elementary verification is left to the reader.

For example, if $\prod \chi_i = \mathbf{1}$, or if n is odd, then we find

$$A = (q)^{\frac{n(n-1)}{2}}.$$

7.4.2. Now consider the case when all $b_i = 1$ and when all $\chi_i = \mathbf{1}$. We denote this sheaf simply

$$\mathrm{Kl}_n(\psi) \stackrel{\text{dfn}}{=} \mathrm{Kl}(\psi; \underbrace{\mathbf{1}, \ldots, \mathbf{1}}_{n \text{ times}}; \underbrace{1, \ldots, 1}_{n \text{ times}}).$$

Applying the results of the previous paragraph, we find a result originally proven by Deligne(cf. [De-3], 7.8 and 7.15.2).

7.4.3. Theorem. *The local monodromy at zero of $\mathrm{Kl}_n(\psi)$ is unipotent, with a single Jordan block of size n, and F_0 acts on $(\mathrm{Kl}_n(\psi))^{I_0}$ as the identity. If $n \geq 2$, then $\det(\mathrm{Kl}_n(\psi))$ is geometrically constant on $\mathbf{G}_m \otimes \mathbf{F}_q$, and for any closed point x of $\mathbf{G}_m \otimes \mathbf{F}_q$ we have*

$$\det(F_x | \mathrm{Kl}_n(\psi)_{\overline{x}}) = \left(q^{\frac{n(n-1)}{2}} \right)^{\deg(x)}.$$

Proof. All the χ_i are trivial, so $\mathrm{drop}(\mathbf{1}) = n$, $\mathrm{drop}(\chi) = 0$ for $\chi \neq \mathbf{1}$, so the first assertion is just 7.3.2, (1) and (2). If $n \geq 2$, the second assertion is the special case $\chi_1 = \ldots \chi_n = \mathbf{1}$ of 7.4.1.3. ∎

7.5. Appendix: The product formula in the general case (d'après O. Gabber)

7.5.1. Let k be an algebraically closed field of characteristic $p > 0$, l a prime number $l \neq p$, E_λ a finite extension of $\mathbf{Q}_l$, $\mathcal{O}_\lambda$ its ring of integers and $\mathbf{F}_\lambda$ its residue field. Let $\mathcal{F}$ be a lisse E_λ-sheaf (resp. $\mathbf{F}_\lambda$-sheaf) on $\mathbf{G}_m \otimes k$ which is tame at zero and totally wild at ∞. Let γ_0 be a topological generator of I_0^{tame}. The Jordan decomposition of γ_0 acting on $\mathcal{F}$ gives a canonical decomposition of $\mathcal{F}$ as I_0^{tame}-representation

$$\mathcal{F} = \left(\mathcal{F}^{I_0-\text{unip.}}\right) \oplus \left(\bigoplus_{\substack{\chi \text{non-triv.contin.} \\ \text{char. of } I_0^{\text{tame}} \text{ to } \overline{E}_\lambda^\times \\ (\text{resp. to } \overline{\mathbf{F}}_\lambda^\times)}} \mathcal{F}^{\chi-\text{unip.}} \right), \tag{7.5.1.1}$$

in which $\gamma_0 - 1$ operates nilpotently on the first factor and invertibly on the second. The cohomology sequence of 7.1.1 gives a short exact sequence

$$0 \to \mathcal{F}^{I_0} \to H_c^1(\mathbf{G}_m \otimes k, \mathcal{F}) \to H^1(\mathbf{P}^1 \otimes k, j_*\mathcal{F}) \to 0, \tag{7.5.1.2}$$

so in particular

$$\dim(\mathcal{F}^{I_0}) \le h_c^1(\mathcal{F}) = \text{Swan}_\infty(\mathcal{F}). \tag{7.5.1.3}$$

Now $\dim(\mathcal{F}^{I_0})$ is precisely the number of Jordan blocks in $\mathcal{F}^{I_0-\text{unip}}$. Therefore if we decree that $\mathcal{F}^{I_0-\text{unip}}$ has $\text{Swan}_\infty(\mathcal{F})$ Jordan blocks in total, of which $\text{Swan}_\infty(\mathcal{F}) - \dim(\mathcal{F}^{I_0})$ are of size zero, we may define the polynomial

$$f(\mathcal{F},T) \in \mathbf{Z}[T]$$

by the recipe

$$f(\mathcal{F},T) = \sum_{\substack{\text{"all" the Swan}_\infty(\mathcal{F}) \\ \text{Jordan blocks of} \\ \mathcal{F}^{I_0-\text{unip}}}} T^{(\text{dim of Jordan block})}.$$

7.5.2. Theorem. (O. Gabber). *For $\mathcal{F}, \mathcal{G}$ as above, we have the product formula*

$$f(\mathcal{F} * \mathcal{G}, T) = f(\mathcal{F},T) f(\mathcal{G},T).$$

7.5.3. The proof is based upon studying the "forget supports" map

$$H_c^1(\mathbf{G}_m \otimes k, \mathcal{F}) \to H^1(\mathbf{G}_m \otimes k, \mathcal{F})$$

for $\mathcal{F}$ and for *all* its twists by tame characters, especially by the *generic* such character. We must first make precise what this means.

The tame (at both zero and ∞) fundamental group of $\mathbf{G}_m \otimes k$ is canonically isomorphic to $\prod_{l \neq p} \mathbf{Z}_l(1) \simeq \varprojlim_{p \nmid N} \boldsymbol{\mu}_N(k)$, via the coverings $[N]$ of $\mathbf{G}_m \otimes k$ by itself. Because these coverings induce precisely all the tame extensions of $k((t))$, namely the fields obtained by extracting N'th roots of t for N prime to p, we see that we have canonically

$$I_0^{\mathrm{tame}} \xrightarrow{\sim} \pi_1(\mathbf{G}_m \otimes k)^{\mathrm{tame}} \xrightarrow{\sim} \prod_{l \neq p} \mathbf{Z}_l(1).$$

For any complete noetherian local ring A with finite residue field of residue characteristic $l \neq p$, and any continuous character

$$\chi : I_0^{\mathrm{tame}} \to A^\times,$$

we denote by $\mathcal{L}_\chi$ the lisse sheaf of free A-modules of rank-one on $\mathbf{G}_m \otimes k$ which is tame at zero and ∞ and which induces χ on I_0. Alternately, If A is finite as well, then χ is necessarily of finite order prime to p, (because $A^\times$ is), so of order dividing $q-1$ for some power q of p. Via the canonical identification $I_0^{\mathrm{tame}} \xrightarrow{\sim} \varprojlim_{p \nmid N} \boldsymbol{\mu}_N(k)$, we may interpret χ as a character of $\boldsymbol{\mu}_{q-1}(k) = \mathbf{F}_q^\times$, and then the $\mathcal{L}_\chi$ in question is obtained from the earlier defined $\mathcal{L}_\chi$ on $\mathbf{G}_m \otimes \mathbf{F}_q$ (obtained from pushing out the Lang torsor, cf. 4.3) by pulling back to $\mathbf{G}_m \otimes k$. Viewing A as the inverse limit of the finite rings $\{A/\mathfrak{m}^n\}_{n \geq 1}$ allows us to define $\mathcal{L}_\chi$ in the general case by passing to the limit.

7.5.4. Let us now fix a topological generator γ_0 of I_0^{tame}, and a prime number $l \neq p$. We denote by

$$\chi^{\mathrm{gen}} : I_0^{\mathrm{tame}} \to (\mathbf{Z}_l[[X]])^\times$$

the unique continuous tame character satisfying

$$\chi^{\mathrm{gen}}(\gamma_0) = (1+X)^{-1}.$$

In terms of the homomorphism

$$t_l : I_0^{\mathrm{tame}} \simeq \prod_{l \neq p} \mathbf{Z}_l(1) \twoheadrightarrow \mathbf{Z}_l(1),$$

we have the formula

$$\chi^{\mathrm{gen}}(\gamma) = (1+X)^{-t_l(\gamma)/t_l(\gamma_0)}.$$

For any complete noetherian local ring A with finite residue field of residue characteristic l, we also denote by

$$\chi^{\mathrm{gen}} : I_0^{\mathrm{tame}} \to (A[[X]])^\times$$
$$\chi^{\mathrm{gen}}(\gamma_0) = (1+X)^{-1}$$

the character obtained by extension of scalars.

For such an A, suppose we are given a lisse sheaf $\mathcal{F}$ on $\mathbf{G}_m \otimes k$ of free A-modules of finite rank which is tame at zero and totally wild at ∞. The long exact cohomology sequence on $\mathbf{P}^1 \otimes k$ for $j : \mathbf{G}_m \hookrightarrow \mathbf{P}^1$ and

$$0 \to j_!\mathcal{F} \to Rj_*\mathcal{F} \to \text{coker} \to 0$$

gives a canonical four-term exact sequence of A-modules

$$0 \to \mathcal{F}^{I_0} \to H^1_c(\mathbf{G}_m \otimes k, \mathcal{F}) \to H^1(\mathbf{G}_m \otimes k, \mathcal{F}) \to \mathcal{F}_{I_0}(-1) \to 0$$

in which the middle two terms are free A-modules of rank $= \text{Swan}_\infty(\mathcal{F})$ (cf. 2.0.7, 2.1.1).

Now extend scalars from A to $A[[X]]$, so that it makes sense to form $\mathcal{F} \underset{A}{\otimes} \mathcal{L}_{\chi^{\text{gen}}}$. The above four-term sequence gives an exact sequence of $A[[X]]$-modules

$$(7.5.4.1)\quad 0 \to \left(\mathcal{F} \otimes \mathcal{L}_{\chi^{\text{gen}}}\right)^{I_0} \to H^1_c\left(\mathbf{G}_m \otimes k, \mathcal{F} \otimes \mathcal{L}_{\chi^{\text{gen}}}\right) \to$$
$$\to H^1\left(\mathbf{G}_m \otimes k, \mathcal{F} \otimes \mathcal{L}_{\chi^{\text{gen}}}\right) \to \left(\mathcal{F} \otimes \mathcal{L}_{\chi^{\text{gen}}}\right)_{I_0}(-1) \to 0,$$

in which the two middle terms are free $A[[X]]$-modules of rank $\text{Swan}_\infty(\mathcal{F})$.

As I_0-representation, $\mathcal{F} \otimes \mathcal{L}_{\chi^{\text{gen}}}$ may be viewed explicitly as follows. Denoting by $\mathcal{F}$ the free A-module of finite rank on which I_0^{tame} acts, say by ρ, we have

$$\begin{cases} \mathcal{F} \otimes \mathcal{L}_{\chi^{\text{gen}}} = \mathcal{F}[[X]] \text{ as free } A[[X]]\text{-module of finite rank,} \\ I_0^{\text{tame}} \text{ acts by } \gamma_0 \to \rho(\gamma_0)(1+X)^{-1}. \end{cases}$$

Thus we have a two-term complex

$$\mathcal{F}[[X]] \xrightarrow{\rho(\gamma_0)(1+X)^{-1}-1} \mathcal{F}[[X]]$$

whose kernel and cokernel are $(\mathcal{F} \otimes \mathcal{L}_{\chi^{\text{gen}}})^{I_0}$ and $(\mathcal{F} \otimes \mathcal{L}_{\chi^{\text{gen}}})_{I_0}$ respectively.

7.5.5. Lemma. *We have*

$$\begin{cases} \left(\mathcal{F} \otimes \mathcal{L}_{\chi^{\text{gen}}}\right)^{I_0} = 0 \\ \left(\mathcal{F} \otimes \mathcal{L}_{\chi^{\text{gen}}}\right)_{I_0} \simeq \mathcal{F}[[X]]/(X - (\rho(\gamma_0) - 1)). \end{cases}$$

Proof. The complex

$$\mathcal{F}[[X]] \xrightarrow{\rho(\gamma_0)(1+X)^{-1}-1} \mathcal{F}[[X]]$$

is isomorphic to the complex

$$\mathcal{F}[[X]] \xrightarrow{\rho(\gamma_0)-1-X} \mathcal{F}[[X]],$$

by the isomorphism $(1, 1+X)$. This second complex is obtained from the complex

$$\mathcal{F}[X] \xrightarrow{\rho(\gamma_0)-1-X} \mathcal{F}[X]$$

of $A[X]$-modules by the flat (A is noetherian) extension of scalars $A[X] \hookrightarrow A[[X]]$. This last complex clearly has its differential $\rho(\gamma_0)-1-X$ injective,

$$\mathcal{F}[X] \xhookrightarrow{\rho(\gamma_0)-1-X} \mathcal{F}[X],$$

as is obvious by looking at the coefficient of the highest power of X in an element of $\mathcal{F}[X]$ suspected of lying in the kernel. By flatness, the map remains injective after extending scalars from $A[X]$ to $A[[X]]$. The assertion concerning its cokernel is obvious. ∎

7.5.6. Let us now insert this data into our earlier four-term exact sequence 7.5.4.1, and remember that the choice of a topological generator γ_0 of I_0^{tame} also provides a generator $t_l(\gamma_0)$ of $\mathbf{Z}_l(1)$, which we use to identify $\mathbf{Z}_l(1)$ to $\mathbf{Z}_l$. We find a short exact sequence of $A[[X]]$-modules

$$(7.5.6.1)\quad 0 \to H^1_c(\mathbf{G}_m \otimes k, \mathcal{F} \otimes \mathcal{L}_{\chi^{\text{gen}}}) \to H^1(\mathbf{G}_m \otimes k, \mathcal{F} \otimes \mathcal{L}_{\chi^{\text{gen}}}) \to$$
$$\to \mathcal{F}[[X]]/(X - (\rho(\gamma_0) - 1)) \to 0,$$

the first two of which are free $A[[X]]$-modules of rank $= \mathrm{Swan}_\infty(\mathcal{F})$.

7.5.7. Before proceeding, it is convenient to axiomatize the sort of structure at which we have arrived. The appropriate notion is a variant of Mazur's "spans," which he introduced in [Maz] in studying Hodge and Newton polygons.

7.5.8. Definition. Let A be an arbitrary ring, and X an indeterminate. An "A-span" $S^\bullet$ is a two-term complex $S^\bullet : M \xrightarrow{\phi} N$, of free $A[[X]]$-modules of the same finite rank, whose differential ϕ is $A[[X]]$-linear and injective. The common rank of M and N is called the rank of the span, denoted $\mathrm{rk}(S^\bullet)$.

If B is a flat A-algebra such that $B[[X]]$ is flat over $A[[X]]$ (e.g., this is automatic if A and B are noetherian, by ([A-K], Chapter V. 3.2 (i) $\longleftrightarrow$ (iv), applied to $A[[X]]$, $I = (X)$, $M = B[[X]]$)) then an A-span $S^\bullet$ gives rise to a B-span, namely

$$S^\bullet \underset{A[[X]]}{\otimes} B[[X]] : \quad M \underset{A[[X]]}{\otimes} B[[X]] \xrightarrow{\phi\otimes\mathrm{id}} N \underset{A[[X]]}{\otimes} B[[X]]$$

(the map $\phi \otimes \mathrm{id}$ remains injective by the supposed flatness).

Given two A-spans $S_1^\bullet$ and $S_2^\bullet$,

$$S_i^\bullet : M_i \xrightarrow{\phi_i} N_i \qquad \text{for } i = 1, 2,$$

their tensor product $S_1^\bullet \otimes S_2^\bullet$ is defined to be

$$S_1^\bullet \otimes S_2^\bullet : M_1 \underset{A[[X]]}{\otimes} M_2 \xrightarrow{\phi_1 \otimes \phi_2} N_1 \underset{A[[X]]}{\otimes} N_2.$$

(To check that this map between free $A[[X]]$-modules of rank $\mathrm{rk}(S_1^\bullet)\,\mathrm{rk}(S_2^\bullet)$ is in fact injective, observe that it factors into two injective maps

$$M_1 \otimes M_2 \xrightarrow{\phi_1 \otimes \mathrm{id}_{M_2}} N_1 \otimes M_2 \xrightarrow{\mathrm{id}_{N_1} \otimes \phi_2} N_1 \otimes N_2.)$$

If A is a *field*, then given an A-span

$$S^\bullet : M \xrightarrow{\phi} N,$$

the theory of elementary divisors shows that, if $\mathrm{rk}(S) = r > 0$ then M and N admit $A[[X]]$-bases in terms of which the matrix of ϕ is a diagonal matrix

$$\begin{pmatrix} X^{n_1} & & \bigcirc \\ & \ddots & \\ \bigcirc & & X^{n_r} \end{pmatrix}$$

with integral non-negative powers of X as diagonal entries. Intrinsically $n_1, \ldots, n_r$ may be recovered as the unique set of r non-negative integers for which there exists an $A[[X]]$-isomorphism

$$N/\phi M \simeq \bigoplus_{i=1}^{r} A[[X]]/(X^{n_i}).$$

The associated polynomial to such a span, denoted

$$f(S^\bullet, T) \in \mathbf{Z}[T]$$

is defined to be

$$f(S^\bullet, T) = \sum_{i=1}^{\mathrm{rk}(S^\bullet)} T^{n_i}.$$

For the zero-span, we define

$$f(0^\bullet, T) = 0.$$

7.5.9. Lemma. *If A is a field, then for any two A-spans $S_1^\bullet$ and $S_2^\bullet$ we have the formula for associated polynomials*

$$f(S_1^\bullet \otimes S_2^\bullet, T) = f(S_1^\bullet, T) f(S_2^\bullet, T).$$

Proof. If either $S_1^\bullet$ or $S_2^\bullet$ are zero, then $S_1^\bullet \otimes S_2^\bullet$ is zero and the assertion is "$0 = 0$ in $\mathbf{Z}[T]$." If both $S_1^\bullet$ and $S_2^\bullet$ are non-zero, then in suitable bases we have matrices for ϕ_1 and ϕ_2 of the form

$$\phi_1 = \begin{pmatrix} X^{n_1} & & \\ & \ddots & \\ & & X^{n_{r_1}} \end{pmatrix}, \qquad \phi_2 = \begin{pmatrix} X^{M_1} & & \\ & \ddots & \\ & & X^{m_{r_2}} \end{pmatrix}$$

whence $\phi_1 \otimes \phi_2$ has matrix

$$\begin{pmatrix} \ddots & & \\ & X^{n_i+m_j} & \\ & & \ddots \end{pmatrix}. \quad \blacksquare$$

7.5.10. Now let us return to the geometric situation at hand. Suppose that A is a complete noetherian local ring with finite residue field $\mathbf{F}_\lambda$ of residue characteristic $l \neq p$, and that $\mathcal{F}$ is a lisse sheaf of free finitely generated A-modules on $\mathbf{G}_m \otimes k$, k algebraically closed of characteristic $p > 0$, which is tame at zero and completely wild at ∞. We define the monodromy span of $\mathcal{F}$, $\mathrm{Span}(\mathcal{F})$, to be the A-span of rank $= \mathrm{Swan}_\infty(\mathcal{F})$ defined by

$$\mathrm{Span}(\mathcal{F}) : H_c^1(\mathbf{G}_m \otimes k, \mathcal{F} \otimes \mathcal{L}_{\chi^{\mathrm{gen}}}) \to$$
$$\xrightarrow{\text{forget supports}} H^1((\mathbf{G}_m \otimes k, \mathcal{F} \otimes \mathcal{L}_{\chi^{\mathrm{gen}}}).$$

7.5.11. Lemma. *For $\mathcal{F}, \mathcal{G}$ sheaves of the above type (i.e. in $\mathcal{C}_{A,k}$), the monodromy spans of $\mathcal{F}, \mathcal{G}$ and $\mathcal{F} * \mathcal{G}$ are related by a canonical isomorphism*

$$\mathrm{Span}(\mathcal{F} * \mathcal{G}) \simeq \mathrm{Span}(\mathcal{F}) \otimes \mathrm{Span}(\mathcal{G}).$$

Proof. This results immediately from parts (6) and (9) of the Convolution Theorem 5.1, applied to $\mathcal{F} \otimes \mathcal{L}_{\chi^{\mathrm{gen}}}$ and $\mathcal{G} \otimes \mathcal{L}_{\chi^{\mathrm{gen}}}$ over $A[[X]]$. $\blacksquare$

7.5.12. We may now conclude the proof of the product formula 7.5.2. Suppose first that $A = \mathbf{F}_\lambda$, a finite extension of $\mathbf{F}_l$. Then for $\mathcal{F} \in \mathcal{C}_{A,k}$, the cokernel of $\mathrm{Span}(\mathcal{F})$ is the $\mathbf{F}_\lambda[[X]]$-module

$$\mathcal{F}[[X]]/(X - (\rho(\gamma_0) - 1)).$$

Write $\mathcal{F}$ as I_0-module as the direct sum

$$\mathcal{F} = \mathcal{F}^{I_0-\mathrm{unip}} \otimes \mathcal{F}^{(\gamma_0 - 1\ \mathrm{invertible})}.$$

Clearly we have

$$(\mathcal{F}^{(\gamma_0-1\ \mathrm{invertible})})[[X]]/(X - (\rho(\gamma_0) - 1)) = 0,$$

(by Nakayama's Lemma), so that the cokernel of $\mathrm{Span}(\mathcal{F})$ is isomorphic to

$$\mathcal{F}^{I_0-\mathrm{unip}}[[X]]/(X-(\rho(\gamma_0)-1)).$$

In terms of the Jordan decomposition of the nilpotent operator $\rho(\gamma_0)-1$ on $\mathcal{F}^{I_0-\mathrm{unip}}$ as a sum of Jordan blocks of sizes $n_1,\ldots,n_k$, we have, tautologously,

$$\mathcal{F}^{I_0-\mathrm{unip}}[[X]]/(X-(\rho(\gamma_0)-1)) \simeq \bigoplus_{i=1}^{k} \mathbf{F}_\lambda[[X]]/(X^{n_i}).$$

Adding on "extra" Jordan blocks of size zero to get a total of $\mathrm{Swan}_\infty(\mathcal{F})$ blocks in "total," we see that

$$f(\mathcal{F},T)=f(\mathrm{Span}(\mathcal{F}),T).$$

From this identity, the product formula

$$f(\mathcal{F}*\mathcal{G},T)=f(\mathcal{F},T)f(\mathcal{G},T)$$

follows immediately from the two properties of $\mathbf{F}_\lambda$-spans

$$\begin{cases} \mathrm{Span}(\mathcal{F}*\mathcal{G})=\mathrm{Span}(\mathcal{F})\otimes\mathrm{Span}(\mathcal{G}) \\ \\ f(S_1^\bullet\otimes S_2^\bullet,T)=f(S_1^\bullet,T)f(S_2^\bullet,T) \end{cases}$$

which we have already established. This concludes the proof of the product formula in the case of $\mathbf{F}_\lambda$-sheaves.

In the case of E_λ-sheaves, we must argue slightly differently. Given a lisse E_λ-sheaf $\mathcal{F}$ on $\mathbf{G}_m\otimes k$ which is tame at zero and totally wild at ∞, pick a lisse $\mathcal{O}_\lambda$-sheaf $\boldsymbol{\mathcal{F}}$ of free $\mathcal{O}_\lambda$-modules of finite rank which gives rise to it. We may consider the $\mathcal{O}_\lambda$-span

$$\mathrm{Span}(\boldsymbol{\mathcal{F}}),$$

whose cokernel is isomorphic to the $\mathcal{O}_\lambda[[X]]$-module

$$\boldsymbol{\mathcal{F}}[[X]]/(X-(\rho(\gamma_0)-1)).$$

By the flatness of $E_\lambda[[X]]$ over $\mathcal{O}_\lambda[[X]]$, it makes sense to form the E_λ-span

$$\mathrm{Span}(\boldsymbol{\mathcal{F}})\underset{\mathcal{O}_\lambda[[X]]}{\otimes}E_\lambda[[X]],$$

whose cokernel is isomorphic to the $E_\lambda[[X]]$-module

$$\mathcal{F}[[X]]/(X-(\rho(\gamma_0)-1)),$$

and just as above this cokernel is itself isomorphic to

$$\mathcal{F}^{I_0-\text{unip}}[[X]]/(X-(\rho(\gamma_0)-1)),$$

so we see as above that

$$f(\mathcal{F},T)=f\left(\text{Span}(\boldsymbol{\mathcal{F}})\underset{\mathcal{O}_\lambda[[X]]}{\otimes}E_\lambda[[X]],T\right).$$

In fact, one can easily check that the E_λ-span

$$\text{Span}(\boldsymbol{\mathcal{F}}\underset{\mathcal{O}_\lambda[[X]]}{\otimes}E_\lambda[[X]]$$

depends only on $\boldsymbol{\mathcal{F}}\underset{\mathcal{O}_\lambda[[X]]}{\otimes}E_\lambda=\mathcal{F}$, and not on the auxiliary choice of $\boldsymbol{\mathcal{F}}$; we denote it simply as $\text{Span}(\mathcal{F})$. The formula for $\mathcal{O}_\lambda$-spans

$$\text{Span}(\boldsymbol{\mathcal{F}}*\boldsymbol{\mathcal{G}})=\text{Span}(\boldsymbol{\mathcal{F}})\otimes\text{Span}(\boldsymbol{\mathcal{G}})$$

yields by extension of scalars $\mathcal{O}_\lambda[[X]]\to E_\lambda[[X]]$ an isomorphism of E_λ-spans

$$\text{Span}(\mathcal{F}*\mathcal{G})=\text{Span}(\mathcal{F})\otimes\text{Span}(\mathcal{G}),$$

and the proof of the product formula concludes exactly as in the $\mathbf{F}_\lambda$-case. ∎

7.6. Appendix: an open problem concerning breaks of a convolution. One problem concerning convolution which we are unable to solve is the following. Given a lisse λ-adic sheaf $\mathcal{F}$ on $\mathbf{G}_m\otimes k$ in our convolution category $\mathcal{C}$ (tame at zero, totally wild at infinity), let us define its Swan polygon to be the polygon whose slopes are the breaks (of $\mathcal{F}$ as representation of I_∞, the inertia group at ∞), with *multiplicity* the multiplicity of that break. (If the distinct breaks are

$$0<\lambda_1<\lambda_2<\cdots<\lambda_r;\quad n_i=\text{ multiplicity of }\lambda_i$$

then $\text{rank}(\mathcal{F})=\sum n_i, \text{Swan}_\infty(\mathcal{F})=\sum n_i\lambda_i$, and the Swan polygon is (for $r=3$).

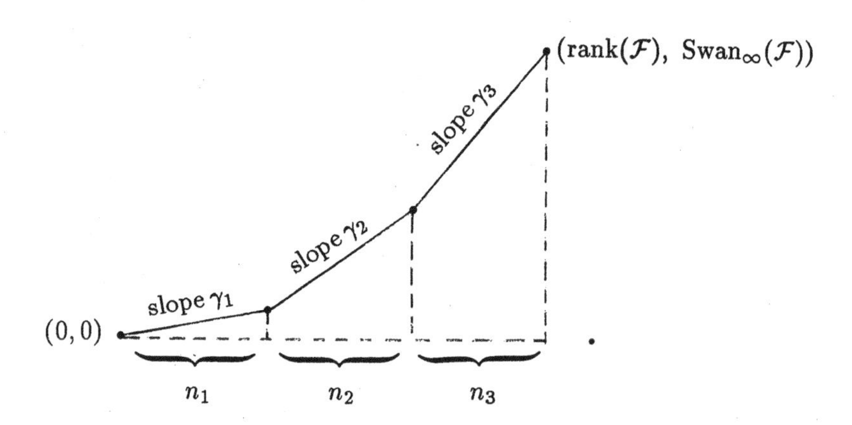

In view of 1.9 (essentially the Hasse–Arf Theorem) the problem is to compute the Swan polygon of a convolution in terms of the Swan polygons of the convolvees. The natural guess is that if

$$\begin{cases} \mathcal{F} \text{ has breaks } \lambda_i \text{ with multiplicities } n_i \\ \mathcal{G} \text{ has breaks } \mu_j \text{ with multiplicities } m_j \end{cases}$$

then their convolution $\mathcal{F} * \mathcal{G}$ has

$$\text{breaks } \frac{1}{(1/\lambda_i) + (1/\mu_j)} \text{ with multiplicities } n_i m_j (\lambda_i + \mu_j).$$

Let us say that $\mathcal{F}$ is "unibreak" if, as I_∞-representation. $\mathcal{F}$ has a single break (with multiplicity equal to rank($\mathcal{F}$)). If the above guess is correct, then we find that for $\mathcal{F}$, $\mathcal{G}$ in $\mathcal{C}$, we have

$$\mathcal{F}, \mathcal{G} \text{ both unibreak} \Longrightarrow \mathcal{F} * \mathcal{G} \text{ unibreak}.$$

Conversely, the universal truth of this last implication would imply the universal truth of the hoped-for Swan polygon formula in the general case, thanks to the existence of "local convolution," established in Chapter 6 following suggestions of Gabber and Laumon.

Here are two reformulations of the Swan polygon formula. Given any "polygon" in the sense of [Ka-4] with slopes $\lambda_i > 0$ and multiplicities n_i, define the "inverse polygon" to be the one with slopes $1/\lambda_i$, and multiplicities $n_i\lambda_i$. Pictorially, the operation of "inversion" of polygons is

$$(x, y) \mapsto \left(\sum n_i\lambda_i - y,\ \sum n_i - x\right);$$

if we horizontally translate all such polygons so that they "exactly" lie in the second quadrant, then inversion is reflection in the line $x + y = 0$. In this language, the hope is that the *inverse* Swan polygon of a convolution $\mathcal{F} * \mathcal{G}$ is the tensor product, in the sense of [Ka-4], of the inverse Swan polygons of $\mathcal{F}$ and $\mathcal{G}$.

A second reformulation is this. If $\mathcal{F}$ has breaks λ_i with multiplicities n_i, define its inverse Swan polynomial to be the $\mathbf{Z}$-polynomial in fractional powers of an indeterminate T

$$\mathrm{inv}\,.\,\mathrm{Swan}(\mathcal{F}, T) = \sum n_i \lambda_i T^{1/\lambda_i}.$$

Then the Swan polygon formula to be proven is the product formula

$$\mathrm{inv}\,.\,\mathrm{Swan}(\mathcal{F} * \mathcal{G}, T) = (\mathrm{inv}\,.\,\mathrm{Swan}(\mathcal{F}, T))(\mathrm{inv}\,.\,\mathrm{Swan}(\mathcal{G}, T)),$$

for $\mathcal{F}, \mathcal{G}$ in $\mathcal{C}$.

CHAPTER 8

Complements on Convolution

8.0. A Cancellation Theorem for Convolution

Let k be an algebraically closed field of characteristic $p > 0$, l a prime number $l \neq p$, and A an l-adic coefficient ring as in 5.0. We have seen (5.2.1, 5.2.3) that for $\mathcal{F} \in \mathcal{C}$ and $\mathcal{G} \in \mathcal{F}_1$, the two convolutions $\mathcal{F}_! * \mathcal{G}$ and $\mathcal{F}_* * \mathcal{G}$ both lie in $\mathcal{T}_1$, and that, *if* $\mathcal{G}$ lies in $\mathcal{C}_1$, then these two convolutions *coincide, and* they lie in $\mathcal{C}_1$.

8.0.1. Cancellation Theorem. *Let $\mathcal{F}$ be a non-zero object of $\mathcal{C}$, and $\mathcal{G}$ an object of $\mathcal{T}_1$. If $\mathcal{F}_! * \mathcal{G}$ lies in $\mathcal{C}_1$, then $\mathcal{G}$ itself lies in $\mathcal{C}_1$.*

Proof. If $A = E_\lambda$, then by picking $\mathcal{O}_\lambda$-forms of $\mathcal{F}$ and $\mathcal{G}$, we may reduce to the case $A = \mathcal{O}_\lambda$. If A is complete noetherian local with finite residue field $\mathbf{F}_\lambda$, we may reduce to the case $A = \mathbf{F}_\lambda$ (formation of $\mathcal{F}_! * \mathcal{G}$ and of the break-decompositions of $\mathcal{F}_! * \mathcal{G}$ and of $\mathcal{G}$ as P_∞-representations commutes with the extension of scalars of coefficients $A \to \mathbf{F}_\lambda$; therefore if $\mathcal{F}_! * \mathcal{G}$ is completely wild at ∞ so is $(\mathcal{F} \otimes_A \mathbf{F}_\lambda)_! * (\mathcal{G} \otimes_A \mathbf{F}_\lambda)$; by the theorem over $\mathbf{F}_\lambda$, $\mathcal{G} \otimes_A \mathbf{F}_\lambda$ is totally wild at ∞; by Nakayama's Lemma applied to the P_∞-invariants of $\mathcal{G}$, we see that $\mathcal{G}$ is itself totally wild at ∞).

8.0.2. Lemma. *For $A = \mathbf{F}_\lambda$, $\mathcal{F}$ in $\mathcal{C}$ and $\mathcal{G}$ in $\mathcal{T}$, we have*

(1) *The sheaves $\mathcal{F}_! * \mathcal{G}$ and $\mathcal{F}_* * \mathcal{G}$ are lisse $\mathbf{F}_\lambda$-sheaves on $\mathbf{G}_m \otimes k$ of the same rank, namely* $\operatorname{rank}(\mathcal{F}) \operatorname{Swan}_\infty(\mathcal{G}) + \operatorname{rank}(\mathcal{G}) \operatorname{Swan}_\infty(\mathcal{F})$.

(2) *The kernel "$Ker(\mathcal{F}, \mathcal{G})$" of the canonical map $\mathcal{F}_! * \mathcal{G} \to \mathcal{F}_* * \mathcal{G}$ is lisse on $\mathbf{G}_m \otimes k$, and tame at both zero and ∞.*

(3) *If $\mathcal{F}_! * \mathcal{G}$ lies in $\mathcal{C}$, then $\mathcal{F}_! * \mathcal{G} \xrightarrow{\sim} \mathcal{F}_* * \mathcal{G}$.*

Proof. For (1), we already know both convolutions are lisse, tame at 0, and that their formation commutes with passage to fibres. The rank is computed by the Euler-Poincaré formula, applied to any fibre (cf. the proof of the same formula for $\mathcal{F}, \mathcal{G}$ both in $\mathcal{C}$).

For (2), the fact that Ker is lisse on $\mathbf{G}_m \otimes k$ and tame at zero is obvious, because it is the kernel of a map of sheaves each of which is lisse, and tame at zero. Granting for a moment that Ker is also tame at ∞, we may easily deduce (3), for if $\mathcal{F}_! * \mathcal{G}$ is totally wild at ∞, then its tame-at-∞ subsheaf Ker must vanish. As $\mathcal{F}_! * \mathcal{G}$ and $\mathcal{F}_* * \mathcal{G}$ are lisse of the same rank over the *field* $\mathbf{F}_\lambda$, the injective map between them must be an isomorphism.

It remains to prove that Ker is tame at ∞, for $\mathcal{F} \in \mathcal{C}$ and $\mathcal{G} \in \mathcal{T}$. We first reduce to the case when $\mathcal{F}$ and $\mathcal{G}$ are *lisse at zero*. Because $\mathcal{F}$ is tame at zero, and we work with finite coefficients $\mathbf{F}_\lambda$, there exists an integer $N \geq 1$ prime to p for which $[N]^*\mathcal{F}$ and $[N]^*(\mathcal{G})$ are lisse at zero, (N.B.: $[N]^*\mathcal{F}$ is still totally wild at ∞). So the spectral sequence (5.2.1.1) and its variant (5.2.1.2) yield isomorphisms

$$[N]^*(\mathcal{F}_! * \mathcal{G}) \xrightarrow{\sim} ([N]^*(\mathcal{F})_! * [N]^*(\mathcal{G}))^{\boldsymbol{\mu}_N(k)}$$

$$[N]^*(\mathcal{F}_* * \mathcal{G}) \xrightarrow{\sim} ([N]^*(\mathcal{F})_* * [N]^*(\mathcal{G}))^{\boldsymbol{\mu}_N(k)},$$

whence an isomorphism

$$[N]^*(\mathrm{Ker}(\mathcal{F}, \mathcal{G})) \xrightarrow{\sim} (\mathrm{Ker}([N]^*\mathcal{F}, [N]^*\mathcal{G}))^{\boldsymbol{\mu}_N}.$$

Therefore it suffices to prove that $\mathrm{Ker}([N]^*\mathcal{F}, [N]^*\mathcal{G})$ is itself tame at ∞, i.e., to prove that $\mathrm{Ker}(\mathcal{F}, \mathcal{G})$ is tame at ∞ under the additional hypothesis that $\mathcal{F}, \mathcal{G}$ are both lisse at zero.

For any $\mathcal{F}, \mathcal{G}$ lisse on $\mathbf{G}_m$, the standard compactification of the multiplication map $\pi : \mathbf{G}_m \times \mathbf{G}_m \to \mathbf{G}_m$ as

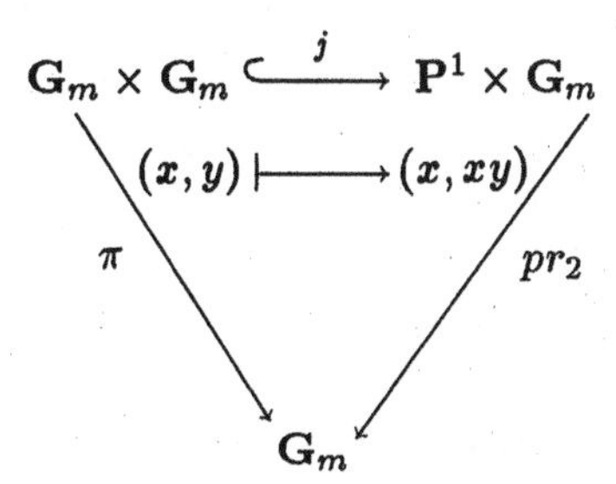

gives

$$R\pi_!(\mathcal{F} \boxtimes \mathcal{G}) = R(pr_2)_*(j_!(\mathcal{F} \boxtimes \mathcal{G}))$$
$$R\pi_*(\mathcal{F} \boxtimes \mathcal{G}) = R(pr_2)_*(Rj_*(\mathcal{F} \boxtimes \mathcal{G})).$$

Because the natural map in $D^b_c(\mathbf{G}_m \otimes k, \mathbf{F}_\lambda)$

$$j_!(\mathcal{F} \boxtimes \mathcal{G}) \to Rj_*(\mathcal{F} \boxtimes \mathcal{G})$$

is an isomorphism on $\mathbf{G}_m \times \mathbf{G}_m$, the cohomology of its mapping cylinder is concentrated on the two sections $\{0\} \times \mathbf{G}_m$ and $\{\infty\} \times \mathbf{G}_m$ of

$pr_2 : \mathbf{P}^1 \times \mathbf{G}_m \to \mathbf{G}_m$. For any sheaf $\mathcal{H}$ on $\mathbf{P}^1 \times \mathbf{G}_m$ which is concentrated on these two sections, we have

$$R^i(pr_2)_*(\mathcal{H}) = \begin{cases} 0 & \text{if } i \neq 0 \\ (\mathcal{H} \mid \{0\} \times \mathbf{G}_m) \bigoplus (\mathcal{H} \mid \{\infty\} \times \mathbf{G}_m) & \text{if } i = 0, \end{cases}$$

where in the last formula both $\{0\} \times \mathbf{G}_m$ and $\{\infty\} \times \mathbf{G}_m$ are viewed as "$\mathbf{G}_m$." Therefore the long exact cohomology sequence (for $R(pr_2)_*$, and the above $j_! \to Rj_*$) is

$$\begin{aligned} 0 \to R^0\pi_!(\mathcal{F} \boxtimes \mathcal{G}) &\to R^0\pi_*(\mathcal{F} \boxtimes \mathcal{G}) \to \\ \to (j_*(\mathcal{F} \boxtimes \mathcal{G}) \mid \{0\} \times \mathbf{G}_m) &\bigoplus (j_*(\mathcal{F} \boxtimes \mathcal{G}) \mid \{\infty\} \times \mathbf{G}_m) \to R^1\pi_!(\mathcal{F} \boxtimes \mathcal{G}) \to \\ &\to R^1\pi_*(\mathcal{F} \boxtimes \mathcal{G}) \to (R^1 j_*(\mathcal{F} \boxtimes \mathcal{G}) \mid \{0\} \times \mathbf{G}_m) \bigoplus \to \\ &\to (R^1 j_*(\mathcal{F} \boxtimes \mathcal{G}) \mid \{\infty\} \times \mathbf{G}_m) \to \\ &\to R^2\pi_!(\mathcal{F} \boxtimes \mathcal{G}) \to R^2\pi_*(\mathcal{F} \boxtimes \mathcal{G}) \to \ldots \end{aligned}$$

Now suppose that $\mathcal{F}$ and $\mathcal{G}$ are each lisse at zero, and denote by $\tilde{\mathcal{F}}, \tilde{\mathcal{G}}$ the unique lisse $\mathbf{F}_\lambda$-sheaves on $\mathbf{A}^1 \otimes k$ extending $\mathcal{F}$ and $\mathcal{G}$ respectively (i.e., in terms of $j_0 : \mathbf{G}_m \hookrightarrow \mathbf{A}^1$, $\tilde{\mathcal{F}} = (j_0)_*\mathcal{F}$, and $\tilde{\mathcal{G}} = (j_0)_*\mathcal{G}$). Then we claim that we have canonical isomorphisms of sheaves

$$\begin{cases} j_*(\mathcal{F} \boxtimes \mathcal{G}) \mid \{0\} \times \mathbf{G}_m \simeq \tilde{\mathcal{F}}(0) \underset{\mathbf{F}_\lambda}{\otimes} (j_*(\mathbf{F}_\lambda \boxtimes \mathcal{G}) \mid \{0\} \times \mathbf{G}_m) \\ j_*(\mathcal{F} \boxtimes \mathcal{G}) \mid \{\infty\} \times \mathbf{G}_m \simeq \tilde{\mathcal{G}}(0) \underset{\mathbf{F}_\lambda}{\otimes} (j_*(\mathcal{F} \boxtimes \mathbf{F}_\lambda) \mid \{\infty\} \times \mathbf{G}_m). \end{cases}$$

Indeed, to calculate $j_*(\mathcal{F} \boxtimes \mathcal{G})$ restricted to $\{0\} \times \mathbf{G}_m$, it suffices to calculate in the Zariski neighborhood $\mathbf{A}^1 \times \mathbf{G}_m$ of $\{0\} \times \mathbf{G}_m$, i.e. to calculate $k_*(\mathcal{F} \boxtimes \mathcal{G})|\{0\} \times \mathbf{G}_m$ where

$$\begin{aligned} k : \mathbf{G}_m \times \mathbf{G}_m &\longrightarrow \mathbf{A}^1 \times \mathbf{G}_m \\ (x, y) &\longmapsto (x, xy). \end{aligned}$$

In terms of this calculation, we have

$$\mathcal{F} \boxtimes \mathcal{G} = k^*(\tilde{\mathcal{F}} \boxtimes \mathbf{F}_\lambda) \otimes (\mathbf{F}_\lambda \boxtimes \mathcal{G}),$$

so the projection formula gives an isomorphism on $\mathbf{A}^1 \times \mathbf{G}_m$

$$R^i k_*(\mathcal{F} \boxtimes \mathcal{G}) \simeq (\tilde{\mathcal{F}} \boxtimes \mathbf{F}_\lambda) \otimes R^i k_*(\mathbf{F}_\lambda \boxtimes \mathcal{G}).$$

Taking $i = 0$ and restricting to $\{0\} \times \mathbf{G}_m$ in $\mathbf{A}^1 \times \mathbf{G}_m$ gives the assertion for $j_*(\mathcal{F} \boxtimes \mathcal{G})|\{0\} \times \mathbf{G}_m$.

To calculate $j_*((\mathcal{F} \boxtimes \mathcal{G})|\{\infty\} \times \mathbf{G}_m$, we may calculate in the Zariski open neighborhood $(\mathbf{P}^1 - \{0\}) \times \mathbf{G}_m$ of $\{\infty\} \times \mathbf{G}_m$ in $\mathbf{P}^1 \times \mathbf{G}_m$.

From the commutative diagram

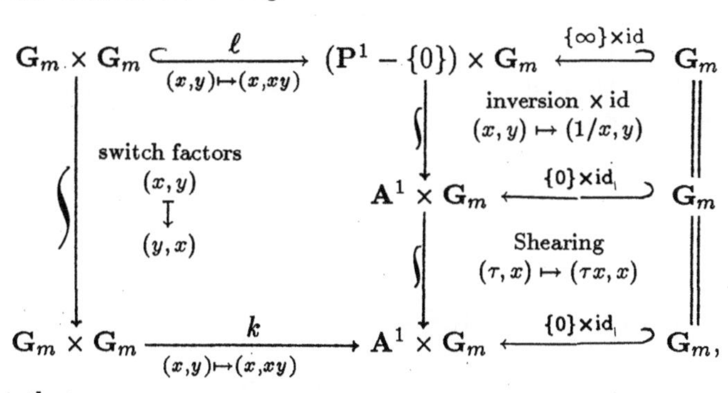

we see that

$$j_*(\mathcal{F} \boxtimes \mathcal{G}) \mid \{\infty\} \times \mathbf{G}_m = l_*(\mathcal{F} \boxtimes \mathcal{G}) \mid \{\infty\} \times \mathbf{G}_m$$
$$\big\downarrow \wr$$
$$k_*(\mathcal{G} \boxtimes \mathcal{F}) \mid \{0\} \times \mathbf{G}_m,$$

and this last is, as we have seen just above, isomorphic to

$$\widetilde{\mathcal{G}}(0) \underset{|\mathbf{F}_\lambda}{\otimes} (k_*(\mathbf{F}_\lambda \boxtimes \mathcal{F}) \mid \{0\} \times \mathbf{G}_m).$$

Reading backwards with $\mathcal{G}$ replaced by $\mathbf{F}_\lambda$, we may rewrite this as

$$\widetilde{\mathcal{G}}(0) \underset{\mathbf{F}_\lambda}{\otimes} (j_*(\mathcal{F} \boxtimes \mathbf{F}_\lambda) \mid \{\infty\} \times \mathbf{G}_m)),$$

as asserted. This concludes the proof of the claim.

Thus we are reduced to showing that if $\mathcal{F}$ and $\mathcal{G}$ are lisse sheaves on $\mathbf{G}_m \otimes k$ which are lisse at zero, then the sheaves

$$(j_*(\mathbf{F}_\lambda \boxtimes \mathcal{G}) \mid \{0\} \times \mathbf{G}_m),$$
$$(j_*(\mathcal{F} \boxtimes \mathbf{F}_\lambda) \mid \{\infty\} \times \mathbf{G}_m),$$

are tame at ∞. But each of these sheaves is a direct factor in a sheaf which, by the above long exact sequence formed with $\mathbf{F}_\lambda$ and $\mathcal{G}$ (resp.,with $\mathcal{F}$ and $\mathbf{F}_\lambda$), is caught between $R^0\pi_*(\mathbf{F}_\lambda \boxtimes \mathcal{G})$ and $R^1\pi_!(\mathcal{F}_\lambda \boxtimes \mathcal{G})$ (resp. between $R^0\pi_*(\mathcal{F} \boxtimes \mathbf{F}_\lambda)$ and $R^1\pi_!(\mathcal{F} \boxtimes \mathbf{F}_\lambda)$). By Lemma 5.2.1.4, we see that all these last sheaves are *constant* sheaves on $\mathbf{G}_m \otimes k$. ∎

We now return to the proof of the theorem with $A = \mathbf{F}_\lambda$, $\mathcal{F}$ non-zero in $\mathcal{C}$, and $\mathcal{G}$ in $\mathcal{T}$ such that (thanks to the lemma)

$$\mathcal{F}_! * \mathcal{G} \xrightarrow{\sim} \mathcal{F}_* * \mathcal{G} \text{ lies in } \mathcal{C}.$$

We wish to prove that $\mathcal{G}$ lies in $\mathcal{C}$, i.e., is totally wild at ∞.

8.0.3. To prove this, we will use Gabber's method of spans (cf. 7.5) to analyze the local monodromy at both zero *and* ∞ of $\mathcal{G}$. For any finite local $\mathbf{F}_\lambda$-algebra A, and any continuous character

$$\chi : \pi_1(\mathbf{G}_m \otimes k)^{\text{tame}} = \prod_{l \neq p} \mathbf{Z}_l(1) \to A^\times,$$

we have

$$\begin{aligned}(\mathcal{F}_! * \mathcal{G}) \underset{\mathbf{F}_\lambda}{\otimes} \mathcal{L}_\chi &= ((\mathcal{F}_! * \mathcal{G}) \underset{\mathbf{F}_\lambda}{\otimes} A) \underset{A}{\otimes} \mathcal{L}_\chi \\ &\simeq ((\mathcal{F} \otimes A)_! * (\mathcal{G} \otimes A)) \underset{A}{\otimes} \mathcal{L}_\chi \\ &\simeq ((\mathcal{F} \otimes A) \underset{A}{\otimes} \mathcal{L}_\chi)_! * ((\mathcal{G} \otimes A) \underset{A}{\otimes} \mathcal{L}_\chi) \\ &= (\mathcal{F} \underset{\mathbf{F}_\lambda}{\otimes} \mathcal{L}_\chi)_! * (\mathcal{G} \underset{\mathbf{F}_\lambda}{\otimes} \mathcal{L}_\chi),\end{aligned}$$

and similarly

$$(\mathcal{F}_* * \mathcal{G}) \underset{\mathbf{F}_\lambda}{\otimes} \mathcal{L}_\chi \simeq (\mathcal{F} \underset{\mathbf{F}_\lambda}{\otimes} \mathcal{L}_\chi)_* * \mathcal{G} \underset{\mathbf{F}_\lambda}{\otimes} \mathcal{L}_\chi).$$

Applying 5.2.2.1, we see that for $i \neq 1$,

$$H_c^i(\mathbf{G}_m \otimes k, \mathcal{G} \underset{\mathbf{F}_\lambda}{\otimes} \mathcal{L}_\chi) = 0 = H^i(\mathbf{G}_m \otimes k, \mathcal{G} \underset{\mathbf{F}_\lambda}{\otimes} \mathcal{L}_\chi);$$

and for $i = 1$ we have isomorphisms

$$\begin{aligned}H_c^1(\text{---}, (\mathcal{F}_! * \mathcal{G}) \underset{\mathbf{F}_\lambda}{\otimes} \mathcal{L}_\chi) &\xrightarrow{\sim} H_c^1(\text{---}, \mathcal{F} \underset{\mathbf{F}_\lambda}{\otimes} \mathcal{L}_\chi) \underset{A}{\otimes} H_c^1(\text{---}, \mathcal{G} \underset{\mathbf{F}_\lambda}{\otimes} \mathcal{L}_\chi) \\ H^1(\text{---}, (\mathcal{F}_! * \mathcal{G}) \underset{\mathbf{F}_\lambda}{\otimes} \mathcal{L}_\chi) &\xrightarrow{\sim} H^1(\text{---}, \mathcal{F} \underset{\mathbf{F}_\lambda}{\otimes} \mathcal{L}_\chi) \underset{A}{\otimes} H^1(\text{---}, \mathcal{G} \underset{\mathbf{F}_\lambda}{\otimes} \mathcal{L}_\chi).\end{aligned}$$

Because both H_c^1 and H^1 of $\mathcal{F} \underset{\mathbf{F}_\lambda}{\otimes} \mathcal{L}_\chi$ are free A-modules of non-zero rank $= \mathrm{Swan}_\infty(\mathcal{F})$, while the H_c^1 and H^1 of $(\mathcal{F}_! * \mathcal{G}) \underset{\mathbf{F}_\lambda}{\otimes} \mathcal{L}_\chi$ and of $(\mathcal{F}_* * \mathcal{G}) \underset{\mathbf{F}_\lambda}{\otimes} \mathcal{L}_\chi$ respectively are themselves free A-modules, (because the sheaves in question lie in $\mathcal{C}$), it follows that

$$H_c^1(\mathbf{G}_m \otimes k, \mathcal{G} \underset{\mathbf{F}_\lambda}{\otimes} \mathcal{L}_\chi) \text{ and } H^1(\mathbf{G}_m \otimes k, \mathcal{G} \underset{\mathbf{F}_\lambda}{\otimes} \mathcal{L}_\chi)$$
$$\text{are each free } A\text{-modules of rank } \mathrm{Swan}_\infty(\mathcal{G}).$$

The long exact sequence for the mapping cylinder of

$$j_!(\mathcal{G} \underset{\mathbf{F}_\lambda}{\otimes} \mathcal{L}_\chi) \to Rj_*(\mathcal{G} \underset{\mathbf{F}_\lambda}{\otimes} \mathcal{L}_\chi),$$

for $j : \mathbf{G}_m \hookrightarrow \mathbf{P}^1$ the inclusion, thus reduces to a four term exact sequence (in view of the vanishing of H^i and H^i_c for $i \neq 1$)

$$0 \to (\mathcal{G} \otimes \mathcal{L}_\chi)^{I_0} \bigoplus (\mathcal{G} \otimes \mathcal{L}_\chi)^{I_\infty} \to H^1_c(\mathbf{G}_m \otimes k, \mathcal{G} \otimes \mathcal{L}_\chi) \to$$
$$\to H^1(\mathbf{G}_m \otimes k, \mathcal{G} \otimes \mathcal{L}_\chi) \to (\mathcal{G} \otimes \mathcal{L}_\chi)_{I_0}(-1) \bigoplus (\mathcal{G} \otimes \mathcal{L}_\chi)_{I_\infty}(-1) \to 0.$$

Applying this with

$$A = \text{ a finite over-field of } \mathbf{F}_\lambda, \text{ say } \mathbf{F}'_\lambda$$
$$\chi = \chi_0 : \pi_1(\mathbf{G}_m \otimes k)^{\text{tame}} \to A^\times \text{ a fixed character,}$$

we see in particular that

$$\dim(\mathcal{G} \otimes \mathcal{L}_{\chi_0})^{I_0} + \dim(\mathcal{G} \otimes \mathcal{L}_{\chi_0})^{I_\infty} \leq \operatorname{Swan}_\infty(\mathcal{G});$$

i.e.

$$\begin{pmatrix}\text{\# of Jordan blocks} \\ \text{in } (\mathcal{G} \otimes \mathcal{L}_{\chi_0})^{I_0\text{-unip}}\end{pmatrix} + \begin{pmatrix}\text{\# of Jordan blocks} \\ \text{in } (\mathcal{G} \otimes \mathcal{L}_{\chi_0})^{I_\infty\text{-unip}}\end{pmatrix} \leq \operatorname{Swan}_\infty(\mathcal{G}).$$

We define

$$f(\mathcal{G} \otimes \mathcal{L}_{\chi_0}, T) \in \mathbf{Z}[T]$$

by the recipe

$$f(\mathcal{G} \otimes \mathcal{L}_{\chi_0}, T) = \sum T^{(\text{dim of Jordan block})}$$

the sum extended to "all" the $\operatorname{Swan}_\infty(\mathcal{G})$ Jordan blocks of $(\mathcal{G} \otimes \mathcal{L}_{\chi_0})^{I_0\text{-unip}}$ at zero *and* of $(\mathcal{G} \otimes \mathcal{L}_{\chi_0})^{I_\infty\text{-unip}}$ at ∞, with the convention that of these, $\operatorname{Swan}_\infty(\mathcal{G}) - \dim(\mathcal{G} \otimes \mathcal{L}_{\chi_0})^{I_0} - \dim(\mathcal{G} \otimes \mathcal{L}_{\chi_0})^{I_\infty}$ are of size zero.

Pass now to

$$\begin{cases} A = \mathbf{F}'_\lambda[[X]] \\ \chi = \chi_0\chi^{\text{univ}}, \text{ where } \chi^{\text{univ}} : \pi_1(\mathbf{G}_m \otimes k)^{\text{tame}} \to A^\times \\ \quad \text{maps a chosen generator } \gamma_0 \text{ of } I_0^{\text{tame}} \text{ to } (1+X)^{-1}, \\ \text{and also maps a generator } \gamma_\infty \text{ of } I_\infty^{\text{tame}} \text{ to } (1+X)^{-1}. \end{cases}$$

(Because the inclusions of both I_0 and I_∞ in $\pi_1(\mathbf{G}_m \otimes k)$ induce isomorphisms

$$\begin{array}{ccc} I_0^{\text{tame}} & \xrightarrow{\sim} & \pi_1(\mathbf{G}_m \otimes k)^{\text{tame}} \\ & & \nearrow \wr \\ I_\infty^{\text{tame}} & & \end{array}$$

the chosen generator γ_0 of I_0^{tame} defines a unique generator γ_∞ of I_∞^{tame} which has the same image as γ_0 in $\pi_1(\mathbf{G}_m \otimes k)^{\mathrm{tame}}$.) Just as in 7.5.6, we obtain a three-term exact sequence

$$0 \to H_c^1(\mathbf{G}_m \otimes k, (\mathcal{G} \otimes \mathcal{L}_{\chi_0}) \otimes \mathcal{L}_{\chi^{\mathrm{univ}}}) \to H^1(\mathbf{G}_m \otimes k, (\mathcal{G} \otimes \mathcal{L}_{\chi_0}) \otimes \mathcal{L}_{\chi^{\mathrm{univ}}}) \to$$
$$\to (\mathcal{G} \otimes \mathcal{L}_{\chi_0} \otimes \mathcal{L}_{\chi^{\mathrm{univ}}})_{I_0}(-1) \bigoplus (\mathcal{G} \otimes \mathcal{L}_{\chi_0} \otimes \mathcal{L}_{\chi^{\mathrm{univ}}})_{I_\infty}(-1) \to 0.$$

Exactly as in 7.5.6, we see that

$$H_c^1(\mathbf{G}_m \otimes k, \mathcal{G} \otimes \mathcal{L}_{\chi_0} \otimes \mathcal{L}_{\chi^{\mathrm{univ}}}) \xrightarrow{\text{forget supports}} H^1(\mathbf{G}_m \otimes k, \mathcal{G} \otimes \mathcal{L}_{\chi_0} \otimes \mathcal{L}_{\chi^{\mathrm{univ}}})$$

is an $\mathbf{F}'_\lambda$-span, denoted $\mathrm{Span}(\mathcal{G} \otimes \mathcal{L}_{\chi_0})$, whose associated polynomial is just $f(\mathcal{G} \otimes \mathcal{L}_{\chi_0}, T)$:

$$f(\mathrm{Span}(\mathcal{G} \otimes \mathcal{L}_{\chi_0}), T) = f(\mathcal{G} \otimes \mathcal{L}_{\chi_0}, T).$$

From the isomorphisms 5.2.2.1, we see that the already-defined monodromy spans of the objects of $\mathcal{C}$ which are $\mathcal{F} \otimes \mathcal{L}_{\chi_0}$ and $(\mathcal{F}_! * \mathcal{G}) \otimes \mathcal{L}_{\chi_0} \simeq (\mathcal{F}_* * \mathcal{G}) \otimes \mathcal{L}_{\chi_0}$ are related to $\mathrm{Span}(\mathcal{G} \otimes \mathcal{L}_{\chi_0})$ by

$$\mathrm{Span}((\mathcal{F}_! * \mathcal{G}) \otimes \mathcal{L}_{\chi_0}) \simeq \mathrm{Span}(\mathcal{F} \otimes \mathcal{L}_{\chi_0}) \bigotimes \mathrm{Span}(\mathcal{G} \otimes \mathcal{L}_{\chi_0}).$$

Passing to associated polynomials, we find a product formula

$$f((\mathcal{F}_! * \mathcal{G}) \otimes \mathcal{L}_{\chi_0}, T) = f(\mathcal{F} \otimes \mathcal{L}_{\chi_0}, T) f(\mathcal{G} \otimes \mathcal{L}_{\chi_0}, T).$$

Differentiating and evaluating at $T = 1$, we find

$$\begin{aligned}\mathrm{rank}(((\mathcal{F}_! * \mathcal{G}) \otimes \mathcal{L}_{\chi_0})^{I_0-\mathrm{unip}}) &= f'(\mathcal{F} \otimes \mathcal{L}_{\chi_0}, 1) f(\mathcal{G} \otimes \mathcal{L}_{\chi_0}, 1) \\ &\quad + f(\mathcal{F} \otimes \mathcal{L}_{\chi_0}, 1) f'(\mathcal{G} \otimes \mathcal{L}_{\chi_0}, 1) \\ &= (\dim(\mathcal{F} \otimes \mathcal{L}_{\chi_0})^{I_0-\mathrm{unip}}) \,\mathrm{Swan}_\infty(\mathcal{G}) \\ &\quad + \mathrm{Swan}_\infty(\mathcal{F}) \left[\dim((\mathcal{G} \otimes \mathcal{L}_{\chi_0})^{I_0-\mathrm{unip}}) + \dim((\mathcal{G} \otimes \mathcal{L}_{\chi_0})^{I_\infty-\mathrm{unip}})\right].\end{aligned}$$

Summing over all χ_0, this yields

$$\begin{aligned}\mathrm{rank}(\mathcal{F}_! * \mathcal{G}) &= \mathrm{rank}(\mathcal{F}) \,\mathrm{Swan}_\infty(\mathcal{G}) + \mathrm{Swan}_\infty(\mathcal{F}) \,\mathrm{rank}(\mathcal{G}) \\ &\quad + \mathrm{Swan}_\infty(\mathcal{F}) \dim((\mathcal{G})^{P_\infty}).\end{aligned}$$

But we know that $\mathrm{rank}(\mathcal{F}_! * \mathcal{G})$ is equal to the sum of the first two terms alone, whence, as $\mathrm{Swan}_\infty(\mathcal{F}) \neq 0$, we conclude that $\dim(\mathcal{G})^{P_\infty} = 0$, i.e., $\mathcal{G}$ is totally wild at ∞. ∎

8.1. Two Variants of the Cancellation Theorem

8.1.1. First Variant. *Let k be algebraically closed of characteristic $p > 0$, l a prime number $l \neq p$, $\mathbf{F}_\lambda$ a finite extension of $\mathbf{F}_l$, A the coefficient ring $\mathbf{F}_\lambda$. Suppose we are given $\mathcal{F} \in \mathcal{C}$, $\mathcal{G} \in \mathcal{T}$ such that*

(1) $\mathcal{F}_! * \mathcal{G} \xrightarrow{\sim} \mathcal{F}_* * \mathcal{G}$

(2) *for all finite extension fields $\mathbf{F}'_\lambda$ of $\mathbf{F}_\lambda$, and all continuous characters $\chi : \pi_1(\mathbf{G}_m \otimes k)^{\text{tame}} \to (\mathbf{F}'_\lambda)^\times$, we have $H^0(\mathbf{G}_m \otimes k, \mathcal{G} \otimes \mathcal{L}_\chi) = 0 = H^2_c(\mathbf{G}_m \otimes k, \mathcal{G} \otimes \mathcal{L}_\chi)$.*

Then we have the formula

$$\dim((\mathcal{F}_! * \mathcal{G})^{P_\infty}) = \operatorname{Swan}_\infty(\mathcal{F}) \dim((\mathcal{G})^{P_\infty}).$$

Proof. The isomorphisms 5.2.2.1 show that the sheaf $\mathcal{F}_! * \mathcal{G} \simeq \mathcal{F}_* * \mathcal{G}$ also satisfies the vanishing properties of (2) above:

$$H^0(\mathbf{G}_m \otimes k, (\mathcal{F}_* * \mathcal{G}) \otimes \mathcal{L}_\chi) = 0 = H^2_c(\mathbf{G}_m \otimes k, (\mathcal{F}_! * \mathcal{G}) \otimes \mathcal{L}_\chi),$$

for all χ as above. Because $\mathcal{F}$ lies in $\mathcal{C}$, it too satisfies these vanishings. We wish to pass to the spans attached to these sheaves. For this, we need the following standard lemma, the second part of which was already proven (cf. 2.2.7). ∎

8.1.2. Lemma. *Let A be a complete noetherian local ring with finite residue field $\mathbf{F}_\lambda$ of characteristic $l \neq p$, $\mathcal{H}$ a lisse sheaf of free A-modules of finite rank on a smooth open connected curve U over k, still supposed algebraically closed of characteristic $\neq l$.*

(1) *If $H^0(U, \mathcal{H} \underset{A}{\otimes} \mathbf{F}_\lambda) = 0$, then $H^0(U, \mathcal{H}) = 0$ and $H^1(U, \mathcal{H})$ is a free A-module whose formation commutes with arbitrary extensions of coefficient rings $A \to A'$.*

(2) *If $H^2_c(U, \mathcal{H} \underset{A}{\otimes} \mathbf{F}_\lambda) = 0$, then $H^2_c(U, \mathcal{H}) = 0$, and $H^1_c(U, \mathcal{H})$ is a free A-module whose formation commutes with arbitrary extensions of coefficient rings $A \to A'$.*

Proof. In both cases, the cohomology groups in question are those of a two-term complex of free A-modules of finite rank, say $M \to N$, (cf. 2.2) which after any extension of coefficient rings $A \to A'$ computes the cohomology groups of $\mathcal{H} \underset{A}{\otimes} A'$. In case (1), we are told that $M \otimes \mathbf{F}_\lambda \hookrightarrow N \otimes \mathbf{F}_\lambda$; therefore $\operatorname{rank}(M) \leq \operatorname{rank}(N)$, and in A-bases of M and N respectively, the matrix of $M \to N$ has a minor of size $\operatorname{rank}(M)$ which is invertible over $\mathbf{F}_\lambda$, so invertible. Therefore M is a direct factor of N, say $M = N \oplus M'$, and the assertion is obvious. In the second case, we are told that $M \to N$ is surjective after $\otimes \mathbf{F}_\lambda$, so surjective. As N is free, there exists a splitting $M \simeq \mathrm{Ker}(M \to N) \oplus N$. ∎

Fix a finite over-field $\mathbf{F}'_\lambda$ of $\mathbf{F}_\lambda$, and a character

$$\chi_0 : \pi_1(\mathbf{G}_m \otimes k)^{\mathrm{tame}} \to (\mathbf{F}'_\lambda)^\times.$$

We will apply the lemma to $U = \mathbf{G}_m \otimes k$, $A = \mathbf{F}'_\lambda[[X]]$, $\mathcal{H}$ the sheaf $\mathcal{K} \underset{\mathbf{F}'_\lambda}{\otimes} \mathcal{L}_{\chi^{\mathrm{univ}}}$ where $\mathcal{K}$ is any of the lisse $\mathbf{F}'_\lambda$-sheaves $\mathcal{F} \otimes \mathcal{L}_{\chi_0}, \mathcal{G} \otimes \mathcal{L}_{\chi_0}$, $(\mathcal{F}_! * \mathcal{G}) \otimes \mathcal{L}_{\chi_0}$. This allows us to form the *spans* of each, and their associated polynomials $f(\mathcal{K}, T)$, given by

$$f(\mathcal{K}, T) = \sum T^{(\mathrm{dim.\ of\ Jordan\ blocks})}$$

where the sum is extended to "all" the $\mathrm{Swan}_\infty(\mathcal{K})$ Jordan blocks of both $(\mathcal{K})^{I_0-\mathrm{unip}}$ as I_0-representation *and* of $(\mathcal{K})^{I_\infty-\mathrm{unip}}$ as I_∞-representation, with the usual convention that $\mathrm{Swan}_\infty(\mathcal{K}) - \dim(\mathcal{K}^{I_0}) - \dim(\mathcal{K}^{I_\infty})$ of these blocks have dimension zero. Just as in 8.0.3 above, we have the tensor product formula

$$\mathrm{Span}((\mathcal{F}_! * \mathcal{G}) \otimes \mathcal{L}_{\chi_0}) = \mathrm{Span}(\mathcal{F} \otimes \mathcal{L}_{\chi_0}) \otimes \mathrm{Span}(\mathcal{G} \otimes \mathcal{L}_{\chi_0})$$

with its resulting product formula

$$f((\mathcal{F}_! * \mathcal{G}) \otimes \mathcal{L}_{\chi_0}, T) = f(\mathcal{F} \otimes \mathcal{L}_{\chi_0}, T) f(\mathcal{G} \otimes \mathcal{L}_{\chi_0}, T).$$

Differentiating and evaluating at 1, and summing over all χ_0, gives the assertion. ∎

8.1.3. Second Variant. *Let k be a finite field of characteristic p, l a prime number $l \neq p$, E_λ a finite extension of $\mathbf{Q}_\lambda$, A the coefficient ring E_λ. Suppose we are given $\mathcal{F} \in \mathcal{C}$, $\mathcal{G} \in \mathcal{T}$ such that*

1. *$\mathcal{F}, \mathcal{G}$ and $\mathcal{F}_! * \mathcal{G}$ are all pure sheaves on $\mathbf{G}_m \otimes k$.*
2. *For any finite extension field E'_λ of E_λ, and any continuous character of finite order $\chi : \pi_1(\mathbf{G}_m \otimes k^{\mathrm{sep}})^{\mathrm{tame}} \to (E'_\lambda)^\times$, we have $H^0(\mathbf{G}_m \otimes k^{\mathrm{sep}}, \mathcal{G} \otimes \mathcal{L}_\chi) = 0 = H^2_c(\mathbf{G}_m \otimes k^{\mathrm{sep}}, \mathcal{G} \otimes \mathcal{L}_\chi)$.*

Then we have the formula

$$\dim((\mathcal{F}_! * \mathcal{G})^{P_\infty}) = \mathrm{Swan}_\infty(\mathcal{F}) \dim((\mathcal{G})^{P_\infty}).$$

Proof. By the isomorphisms 5.2.2.1, (resp., the fact that $\mathcal{F} \in \mathcal{C}$), we see that $\mathcal{F}_! * \mathcal{G}$ (resp. $\mathcal{F}$) also satisfies the vanishing properties (2) above. For

$\mathcal{K}$ any of the sheaves $\mathcal{F}\otimes\mathcal{L}_\chi$, $\mathcal{G}\otimes\mathcal{L}_\chi$, $(\mathcal{F}_!*\mathcal{G})\otimes\mathcal{L}_\chi$, we have, denoting by $j:\mathbf{G}_m\hookrightarrow\mathbf{P}^1$ the inclusion, a three-term exact sequence

$$0\to\mathcal{K}^{I_0}\bigoplus\mathcal{K}^{I_\infty}\to H^1_c(\mathbf{G}_m\otimes k^{\text{sep}},\mathcal{K})\to H^1(\mathbf{P}^1\otimes k^{\text{sep}},j_*\mathcal{K})\to 0.$$

We may apply the weight analysis of 7.1.1, to find that $f(\mathcal{K},T)$ as defined in the first variant is equal to

$$\sum_{\substack{\alpha\text{ eigenvalue}\\ \text{of }F\text{ on}\\ H^1_c(\mathbf{G}_m\otimes k^{\text{sep}},\mathcal{K})}} T^{(1+\text{weight}(\mathcal{K})-\text{weight}(\alpha))}.$$

From this, and the isomorphism

$$\begin{aligned}H^1_c(\mathbf{G}_m\otimes k^{\text{sep}},\mathcal{F}\otimes\mathcal{L}_\chi)\otimes H^1_c(\mathbf{G}_m\otimes k^{\text{sep}},\mathcal{G}\otimes\mathcal{L}_\chi)&\\ \simeq H^1_c(\mathbf{G}_m\otimes k^{\text{sep}},(\mathcal{F}_!*\mathcal{G})\otimes\mathcal{L}_\chi),&\end{aligned}$$

it follows (cf. 7.1.4) that we have the product formula

$$f((\mathcal{F}_!*\mathcal{G})\otimes\mathcal{L}_\chi,T)=f(\mathcal{F}\otimes\mathcal{L}_\chi,T)\cdot(\mathcal{G}\otimes\mathcal{L}_\chi,T).$$

Differentiating, evaluating at $T=1$, and summing over all χ of finite order gives the assertion. ∎

8.1.4. Remark. For $\mathcal{F}$ and $\mathcal{G}$ as in the first variant, if we *do not* use the fact that $\mathcal{F}_!*\mathcal{G}$ is tame at zero, the calculation ending the proof will give, for each χ_0,

$$\begin{aligned}&\dim((\mathcal{F}_!*\mathcal{G})\otimes\mathcal{L}_{\chi_0})^{I_0-\text{unip}}+\dim((\mathcal{F}_!*\mathcal{G})\otimes\mathcal{L}_{\chi_0})^{I_\infty-\text{unip}}\\ &=\text{Swan}_\infty(\mathcal{F})\big[\dim(\mathcal{G}\otimes\mathcal{L}_{\chi_0})^{I_0-\text{unip}}+\dim(\mathcal{G}\otimes\mathcal{L}_{\chi_0})^{I_0-\text{unip}}\big]\\ &\qquad+\text{Swan}_\infty(\mathcal{G})\big[\dim(\mathcal{F}\otimes\mathcal{L}_{\chi_0})^{I_0-\text{unip}}\big].\end{aligned}$$

Summing over all χ_0, we find

$$\begin{aligned}&\dim((\mathcal{F}_!*\mathcal{G})^{P_0})+\dim((\mathcal{F}_!*\mathcal{G})^{P_\infty})\\ &=\text{Swan}_\infty(\mathcal{F})[\text{rank}(\mathcal{G})+\dim(\mathcal{G}^{P_\infty})]+\text{Swan}_\infty(\mathcal{G})\,\text{rank}(\mathcal{F}).\end{aligned}$$

But the Euler-Poincaré formula gives

$$\text{rank}(\mathcal{F}_!*\mathcal{G})=\text{Swan}_\infty(\mathcal{F})\,\text{rank}(\mathcal{G})+\text{Swan}_\infty(\mathcal{G})\,\text{rank}(\mathcal{F}),$$

so we have

$$\begin{aligned}\dim((\mathcal{F}_!*\mathcal{G})^{\text{wild at }0})+\dim((\mathcal{F}_!*\mathcal{G})^{\text{tame at }\infty})&\\ =\text{Swan}_\infty(\mathcal{F})\dim(\mathcal{G}^{\text{tame at }\infty}).&\end{aligned}$$

In particular, if $\mathcal{F}$ and $\mathcal{G}$ *both* lie in $\mathcal{C}$, we have

$$\dim(\mathcal{F}_!*\mathcal{G})^{\text{wild at }0}+\dim(\mathcal{F}_!*\mathcal{G})^{\text{tame at }\infty}=0.$$

This gives *another* proof, independent of 5.2.1(3) and 5.2.3, of the fact that

$$\mathcal{F}, \mathcal{G} \text{ both in } \mathcal{C} \Longrightarrow \mathcal{F} * \mathcal{G} \text{ lies in } \mathcal{C}.$$

8.2. Interlude: Naive Fourier Transform (cf. [Lau-2])

8.2.1. Let k be an algebraically closed field of characteristic $p > 0$, l a prime number $l \neq p$, E_λ a finite extension of $\mathbf{Q}_\ell$, $\mathcal{O}_\lambda$ its ring of integers, $\mathbf{F}_\lambda$ its residue field. Fix a finite subfield $\mathbf{F}_q \subset k$, and a *non-trivial* additive character

$$\psi : (\mathbf{F}_q, +) \to \mathcal{O}_\lambda^\times.$$

(Such a ψ exists if and only if the residue field $\mathbf{F}_\lambda$ contains a non-trivial p'th root of unity, i.e., if and only if $\#(\mathbf{F}_\lambda) \equiv 1 \mod p$.)

We have already defined $\mathcal{L}_\psi$ as a lisse, rank one $\mathcal{O}_\lambda$-sheaf on $\mathbf{A}^1 \otimes \mathbf{F}_q$, obtained by suitably pushing out the Lang torsor. By inverse image, we obtain $\mathcal{L}_\psi$ on $\mathbf{A}^1 \otimes k$. For any $a \in k = \mathbf{A}^1(k)$, we denote by

$$\mathcal{L}_{\psi(ax)} \overset{\text{dfn}}{=} (x \mapsto ax)^*(\mathcal{L}_\psi),$$

another lisse rank one $\mathcal{O}_\lambda$-sheaf on $\mathbf{A}^1 \otimes k$. For $a = 0$, $\mathcal{L}_{\psi(ax)}$ is the constant sheaf $\mathcal{O}_\lambda$; for $a \neq 0$, $\mathcal{L}_{\psi(ax)}$ has $\text{Swan}_\infty = 1$.

8.2.1.1. Definition. For A either of the coefficient fields $\mathbf{F}_\lambda$, E_λ, a constructible A-sheaf $\mathcal{F}$ on $\mathbf{A}^1 \otimes k$ is called "*elementary*" if it satisfies the two following conditions:

Elem (1) $\mathcal{F}$ has no non-zero punctual sections, i.e., if $j : U \hookrightarrow \mathbf{A}^1 \otimes k$ is the inclusion of any non-empty ouvert de lissité of $\mathcal{F}$, the canonical map $\mathcal{F} \longrightarrow j_* j^* \mathcal{F}$ is injective.

Elem (2) For all $a \in k$, $H^2_c(\mathbf{A}^1 \otimes k, \mathcal{F} \otimes \mathcal{L}_{\psi(ax)}) = 0$.

Because each $\mathcal{L}_{\psi(ax)}$ is lisse of rank one on $\mathbf{A}^1 \otimes k$, condition (1) is equivalent to

$$\text{Elem(1)bis : for all } a \in k,\ H^0_c(\mathbf{A}^1 \otimes k, \mathcal{F} \otimes \mathcal{L}_{\psi(ax)}) = 0.$$

8.2.1.2. Definition. A constructible A-sheaf $\mathcal{F}$ on $\mathbf{A}^1 \otimes k$ is called a Fourier sheaf if it is elementary and it satisfies the following two additional conditions:

Fourier (1) For $j : U \hookrightarrow \mathbf{A}^1 \otimes k$ the inclusion of any non-empty ouvert de lissité of $\mathcal{F}$, we have $\mathcal{F} \xrightarrow{\sim} j_* j^* \mathcal{F}$

Fourier (2) For any $a \in k$,

$$H^0(\mathbf{A}^1 \otimes k, \mathcal{F} \otimes \mathcal{L}_{\psi(ax)}) = 0.$$

Notice that the notion of being an "elementary" or a "Fourier"sheaf is independent of the auxiliary choice of $(\mathbf{F}_q, \psi)$, appearing in its definition. Indeed for any other choice $(\mathbf{F}_{q_1}, \psi_1)$, pick a common finite over-field $\mathbf{F}_{q_2}$ of both $\mathbf{F}_q$ and $\mathbf{F}_{q_1}$, and denote by $\tilde{\psi}$ and $\tilde{\psi}_1$ the non-trivial characters of $\mathbf{F}_{q_2}$ obtained from ψ and ψ_1 respectively by composing with the trace from $\mathbf{F}_{q_2}$ to $\mathbf{F}_q$ and to $\mathbf{F}_{q_1}$ respectively. Then there is a unique $b \in (\mathbf{F}_{q_2})^\times$ such that $\tilde{\psi}_1(x) = \tilde{\psi}(bx)$ for all $x \in \mathbf{F}_{q_2}$, and for this b we have an isomorphism of sheaves on $\mathbf{A}^1 \otimes k$

$$\mathcal{L}_{\psi_1(ax)} \simeq \mathcal{L}_{\psi(abx)}.$$

8.2.1.3. Lemma. *Let $\mathcal{F}$ be a constructible A-sheaf on $\mathbf{A}^1 \otimes k$, $j : U \hookrightarrow \mathbf{A}^1 \otimes k$ the inclusion of a non-empty "ouvert de lissité" of $\mathcal{F}$. Denote by $(j^*\mathcal{F})^\vee$ the A-linear dual of $j^*\mathcal{F}$ (contragradient representation of $\pi_1(U, \bar{\eta})$). Then we have the following implications:*

(1) *If $\mathcal{F}$ is Fourier, then $\mathcal{F}$ is elementary.*
(2) *If $\mathcal{F} \xrightarrow{\sim} j_*j^*\mathcal{F}$, and if $(j^*\mathcal{F})_{\bar{\eta}}$ is a completely reducible representation of $\pi_1(U, \bar{\eta})$, then $\mathcal{F}$ is elementary if and only if $\mathcal{F}$ is Fourier, indeed* Elem(2) $\Longleftrightarrow$ Fourier(2) *for such an $\mathcal{F}$.*
(3) *If $\mathcal{F} \xrightarrow{\sim} j_*j^*\mathcal{F}$, then $\mathcal{F}$ is Fourier if and only if both $\mathcal{F}$ and $j_*((j^*\mathcal{F})^\vee)$ are elementary.*
(4) *If $\mathcal{F}$ is Fourier, then $j_*((j^*\mathcal{F})^\vee)$ is Fourier.*

Proof. (1) is obvious. (2) holds because Elem(2) is equivalent to requiring that $j^*\mathcal{F}$ as $\pi_1(U, \bar{\eta})$ representation have *no quotient* isomorphic to any of the one-dimensional representations $j^*\mathcal{L}_{\psi(ax)}$, for any $a \in k$, while Fourier(2) requires in addition that $j^*\mathcal{F}$ as $\pi_1(U, \bar{\eta})$ have *no subrepresentation* isomorphic to any of the $j^*\mathcal{L}_{\psi(ax)}$, $a \in k$. (3) is obvious from the fact that $H^0(U, (j^*\mathcal{F})^\vee \otimes \mathcal{L}_{\psi(-ax)})$ is the $A(-1)$-dual of $H^2_c(U, j^*\mathcal{F} \otimes \mathcal{L}_{\psi(ax)})$, and similarly with $\mathcal{F}$ replaced by $\mathcal{F}^\vee$. (4) is obvious from (3). ∎

8.2.1.4. Lemma. *Let K be an algebraically closed over-field of k, $\mathcal{F}$ a constructible A-sheaf on $\mathbf{A}^1 \otimes k$, $\mathcal{F}_K$ its inverse image on $\mathbf{A}^1 \otimes K$. Then $\mathcal{F}$ is elementary (resp. Fourier) if and only if $\mathcal{F}_K$ is elementary (resp. Fourier).*

Proof. Consider the diagram

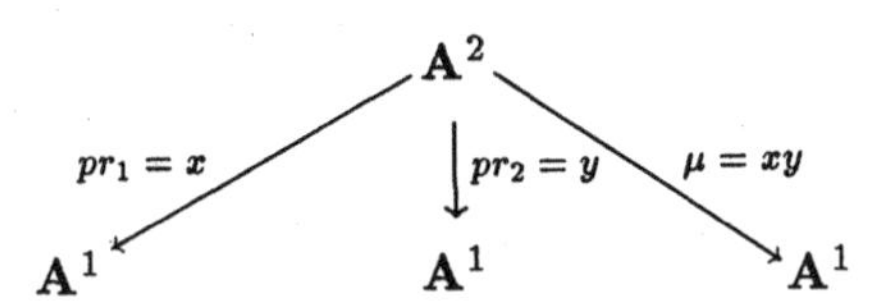

Then $\mathcal{F}$ on $\mathbf{A}^1 \otimes k$ is elementary if and only if

$$R^i(pr_2)_!(pr_1^*(\mathcal{F}) \otimes \mu^*(\mathcal{L}_\psi)) = 0 \quad \text{for} \quad i \neq 1,$$

so the asserted invariance of the property of being elementary results from the smooth base change theorem (SGA IV, Exp XVI, 1.1 and 1.5). By part (3) of the previous lemma, the invariance of "Fourierness" follows from that of elementaryness. ∎

8.2.2. Definition. For $A = \mathbf{F}_\lambda$ or E_λ as above, and k *any* field of characteristic p, a constructible A-sheaf $\mathcal{F}$ on $\mathbf{A}^1 \otimes k$ is said to be elementary, respectively Fourier, if for some (or equivalently, for any) algebraically closed over-field K of k, the inverse image of $\mathcal{F}$ on $\mathbf{A}^1 \otimes K$ is elementary (resp. Fourier).

8.2.3. Definition. Let $\mathcal{F}$ be a constructible A-sheaf on $\mathbf{A}^1 \otimes k$, k *any* over-field of $\mathbf{F}_q$. Its naive Fourier transform, denote $\mathrm{NFT}_\psi(\mathcal{F})$, is the constructible A-sheaf on $\mathbf{A}^1 \otimes k$ defined by

$$\mathrm{NFT}_\psi(\mathcal{F}) = R^1(pr_2)_!(pr_!^*(\mathcal{F}) \otimes \mu^*(\mathcal{L}_\psi)),$$

where pr_1, pr_2, μ are the three maps $\mathbf{A}^2 \to \mathbf{A}^1$ in 8.2.1.4 above.

8.2.4. Warning. This definition is reasonable *only* for elementary $\mathcal{F}$'s, as the "correct" Fourier transform of $\mathcal{F}$ is the object $R(pr_2)_!(pr_1^*(\mathcal{F}) \otimes \mu^*(\mathcal{L}_\psi))[1]$ in $D^b_c(\mathbf{A}^1 \otimes k, A)$.The elementary sheaves are precisely those for which this object has *only* its zeroeth cohomology sheaf non-zero, and our "naive" Fourier transform is precisely this cohomology sheaf.

8.2.5. Main Theorem on Fourier Transform (Laumon, Deligne).

(1) *If $\mathcal{F}$ is elementary, then* $\mathrm{NFT}_\psi(\mathcal{F})$ *is elementary, and if we denote by $\overline{\psi}$ the inverse (complex conjugate) character to ψ, we have a canonical isomorphism, functorial in $\mathcal{F}$,*

$$\mathrm{NFT}_{\overline{\psi}}(\mathrm{NFT}_\psi(\mathcal{F})) \xrightarrow{\sim} \mathcal{F}(-1).$$

(2) *If $\mathcal{F}$ is Fourier, then* $\mathrm{NFT}_\psi(\mathcal{F})$ *is Fourier. More precisely, for $\mathcal{F}$ elementary, and $i \neq j$ in $\{1,2\}$, then $\mathcal{F}$ satisfies Fourier(i) $\Longleftrightarrow$* $\mathrm{NFT}_\psi(\mathcal{F})$ *satisfies Fourier(j).*

(3) *If k is a finite field, $A = E_\lambda$, and $\mathcal{F}$ is a Fourier sheaf which is pure of some weight w (meaning that $j^*\mathcal{F}$ is punctually pure of weight w on any non-void ouvert de lissité), then* $\mathrm{NFT}_\psi(\mathcal{F})$ *is a Fourier sheaf which is pure of weight $w+1$.*

Proof. Assertion (1) is the involutivity of the "correct" Fourier transform. Assertion (2) is geometric, so we may assume k algebraically closed. For an elementary $\mathcal{F}$, and $j : U \hookrightarrow \mathbf{A}^1 \otimes k$ the inclusion of a non-empty ouvert de lissité of $\mathcal{F}$, we have a short exact sequence

$$0 \to \mathcal{F} \to j_* j^* \mathcal{F} \to \bigoplus_{a \in k} (\mathcal{F}^{I_a}/\mathcal{F}_a) \to 0.$$

Passing to the long cohomology sequence of the functors

$$R^i(pr_2)_!(pr_1^*(\text{——}) \otimes \mu^*(\mathcal{L}_\psi))$$

we get a short exact sequence of sheaves on $\mathbf{A}^1 \otimes k$

$$0 \to \bigoplus_{a \in k} (\mathcal{F}^{I_a}/\mathcal{F}_a) \otimes \mathcal{L}_{\psi(ax)} \to \mathrm{NFT}_\psi(\mathcal{F}) \to$$
$$\to \mathrm{NFT}_\psi(j_* j^* \mathcal{F}) \to 0.$$

Tensoring this with $\mathcal{L}_{\psi(-ax)}$ and passing to cohomology on $\mathbf{A}^1 \otimes k$, we get, for every $a \in k$,

$$0 \to (\mathcal{F}^{I_a}/\mathcal{F}_a) \to H^0(\mathbf{A}^1 \otimes k, \mathrm{NFT}_\psi(\mathcal{F}) \otimes \mathcal{L}_{\psi(-ax)}).$$

Thus we see that, for $\mathcal{F}$ elementary, we have

(8.2.5.4) $\quad \mathrm{NFT}_\psi(\mathcal{F})$ satisfies Fourier(2) $\Longrightarrow \mathcal{F}$ satisfies Fourier(1).

Now consider all the $\mathcal{L}_{\psi(ax)}$-subsheaves of a given elementary $\mathcal{F}$. Because $H^i_c(\mathbf{A}^1 \otimes k, \mathcal{L}_{\psi(ax)}) = 0$ for $i \neq 2$ and for all $a \in k$, we see that if $\mathcal{F}$ is elementary, and if we are given an injective map $\mathcal{L}_{\psi(ax)} \hookrightarrow \mathcal{F}$, then the cokernel is again elementary, for if

$$0 \to \mathcal{L}_{\psi(ax)} \to \mathcal{F} \to \mathcal{G} \to 0,$$

then after twisting by any $\mathcal{L}_{\psi(bx)}$ we have

$$H^i_c(\mathbf{A}^1 \otimes k, \mathcal{F} \otimes \mathcal{L}_{\psi(bx)}) \twoheadrightarrow H^i_c(\mathbf{A}^1 \otimes k, \mathcal{G} \otimes \mathcal{L}_{\psi(bx)})$$

for $i \neq 1$.

Because the $\mathcal{L}_{\psi(ax)}$ for variable a are pairwise non-isomorphic, we have, for any elementary $\mathcal{F}$, a short exact sequence

$$0 \to \bigoplus_{a \in k} H^0(\mathbf{A}^1 \otimes k, \mathcal{F} \otimes \mathcal{L}_{\psi(ax)}) \otimes \mathcal{L}_{\psi(-ax)} \to \mathcal{F} \to \mathcal{G} \to 0$$

in which $\mathcal{F}$ is again elementary, and in which

$$H^0(\mathbf{A}^1 \otimes k, \mathcal{G} \otimes \mathcal{L}_{\psi(ax)}) = 0 \quad \text{for all } a \in k.$$

(One sees this last vanishing by tensoring the above exact sequence with $\mathcal{L}_{\psi(ax)}$, and passing to the long exact cohomology sequence.)

Now consider the long exact cohomology sequence of the functors

$$R^i(pr_2)_!((pr_1)^*(\text{---}) \otimes \mu^*(\mathcal{L}_\psi)):$$

$$0 \to \mathrm{NFT}_\psi(\mathcal{F}) \to \mathrm{NFT}_\psi(\mathcal{G}) \to \bigoplus_{a\in k} \begin{bmatrix} H^0(\mathbf{A}^1 \otimes k, \mathcal{F} \otimes \mathcal{L}_{\psi(ax)}) \\ \text{conc. at } a \end{bmatrix} \to 0.$$

But both $\mathcal{F}$ and $\mathcal{G}$ being elementary, their NFT_ψ's have no punctual sections. Because $\mathrm{NFT}_\psi(\mathcal{F})$ is, by the above sequence, a subsheaf of $\mathrm{NFT}_\psi(\mathcal{G})$ with punctual quotient, we see that, denoting by $j: U \hookrightarrow \mathbf{A}^1 \otimes k$ the inclusion of a non-empty ouvert de lissité of both $\mathcal{F}$ and $\mathcal{G}$, we have

$$\mathrm{NFT}_\psi(\mathcal{F}) \subset \mathrm{NFT}_\psi(\mathcal{G}) \subset j_*j^*\mathrm{NFT}_\psi(\mathcal{F}) = j_*j^*\mathrm{NFT}_\psi(\mathcal{G}),$$

whence an injective map

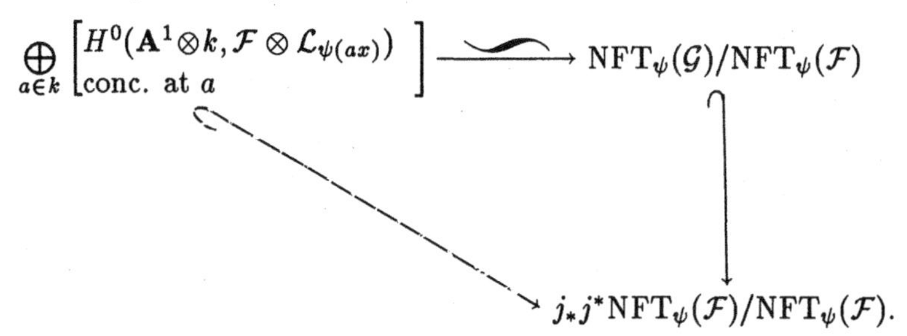

$$\bigoplus_{a\in k} \begin{bmatrix} H^0(\mathbf{A}^1 \otimes k, \mathcal{F} \otimes \mathcal{L}_{\psi(ax)}) \\ \text{conc. at } a \end{bmatrix} \xrightarrow{\sim} \mathrm{NFT}_\psi(\mathcal{G})/\mathrm{NFT}_\psi(\mathcal{F}) \hookrightarrow j_*j^*\mathrm{NFT}_\psi(\mathcal{F})/\mathrm{NFT}_\psi(\mathcal{F}).$$

Therefore we have, for $\mathcal{F}$ elementary,

$$\mathrm{NFT}_\psi(\mathcal{F}) \text{ satisfies Fourier(1)} \Longrightarrow \mathcal{F} \text{ satisfies Fourier(2)}.$$

Combining this with (8.2.5.4), we thus obtain, for $\mathcal{F}$ elementary, and $i \neq j$ in $\{1,2\}$,

$$\mathrm{NFT}_\psi(\mathcal{F}) \text{ satisfies Fourier(i)} \Longrightarrow \mathcal{F} \text{ satisfies Fourier(j)}$$

as required. By the involutivity (1) of NFT, these implications are in fact equivalences: for $\mathcal{F}$ elementary and $i \neq j$ in $\{1,2\}$,

$$\mathrm{NFT}_\psi(\mathcal{F}) \text{ satisfies Fourier(i)} \Longleftrightarrow \mathcal{F} \text{ satisfies Fourier(j)}.$$

Assertion (3) is a special case of ([K-L], 2.2.1). ∎

8.3. Basic examples of Fourier sheaves

8.3.1. Lemma. *Let k be an algebraically closed field of characteristic p, $j: U \hookrightarrow \mathbf{A}^1 \otimes k$ be the inclusion of a non-empty open set, $\mathcal{F}$ a lisse*

A-sheaf on U, for $A = E_\lambda$ or $\mathbf{F}_\lambda$. Then $j_*\mathcal{F}$ is a Fourier sheaf on $\mathbf{A}^1 \otimes k$ if any of the following conditions hold:

(1) *$\mathcal{F}$ is totally wild at ∞ but has no break equal to 1 (i.e., as P_∞-representation, all break of $\mathcal{F}$ are > 0 but $\neq 1$).*

(2) *At some point $a \in \mathbf{A}^1 \otimes k - U$, $\mathcal{F}^{I_a} = 0$ (i.e., $j_*\mathcal{F}$ on $\mathbf{A}^1 \otimes k$ has stalk zero at some point).*

(3) *As representation of $\pi_1(U, \bar{\eta})$, $\mathcal{F}$ is irreducible of rank ≥ 2.*

(4) *As representation of $\pi_1(U, \bar{\eta})$, every irreducible constitutent satisfies one of (1), (2) or (3) above.*

Proof. What must be shown is that $j_*\mathcal{F}$ contains no subsheaf and no quotient sheaf of the form $\mathcal{L}_{\psi(ax)}$ for any $a \in k$, and this obviously holds in all three cases. ∎

8.4. Irreducible Fourier Sheaves

Let k be an algebraically closed field of characteristic $p > 0$, $l \neq p$, $A = \mathbf{F}_\lambda$ or E_λ, $\mathcal{F}$ a Fourier (A-)sheaf on $\mathbf{A}^1 \otimes k$. We say that $\mathcal{F}$ is an irreducible Fourier sheaf if for some (equivalently: for any) non-empty ouvert de lissité $j : U \to \mathbf{A}^1 \otimes k$, $j^*\mathcal{F}$ is *irreducible* as an A-linear representation of $\pi_1(U, \bar{\eta})$. If K/k is an algebraically closed over-field, then a Fourier sheaf $\mathcal{F}$ on $\mathbf{A}^1 \otimes k$ is irreducible if and only if its inverse image on $\mathbf{A}^1 \otimes K$ is irreducible, simply because $\pi_1(U \otimes_k K) \twoheadrightarrow \pi(U)$.

8.4.1. Theorem (Brylinski). *If $\mathcal{F}$ is an irreducible Fourier sheaf, then $\mathrm{NFT}_\psi(\mathcal{F})$ is an irreducible Fourier sheaf.*

Proof. By Fourier inversion, it is equivalent to show that if a Fourier sheaf $\mathcal{F}$ is *not* irreducible, then neither is its NFT. So let $\mathcal{F}$ be a Fourier sheaf, $j : U \hookrightarrow \mathbf{A}^1 \otimes k$ a non-empty ouvert de lissité, and $\mathcal{F}_1 \subset j^*\mathcal{F}$ a lisse subsheaf corresponding to a non-zero *irreducible* representation of $\pi_1(U, \bar{\eta})$.We assume that $\mathcal{F}_1 \neq j^*\mathcal{F}$. Then we have a short exact sequence of lisse non-zero sheaves on U

$$0 \to \mathcal{F}_1 \to j^*\mathcal{F} \to \mathcal{F}_2 \to 0.$$

Apply j_*, we get an exact sequence

$$0 \to j_*\mathcal{F}_1 \to \mathcal{F} \to j_*\mathcal{F}_2 \to \begin{pmatrix}\text{punctual sheaf sup-} \\ \text{ported in } \mathbf{A}^1 \otimes k - U\end{pmatrix} \to 0.$$

Let us define a sheaf $\mathcal{F}_3$ on $\mathbf{A}^1 \otimes k$ by

$$\mathcal{F}_3 = \text{ image of } \mathcal{F} \text{ in } j_*\mathcal{F}_2.$$

We first claim that $\mathcal{F}_3$ is elementary. The exact sequence

$$0 \to \mathcal{F}_3 \to j_*\mathcal{F}_2 \to \begin{pmatrix}\text{punctual, supported in} \\ \mathbf{A}^1 \otimes k - U\end{pmatrix} \to 0$$

shows, applying first j^* and then j_*, that

$$j_*j^*\mathcal{F}_3 \xrightarrow{\sim} j_*\mathcal{F}_2,$$

so that $\mathcal{F}_3 \hookrightarrow j_*j^*\mathcal{F}_3$ satisfies Elem(1). That $\mathcal{F}_3$ satisfies Elem(2) results from the fact that $\mathcal{F}_3$ is a quotient of $\mathcal{F}$, which itself satisfies Elem(2), and the fact that *any* quotient of a sheaf satisfying Elem(2) again satisfies Elem(2).

We next claim that $j_*\mathcal{F}_1$ is Fourier. It obviously satisfies Fourier(1), and, being a subsheaf of $\mathcal{F}$, it satisfies Fourier(2) (because $\mathcal{F}$ does, and Fourier(2) passes to sub-sheaves). Because $j^*(j_*\mathcal{F}_1) = \mathcal{F}_1$ is irreducible, and hence semi-simple, it also satisfies Elem(2), because Elem(2) $\Longleftrightarrow$ Fourier(2) for a completely reducible $\mathcal{F}_1$.

Now apply NFT_ψ to the exact sequence

$$0 \to j_*\mathcal{F}_1 \to \mathcal{F} \to \mathcal{F}_3 \to 0.$$

Because $\mathcal{F}_3$ satisfies Elem(1)bis and $j_*\mathcal{F}_1$ satisfies Elem(2), (both being elementary), we have a short exact sequence

$$0 \to \mathrm{NFT}_\psi(j_*\mathcal{F}_1) \xrightarrow{(a)} \mathrm{NFT}_\psi(\mathcal{F}) \xrightarrow{(b)} \mathrm{NFT}_\psi(\mathcal{F}_3) \to 0.$$

Because $j_*\mathcal{F}_1$ and $\mathcal{F}$ are Fourier, so are $\mathrm{NFT}_\psi(j_*\mathcal{F}_1)$ and $\mathrm{NFT}_\psi(\mathcal{F})$. Therefore for any $k : V \hookrightarrow \mathbf{A}^1 \otimes k$ a non-empty ouvert de lissité of both $\mathrm{NFT}_\psi(j_*\mathcal{F}_1)$ and of $\mathrm{NFT}_\psi(\mathcal{F})$, the given map (a) is equal to $k_*k^*((a))$.

We claim that $k^*\mathrm{NFT}_\psi(\mathcal{F})$ is *not* irreducible. For if it were, then either $k^*((a)) = 0$ or $k^*((a)) = \text{isom}$; because $(a) = k_*k^*((a))$, this implies either $(a) = 0$ or $(a) = \text{isom}$, i.e., either $(a) = 0$ or $(b) = 0$. By Fourier inversion, either $j_*\mathcal{F}_1 = 0$ or $\mathcal{F}_3 = 0$. By construction, $j_*\mathcal{F}_1$ is $\neq 0$, so we must have $\mathcal{F}_3 = 0$, in which case $j_*\mathcal{F}_1 \xrightarrow{\sim} \mathcal{F}$ is irreducible, contradiction. ∎

8.5. Numerology of Fourier Transform

8.5.1. Let k be an algebraically closed field of characteristic p, $A = \mathbf{F}_\lambda$ or E_λ, $\psi : (\mathbf{F}_q, +) \to A^\times$ a fixed non-trivial additive character of a finite subfield. For $\mathcal{F}$ a constructible A-sheaf on $\mathbf{A}^1 \otimes k$, and $a \in k = \mathbf{A}^1(k)$, we define an integer $\mathrm{drop}(\mathcal{F}, a)$ by the formula

$$\mathrm{drop}(\mathcal{F}, a) = \mathrm{rank}(\mathcal{F}_{\bar\eta}) - \mathrm{rank}(\mathcal{F}_a).$$

If $j : U \hookrightarrow \mathbf{A}^1 \otimes k$ is the inclusion of a non-empty ouvert de lissité of $\mathcal{F}$, then $\mathrm{drop}(\mathcal{F}, a) = 0$ for $a \in U$. If $\mathcal{F}$ satisfies Elem(1), i.e., if $\mathcal{F} \hookrightarrow j_* j^* \mathcal{F}$, then the *largest* ouvert de lissité of $\mathcal{F}$ is *precisely* the set

$$\{a \in \mathbf{A}^1(k) \quad \text{such that} \quad \mathrm{drop}(\mathcal{F}, a) = 0\}.$$

8.5.2. The Euler-Poincaré formula for a constructible $\mathcal{F}$ on $\mathbf{A}^1 \otimes k$ gives

$$\chi_c(\mathbf{A}^1 \otimes k, \mathcal{F}) = \mathrm{rank}(\mathcal{F}_{\bar{\eta}}) - \sum_{a \in k} \mathrm{drop}(\mathcal{F}, a) - \sum_{a \in k \cup \infty} \mathrm{Swan}_a(\mathcal{F}).$$

8.5.3. If $\mathcal{F}$ is an elementary sheaf on $\mathbf{A}^1 \otimes k$, then $\mathrm{NFT}_\psi(\mathcal{F})$ is elementary, and so the largest ouvert de lissité of $\mathrm{NFT}_\psi(\mathcal{F})$ is the set

$$\{y \in \mathbf{A}^1(k) \quad \text{such that} \quad \mathrm{drop}(\mathrm{NFT}_\psi(\mathcal{F}), y) = 0\}.$$

Because $\mathcal{F}$ is elementary, for any $y \in \mathbf{A}^1(k)$ we have

$$\begin{aligned} \mathrm{rank}((\mathrm{NFT}_\psi(\mathcal{F}))_y) &= h_c^1(\mathbf{A}^1 \otimes k, \mathcal{F} \otimes \mathcal{L}_{\psi(yx)}) \\ &= -\chi_c(\mathbf{A}^1 \otimes k, \mathcal{F} \otimes \mathcal{L}_{\psi(yx)}) = -\mathrm{rank}(\mathcal{F}_{\bar{\eta}}) \\ &\quad + \sum_{a \in k} (\mathrm{drop}(\mathcal{F} \otimes \mathcal{L}_{\psi(yx)}, a) + \mathrm{Swan}_a(\mathcal{F} \otimes \mathcal{L}_{\psi(yx)})) \\ &\quad + \mathrm{Swan}_\infty(\mathcal{F} \otimes \mathcal{L}_{\psi(yx)}). \end{aligned}$$

In this last sum, the terms

$$\mathrm{drop}(\mathcal{F} \otimes \mathcal{L}_{\psi(yx)}, a), \quad \mathrm{Swan}_a(\mathcal{F} \otimes \mathcal{L}_{\psi(yx)})$$

are, for each fixed $a \in k$, constant functions of y (simply because $\mathcal{L}_{\psi(yx)}$ is lisse of rank one on $\mathbf{A}^1 \otimes k$ for every fixed y). Therefore we find, for $\mathcal{F}$ elementary, the formula

$$\mathrm{rank}(\mathrm{NFT}_\psi(\mathcal{F})_y) - \mathrm{rank}(\mathrm{NFT}_\psi(\mathcal{F})_0) = \mathrm{Swan}_\infty(\mathcal{F} \otimes \mathcal{L}_{\psi(yx)}) - \mathrm{Swan}_\infty(\mathcal{F}).$$

In particular, the largest ouvert de lissité of $\mathrm{NFT}_\psi(\mathcal{F})$ is the set of $y \in \mathbf{A}^1(k)$ at which the function

$$\begin{aligned} \mathbf{A}^1(k) &\longrightarrow \mathbf{Z} \\ \cup\!\!\!\!\!\!\;\; & \\ y &\longmapsto \mathrm{Swan}_\infty(\mathcal{F} \otimes \mathcal{L}_{\psi(yx)}) \end{aligned}$$

assumes its *maximum* value.

8.5.4. Lemma. *Let k be an algebraically closed field of characteristic p, $A = \mathbf{F}_\lambda$ or E_λ, $\psi : (\mathbf{F}_q, +) \to A^\times$ a fixed non-trivial additive character of a finite subfield. Let M be a non-zero finite-dimensional A-vector* *space on which P_∞ operates irreducibly, and let t be the break of M, i.e. $M = M(t)$ in the notation of 1.1. Then*

(1) *If $t < 1$, then for all $a \in k^\times$, $M \otimes \mathcal{L}_{\psi(ax)}$ has all its P_∞-breaks = 1. (For $a = 0$, $M \otimes \mathcal{L}_{\psi(ax)} = M$ has all breaks t.)*

(2) *If $t > 1$, then for all $a \in k$, $M \otimes \mathcal{L}_{\psi(ax)}$ has all its P_∞-breaks equal to t.*

(3) *If $t = 1$, then for all but at most one value of $a \in k$, $M \otimes \mathcal{L}_{\psi(ax)}$ has all its P_∞-breaks equal to 1.*

Proof. Assertions (1) and (2) follow immediately from (1.3), and the fact that $\mathcal{L}_{\psi(ax)}$ has P_∞-break 1 for $a \neq 0$, and P_∞-break zero for $a = 0$. For (3), if $a \in k$ then $M \otimes \mathcal{L}_{\psi(ax)}$ is still P_∞-irreducible, and all its breaks are ≤ 1, so either all are 1 or all are some $t < 1$. If $M \otimes \mathcal{L}_{\psi(ax)}$ has all breaks $t < 1$, then for $b \neq a$ we have

$$M \otimes \mathcal{L}_{\psi(bx)} = \underbrace{(M \otimes \mathcal{L}_{\psi(ax)})}_{\text{all breaks} t<1} \otimes \underbrace{\mathcal{L}_{\psi((b-a)x)}}_{\text{break} 1},$$

so, by case (1), $M \otimes \mathcal{L}_{\psi(bx)}$ has all breaks 1 for $b \neq a$. ∎

8.5.5. Corollary. *For $k, A, \mathbf{F}_q, \psi$ as in the previous lemma, let M be a non-zero finite-dimensional A-vector space on which P_∞ acts continuously, with breaks $\lambda_1, \ldots, \lambda_n$, $n = \dim(M)$. Then*

(1) *For all but at most $n = \dim(M)$ values of $a \in k$, the breaks of $M \otimes \mathcal{L}_{\psi(ax)}$ are*

$$\max(1, \lambda_1), \ldots, \max(1, \lambda_n),$$

and

$$\text{Swan}_\infty(M \otimes \mathcal{L}_{\psi(ax)}) = \sum_{i-1}^{n} \max(1, \lambda_i).$$

(2) *For all $a \in k$, we have*

$$\text{Swan}_\infty(M \otimes \mathcal{L}_{\psi(ax)}) \leq \sum \max(1, \lambda_i).$$

8.5.6. Corollary. *Let $\mathcal{F}$ be a non-zero elementary sheaf on $\mathbf{A}^1 \otimes k$, whose P_∞-breaks are $\lambda_1, \ldots, \lambda_n$, where $n \geq 1$ is the generic rank of $\mathcal{F}$. The largest ouvert de lissité of* $\text{NFT}_\psi(\mathcal{F})$ *is the set*

$$\{a \in k \text{ such that } \text{Swan}_\infty(\mathcal{F} \otimes \mathcal{L}_{\psi(ax)}) = \sum \max(1, \lambda_i)\}.$$

8.5.7. Break-Depression Lemma. *Let k be an algebraically closed field of characteristic p, $A = \mathbf{F}_\lambda$ or E_λ, $\psi : (\mathbf{F}_q, +) \to A^\times$ a fixed non-trivial* *additive character of a finite subfield. Let $M \neq 0$ be a finite-dimensional A-vector space on which I_∞ (the inertia group at $\infty \in \mathbf{P}^1_k$) operates continuously, A-linearly, and irreducibly. Suppose that all the breaks of M are equal to 1. Then there exists a unique $a \in k$ such that $M \otimes \mathcal{L}_{\psi(ax)}$ has all its breaks < 1, this a is $\neq 0$, and for any $b \in k$ with $b \neq a$, $M \otimes \mathcal{L}_{\psi(bx)}$ has all its break equal to 1.*

Proof. The uniqueness is easy. Any $M \otimes \mathcal{L}_{\psi(ax)}$ is still I_∞-irreducible, so has all its breaks equal. So either all its breaks are 1, or all are < 1. If $M \otimes \mathcal{L}_{\psi(ax)}$ has all breaks < 1, then for any $b \neq a$ we have

$$M \otimes \mathcal{L}_{\psi(bx)} = \underbrace{(M \otimes \mathcal{L}_{\psi(ax)})}_{\text{all breaks } <1} \otimes \underbrace{\mathcal{L}_{\psi((b-a)x)}}_{\text{break } =1},$$

which shows that $M \otimes \mathcal{L}_{\psi(bx)}$ has all breaks 1 for $b \neq a$.

To show the *existence* of such a "break-depressing" $a \in k$, we resort to a global argument. According to [Ka-2], there exists a lisse A-sheaf $\mathcal{F}$ on $\mathbf{G}_m \otimes k$ which is tame at zero and which is isomorphic to M as an I_∞-representation. Suppose that for *all* $a \in k$, $\mathcal{F} \otimes \mathcal{L}_{\psi(ax)}$ has all breaks 1 at ∞. Then denoting by $j : \mathbf{G}_m \otimes k \hookrightarrow \mathbf{A}^1 \otimes k$ the inclusion, we have that $j_!\mathcal{F}$ is an elementary sheaf: Elem(1) is obvious and Elem(2) holds because

$$H^2_c(\mathbf{A}^1 \otimes k, j_!\mathcal{F} \otimes \mathcal{L}_{\psi(ax)}) \longleftarrow (\mathcal{F} \otimes \mathcal{L}_{\psi(ax)})^{P_\infty} = 0.$$

Consider the sheaf $\mathrm{NFT}_\psi(j_!\mathcal{F})$ on $\mathbf{A}^1 \otimes k$. It is an elementary sheaf on $\mathbf{A}^1 \otimes k$, whose stalk at $a \in k = \mathbf{A}^1(k)$ is

$$H^1_c(\mathbf{G}_m, \mathcal{F} \otimes \mathcal{L}_{\psi(ax)}).$$

By the Euler–Poincaré formula, this stalk has dimension

$$\mathrm{Swan}_\infty(\mathcal{F} \otimes \mathcal{L}_{\psi(ax)}) = \mathrm{rank}(\mathcal{F}),$$

independent of a. Because $\mathrm{NFT}_\psi(j_!\mathcal{F})$ satisfies Elem(1), this constancy of rank means precisely that

$$\mathrm{NFT}_\psi(j_!\mathcal{F}) \quad \text{is lisse on } \mathbf{A}^1 \otimes k, \quad \text{of rank} = \mathrm{rank}(\mathcal{F}).$$

Let us temporarily denote $\mathrm{NFT}_\psi(j_!\mathcal{F})$ by $\mathcal{G}$. By Fourier inversion, we have

$$\mathrm{NFT}_{\overline{\psi}}(\mathcal{G}) \xrightarrow{\sim} j_!\mathcal{F}(-1);$$

because $\mathcal{G}$ is elementary, we see that for $a \in k = \mathbf{A}^1(k)$,

$$-\chi_c(\mathbf{A}^1 \otimes k, \mathcal{G} \otimes \mathcal{L}_{\psi(-ax)}) = \dim(j_!\mathcal{F}(-1))_a = \begin{cases} \operatorname{rank}(\mathcal{F}), & \text{if } a \neq 0 \\ 0 & \text{, if } a = 0. \end{cases}$$

Because $\mathcal{G}$ is lisse on $\mathbf{A}^1 \otimes k$, the Euler–Poincaré Formula gives

$$\begin{aligned} \chi_c(\mathbf{A}^1 \otimes k, \mathcal{G} \otimes \mathcal{L}_{\psi(-ax)}) &= \chi_c(\mathbf{A}^1 \otimes k) \operatorname{rank}(\mathcal{G}) - \operatorname{Swan}_\infty(\mathcal{G} \otimes \mathcal{L}_{\psi(-ax)}) \\ &= \operatorname{rank}(\mathcal{G}) - \operatorname{Swan}_\infty(\mathcal{G} \otimes \mathcal{L}_{\psi(-ax)}). \end{aligned}$$

Taking $a = 0$, this gives

$$\operatorname{Swan}_\infty(\mathcal{G}) = \operatorname{rank}(\mathcal{G}).$$

Taking $-a \neq 0$ and remembering that $\operatorname{rank}(\mathcal{F}) = \operatorname{rank}(\mathcal{G})$, we find

$$\operatorname{Swan}_\infty(\mathcal{G} \otimes \mathcal{L}_{\psi(ax)}) = 2 \operatorname{rank}(\mathcal{G}) \quad \text{for} \quad a \neq 0.$$

Let $n \geq 1$ be the rank of $\mathcal{G} = \operatorname{rank}(\mathcal{F}) = \dim(M)$, and let $\lambda_1, \ldots, \lambda_n$ be the breaks of $\mathcal{G}$ at ∞.

If $a \in k$ remains outside a finite ($\leq \operatorname{rank}(\mathcal{G})$) set of break-depressing values, we have (cf. 8.5.5)

$$\operatorname{Swan}_\infty(\mathcal{G} \otimes \mathcal{L}_{\psi(ax)}) = \sum_{i=1}^{n} \max(1, \lambda_i).$$

Thus we obtain the two equalities

$$\begin{cases} \sum_{i=1}^{n} \lambda_i = n = \operatorname{rank}(\mathcal{G}) \\ \sum_{i=1}^{n} \max(1, \lambda_i) = 2n = 2 \operatorname{rank}(\mathcal{G}), \end{cases}$$

whence

$$\sum_{i=1}^{n} (\max(1, \lambda_i) - \lambda_i - 1) = 0.$$

Each term in the sum is ≤ 0, so we infer that

$$\max(1, \lambda_i) = 1 + \lambda_i, \quad \text{for} \quad i = 1, \ldots, n;$$

and this is only possible if all $\lambda_i = 0$. But $n = \sum \lambda_i$, whence $n = 0$, whence $M = 0$, contradiction. ∎

8.5.7.1. Variant. *Let M be a finite-dimensional A-vector space on which I_∞ operates continuously, A-linearly, and irreducibly. Suppose that all the* *breaks of* M *are equal to an integer* $n \geq 1$ *which is prime to* p. *Then there exists a unique* $a \in k$ *such that* $M \otimes \mathcal{L}_{\psi(ax^n)}$ *has all its breaks* $< n$, *this* a *is* $\neq 0$, *and for any* $b \in k$ *with* $b \neq a$, $M \otimes \mathcal{L}_{\psi(bx^n)}$ *has all its breaks equal to* n.

Proof. Just as in 8.5.7, the only non-obvious assertion is that such a break-depressing a exists. To prove existence, we argue as follows.

By the projection formula, we have

$$[n]_*(M \otimes \mathcal{L}_{\psi(ax^n)}) = ([n]_*(M)) \otimes \mathcal{L}_{\psi(ax)}$$

and by 1.13, $[n]_*(M)$ has all its breaks equal to 1. Applying 8.5.7 to any irreducible constituent of $[n]_*(M)$, we see that there exists an $a \in k$ for which $([n]_*(M)) \otimes \mathcal{L}_{\psi(ax)}$ has some break < 1. By the projection formula and 1.13, we see that, for this a, $M \otimes \mathcal{L}_{\psi(ax^n)}$ has some break $< n$. By irreducibility, all its breaks are equal. ∎

8.5.8. Corollary. *Let* $\mathcal{F}$ *be a non-zero elementary sheaf on* $\mathbf{A}^1 \otimes k$. *Then*

(1) $\mathrm{NFT}_\psi(\mathcal{F})$ *is lisse on* $\mathbf{A}^1 \otimes k$ *if and only if all* P_∞*-breaks of* $\mathcal{F}$ *are* > 1.

(2) $\mathrm{NFT}_\psi(\mathcal{F})$ *is lisse on* $\mathbf{G}_m \otimes k$ *if and only if all* P_∞*-breaks of* $\mathcal{F}$ *are* $\neq 1$.

(3) $\mathrm{NFT}_\psi(\mathcal{F})$ *is lisse at zero if and only if all* P_∞*-breaks of* $\mathcal{F}$ *are* ≥ 1.

Proof. We know that in terms of the ∞-breaks $\lambda_1, \ldots, \lambda_n$ of $\mathcal{F}$, $\mathrm{NFT}_\psi(\mathcal{F})$ is lisse at $a \in \mathbf{A}^1(k)$ if and only if

$$\mathrm{Swan}_\infty(\mathcal{F} \otimes \mathcal{L}_{\psi(ax)}) = \sum \max(1, \lambda_i),$$

and that in general the inequality $\leq$ holds.

Semisimplify $\mathcal{F}$ as I_∞-representation, say

$$\mathcal{F} \sim \sum m_j \mathcal{F}_j,$$

where $\mathcal{F}_j$ has break t_j (with multiplicity rank $(\mathcal{F}_j)$). Then $\mathrm{NFT}_\psi(\mathcal{F})$ is lisse at a if and only if for *each* j,

$$\mathcal{F}_j \otimes \mathcal{L}_{\psi(ax)} \quad \text{has all breaks } = \max(1, t_j).$$

Therefore any $t_j < 1$ introduces a "singularity" of $\mathrm{NFT}_\psi(\mathcal{F})$ at zero, and any $t_j = 1$ introduces a "singularity" at some point $a \in k^\times$. ∎

8.6. Convolution with $\mathcal{L}_\psi$ as Fourier transform

Let k be an over-field of $\mathbf{F}_q$, $l \neq p$, $A = E_\lambda$ or $\mathbf{F}_\lambda$, and $\psi : (\mathbf{F}_q, +) \to A^\times$ a non-trivial additive character. We denote by

$$\mathrm{inv} : \mathbf{G}_m \xrightarrow{\sim} \mathbf{G}_m$$

$$x \mapsto 1/x$$

the multiplicative inversion, and by

$$j : \mathbf{G} \hookrightarrow \mathbf{A}^1$$

the inclusion.

8.6.1. Proposition. *Notations as above. Let $\mathcal{G}$ be a lisse A-sheaf on $\mathbf{G}_m \otimes k$ which is tame at zero (i.e., $\mathcal{G} \in \mathcal{T}$). Then we have a canonical isomorphism*

$$(*) \qquad (\mathcal{L}_\psi)_! * \mathcal{G} \xrightarrow{\sim} j^*\mathrm{NFT}_\psi(j_!\,\mathrm{inv}^*(\mathcal{G})).$$

If in addition $\mathcal{G}^{I_\infty} = 0$, then

(1) $j_!\,\mathrm{inv}^*(\mathcal{G})$ *is a Fourier sheaf on* $\mathbf{A}^1 \otimes k$.
(2) $j_*((\mathcal{L}_\psi)_! * \mathcal{G}) \xrightarrow{\sim} \mathrm{NFT}_\psi(j_!\,\mathrm{inv}^*(\mathcal{G}))$ *is a Fourier sheaf on* $\mathbf{A}^1 \otimes k$.
(3) $\mathrm{NFT}_{\overline{\psi}}(j_*((\mathcal{L}_\psi)_! * \mathcal{G})) \xrightarrow{\sim} j_!\,\mathrm{inv}^*(\mathcal{G})(-1)$.
(4) $(\mathcal{L}_\psi)_! * \mathcal{G} \xrightarrow{\sim} (\mathcal{L}_\psi)_* * \mathcal{G}$.

Construction-proof. In terms of coordinates x, y on $\mathbf{G}_m \times \mathbf{G}_m$, and the new coordinates $A = 1/y$, $B = xy$ on $\mathbf{G}_m \times \mathbf{G}_m$, we have, for any $\mathcal{F}, \mathcal{G}$ on $\mathbf{G}_m \otimes k$,

$$\begin{aligned} \mathcal{F}_! * \mathcal{G} &= R^1(xy)_!(\mathcal{F}_x \otimes \mathcal{G}_y) \\ &= R^1(B)_!(\mathcal{F}_{AB} \otimes \mathcal{G}_{A^{-1}}) \\ &= R^1(pr_2)_!(pr_1^*(\mathrm{inv}^*(\mathcal{G})) \otimes \mu^*(\mathcal{F})). \end{aligned}$$

"Compactifying" $\mathbf{G}_m \times \mathbf{G}_m$ to $\mathbf{A}^1 \otimes \mathbf{A}^1$ with coordinates (A, B), we have

$$\mathcal{F}_! * \mathcal{G} = j^*(R^1(pr_2)_!(pr_1^*(j_!(\mathrm{inv}^*(\mathcal{G})) \otimes \mu^*(j_*\mathcal{F}))),$$

(where now pr_1, pr_2, μ are the maps A, B, AB of $\mathbf{A}^2 \to \mathbf{A}^1$). In particular, taking $\mathcal{F} = j^*\mathcal{L}_\psi$, we obtain a canonical identification

$$(\mathcal{L}_\psi)_! * \mathcal{G} = j^*(\mathrm{NFT}_\psi(j_!\,\mathrm{inv}^*(\mathcal{G}))).$$

Suppose now that $\mathcal{G}^{I_\infty} = 0$. Then $(\mathrm{inv}^*(\mathcal{G}))^{I_0} = 0$, and $\mathrm{inv}^*(\mathcal{G})$ is lisse on $\mathbf{G}_m \otimes k$. Therefore $j_!\,\mathrm{inv}^*(\mathcal{G}) \xrightarrow{\sim} j_*\,\mathrm{inv}^*(\mathcal{G})$ is a Fourier sheaf on $\mathbf{A}^1 \otimes k$ (basic example (2) of 8.3). Therefore $\mathrm{NFT}_\psi(j_!\,\mathrm{inv}^*(\mathcal{G}))$ is a Fourier sheaf (8.2.5 (2)), and so the isomorphism (2) on $\mathbf{A}^1 \otimes k$ is j_* (the isom. $(*)$ on

$\mathbf{G}_m \otimes k$). Assertion (3) is Fourier inversion. Assertion (4) may be checked at a single fibre, say over 1, where it becomes the assertion that $\text{inv}^*(\mathcal{G}) \otimes \mathcal{L}_\psi$ on $\mathbf{G}_m \otimes k$ has $H^1_c \xrightarrow{\sim} H^1$, which holds because the sheaf in question has no non-zero invariants under either I_0 or I_∞. ∎

8.6.2. Proposition. *Notations as above, let $\mathcal{G} \in \mathcal{T}$ have $\mathcal{G}^{I_\infty} = 0$. Then*

(1) $(\mathcal{L}_\psi)_! * \mathcal{G} \in \mathcal{T}$, *and has* $\text{rank} = \text{rank}(\mathcal{G}) + \text{Swan}_\infty(\mathcal{G})$

(2) $((\mathcal{L}_\psi)_! * \mathcal{G})^{I_0}$ *has dimension* $= \text{Swan}_\infty(\mathcal{G})$

(3) *All P_∞-breaks of $(\mathcal{L}_\psi)_! * \mathcal{G}$ are < 1.*

Proof. Assertion (1) is just "mise pour memoire." By the previous proposition, we have

$$j_*((\mathcal{L}_\psi)_! * \mathcal{G}) \xrightarrow{\sim} \text{NFT}_\psi(j_! \,\text{inv}^*(\mathcal{G})).$$

Comparing stalks at zero, we find

$$\begin{aligned}((\mathcal{L}_\psi)_! * \mathcal{G})^{I_0} &\xrightarrow{\sim} H^1_c(\mathbf{A}^1 \otimes \bar{k}, j_! \,\text{inv}^*(\mathcal{G})) = H^1_c(\mathbf{G}_m \otimes \bar{k}, \text{inv}^*(\mathcal{G})) \\ &\xrightarrow{\sim} H^1_c(\mathbf{G}_m \otimes \bar{k}, \mathcal{G}),\end{aligned}$$

while, as $j_! \,\text{inv}^*(\mathcal{G})$ is Fourier, its $H^2_c = 0$, so the Euler-Poincaré formula gives (2).

For (3), we argue as follows. By the previous proposition, $j_*((\mathcal{L}_\psi)_! * \mathcal{G})$ is elementary, and its $\text{NFT}_{\bar{\psi}}(\text{——})(-1)$ is *lisse* on $\mathbf{G}_m$, being $j_! \,\text{inv}^*(\mathcal{G})$. So by 8.5.8 (2), all P_∞-breaks of $j_*((\mathcal{L}_\psi)_! * \mathcal{G})$ are $\neq 1$. So it suffices to show that $j_*((\mathcal{L}_\psi)_! * \mathcal{G})$ has all P_∞-breaks ≤ 1.

We know that

$$\text{NFT}_{\bar{\psi}}(j_*((\mathcal{L}_\psi)_! * \mathcal{G})) \xrightarrow{\sim} j_! \,\text{inv}^*(\mathcal{G})(-1),$$

and we know that $j_*((\mathcal{L}_\psi)_! * \mathcal{G})$ is lisse on $\mathbf{G}_m$, and tame at zero, with, by part (1),

$$\text{drop}(j_*(\ldots, *\mathcal{G}), 0) = \text{rank } \mathcal{G}.$$

Therefore the rank formula for $\text{NFT}_{\bar{\psi}}$ gives, for $a \in \mathbf{A}^1(\bar{k})$,

$$\begin{aligned}\text{rank}(j_! \,\text{inv}^*(\mathcal{G}))_a &= \text{rank } \text{NFT}_{\bar{\psi}}(j_*((\mathcal{L}_\psi)_! * \mathcal{G}))_a \\ &= -\text{rank}((\mathcal{L}_\psi)_! * \mathcal{G}) + \text{drop}(j_*((\mathcal{L}_\psi)_! * \mathcal{G}), 0) \\ &\quad + \text{Swan}_\infty(((\mathcal{L}_\psi)_! * \mathcal{G}) \otimes \mathcal{L}_{\psi(ax)}) \\ &= -(\text{rank}(\mathcal{G}) + \text{Swan}_\infty(\mathcal{G})) + \text{rank}(\mathcal{G}) \\ &\quad + \text{Swan}_\infty(((\mathcal{L}_\psi)_! * \mathcal{G}) \otimes \mathcal{L}_{\psi(ax)}) \\ &= \text{Swan}_\infty(((\mathcal{L}_\psi)_! * \mathcal{G}) \otimes \mathcal{L}_{\psi(ax)}) - \text{Swan}_\infty(\mathcal{G}).\end{aligned}$$

Taking $a \neq 0$, we find

$$\operatorname{rank} \mathcal{G} = \operatorname{Swan}_\infty(((\mathcal{L}_\psi)_! * \mathcal{G}) \otimes \mathcal{L}_{\psi(ax)}) - \operatorname{Swan}_\infty(\mathcal{G}),$$

which we may rewrite as

$$\operatorname{rank}((\mathcal{L}_\psi)_! * \mathcal{G}) = \operatorname{Swan}_\infty(((\mathcal{L}_\psi)_! * \mathcal{G}) \otimes \mathcal{L}_{\psi(ax)})$$

for all $a \neq 0$. Let $\lambda_1, \ldots, \lambda_n$ denote the P_∞-breaks of $((\mathcal{L}_\psi)_! * \mathcal{G})$.

For a sufficiently general, this $\operatorname{Swan}_\infty = \sum \max(1, \lambda_i)$, so the above equality gives

$$n = \operatorname{rank}((\mathcal{L}_\psi)_! * \mathcal{G}) = \sum_{i=1}^{n} \max(1, \lambda_i),$$

i.e.,

$$\sum_{i=1}^{n} (\max(1, \lambda_i) - 1) = 0.$$

As each term in this $\sum$ is ≥ 0, each term vanishes, whence $\lambda_i \leq 1$ for all i. ∎

8.6.3. Proposition. *Notations as above, let $\mathcal{H} \in \mathcal{T}$ satisfy*

(A) $\dim \mathcal{H}^{I_0} = \operatorname{Swan}_\infty(\mathcal{H})$
(B) *all P_∞-breaks of $\mathcal{H}$ are < 1*
(C) *$j_*\mathcal{H}$ is a Fourier sheaf.*

Then we have

(1) $\mathrm{NFT}_{\overline{\psi}}(j_*\mathcal{H})$ *is a Fourier sheaf*
(2) $\mathrm{NFT}_{\overline{\psi}}(j_*\mathcal{H})$ *is lisse on $\mathbf{G}_m$ of* rank = $\operatorname{rank}(\mathcal{H}) - \operatorname{Swan}_\infty(\mathcal{H})$, *vanishes at zero, and is tame at infinity.*

Proof. As always, (C) $\Longrightarrow$ (1). By (B), $\mathrm{NFT}_{\overline{\psi}}(j_*\mathcal{H})$ is lisse on $\mathbf{G}_m$. At zero, the stalk of $\mathrm{NFT}_{\overline{\psi}}(j_*\mathcal{H})$ is

$$H^1_c(\mathbf{A}^1 \otimes \bar{k}, j_*\mathcal{H});$$

By (C), its dimension is $-\chi_c(\mathbf{A}^1 \otimes \bar{k}, j_*\mathcal{H}) = -\dim \mathcal{H}^{I_0} - \chi_c(\mathbf{G}_m \otimes \bar{k}, \mathcal{H}) = -\dim \mathcal{H}^{I_0} + \operatorname{Swan}_\infty(\mathcal{H}) = 0$ (by (A)). Thus we have

$$\mathrm{NFT}_{\overline{\psi}}(j_*\mathcal{H}) = j_!\mathcal{G} = j_*\mathcal{G}$$

for some lisse $\mathcal{G}$ on $\mathbf{G}_m \otimes k$ with $\mathcal{G}^{I_0} = 0$.

For $a \neq 0$,

$$\mathcal{G}_a = H^1_c(\mathbf{A}^1 \otimes \bar{k}, j_*\mathcal{H} \otimes \mathcal{L}_{\overline{\psi}(ax)})$$

has dimension $\mathrm{Swan}_\infty(\mathcal{H} \otimes \mathcal{L}_{\overline{\psi}(ax)}) - \dim \mathcal{H}^{I_0}$, and by (B),

$$\mathrm{Swan}_\infty(\mathcal{H} \otimes \mathcal{L}_{\overline{\psi}(ax)}) = \mathrm{rank}(\mathcal{H}),$$

so rank $\mathcal{G}$ = rank $\mathcal{H} - \mathrm{Swan}_\infty(\mathcal{H})$. Let $\lambda_1, \ldots, \lambda_n$ be the P_∞-breaks of $\mathcal{G}$. We must show that all $\lambda_i = 0$.

We know (Fourier inversion) that

$$\mathrm{NFT}_\psi(j_! \mathcal{G}) \xrightarrow{\sim} j_* \mathcal{H}(-1).$$

For $a \in \mathbf{A}^1(\bar{k})$, we have

$$\begin{aligned} \mathrm{rank}(\mathrm{NFT}_\psi(j_! \mathcal{G}))_a &= -\chi_c(\mathbf{A}^1 \otimes \bar{k}, j_! \mathcal{G} \otimes \mathcal{L}_{\psi(ax)}) \\ &= \mathrm{Swan}_\infty(\mathcal{G} \otimes \mathcal{L}_{\psi(ax)}). \end{aligned}$$

Therefore, comparing dimensions we find

$$\mathrm{Swan}_\infty(\mathcal{G} \otimes \mathcal{L}_{\psi(ax)}) = \begin{cases} \mathrm{rank}(\mathcal{H}) & \text{if } a \neq 0 \\ \\ \mathrm{Swan}_\infty(\mathcal{H}) & \text{if } a = 0. \end{cases}$$

Because $\mathrm{NFT}_\psi(j_! \mathcal{G})$ is *lisse* on $\mathbf{G}_m$, we have, for $a \neq 0$ in $\bar{k}$,

$$(\alpha) \qquad \mathrm{Swan}_\infty(\mathcal{G} \otimes \mathcal{L}_{\psi(ax)}) = \sum_i \max(1, \lambda_i) = \mathrm{rank}\, \mathcal{H}.$$

For $a = 0$,

$$(\beta) \qquad \mathrm{Swan}_\infty(\mathcal{G}) = \sum \lambda_i = \mathrm{Swan}_\infty(\mathcal{H}).$$

As already noted,

$$(\gamma) \qquad \mathrm{rank}(\mathcal{G}) = \mathrm{rank}\, \mathcal{H} - \mathrm{Swan}_\infty(\mathcal{H}).$$

Subtracting $(\alpha) - (\beta)$ and using (γ), we get

$$\sum_{i=1}^{\mathrm{rk}(\mathcal{G})} \max(1, \lambda_i) - \sum_{i=1}^{\mathrm{rk}(\mathcal{G})} \lambda_i = \mathrm{rk}(\mathcal{G}),$$

i.e.,

$$\sum_i [\max(1, \lambda_i) - 1 - \lambda_i] = 0.$$

As each term in this sum is ≤ 0, we have

$$\max(1, \lambda_i) = 1 + \lambda_i,$$

whence $\lambda_i = 0$, as required. ∎

8.6.4. Theorem. *The constructions*

$$\mathcal{G} \longmapsto (\mathcal{L}_\psi)_! * \mathcal{G}$$

$$\mathrm{inv}^* j^* \mathrm{NFT}_{\overline{\psi}}(j_* \mathcal{H})(1) \longleftarrow \mathcal{H}$$

define quasi-inverse equivalences of categories

$$\{\mathcal{G} \in \mathcal{T} \text{ with } \mathcal{G}^{I_\infty} = 0\} \rightleftarrows$$

$$\left\{ \mathcal{H} \in \mathcal{C} \text{ with } \begin{cases} \text{all } P_\infty\text{-breaks} < 1 \\ \dim \mathcal{H}^{I_0} = \mathrm{Swan}_\infty(\mathcal{H}) \\ j_*\mathcal{H} \text{ is Fourier} \end{cases} \right\}$$

we have for $\mathcal{G} \longleftrightarrow \mathcal{H}$ *in these categories*

(1) $\mathrm{Swan}_\infty(\mathcal{G}) = \mathrm{Swan}_\infty(\mathcal{H})$.
(2) $\mathrm{rank}\,\mathcal{G} + \mathrm{Swan}_\infty(\mathcal{G}) = \mathrm{rank}(\mathcal{H})$.
(3) $\mathcal{G} \in \mathcal{C} \Longleftrightarrow \mathcal{H}$ *lies in* $\mathcal{C}$.

In particular, these constructions define quasi-inverse equivalences

$$\mathcal{C} \rightleftarrows \left\{ \mathcal{H} \in \mathcal{C} \text{ with } \begin{cases} \text{all } P_\infty\text{-breaks} < 1 \\ \dim \mathcal{H}^{I_0} = \mathrm{Swan}_\infty(\mathcal{H}) \end{cases} \right\}.$$

Proof. That we obtain quasi-inverse equivalences of the categories in question is the content of the previous two propositions, in which (1) and (2) are also verified. Assertion (3) is the special case $\mathcal{F} = \mathcal{L}_\psi$ of the cancellation theorem 8.0.1. For the final assertion, if $\mathcal{G} \in \mathcal{C}$ then automatically $\mathcal{G}^{I_\infty} = 0$, and if $\mathcal{H}$ in $\mathcal{C}$ has all breaks in $(0, 1)$, then, by 8.3 (1), $j_*\mathcal{H}$ is automatically Fourier. ∎

8.7. Ubiquity of Kloosterman Sheaves

Let k be a perfect over-field of $\mathbf{F}_q$, $l \neq p$, $A = E_\lambda$ or $\mathbf{F}_\lambda$, and $\psi : (\mathbf{F}_q, +) \to A^\times$ a non-trivial additive character.

8.7.1. Theorem. *Let* $\mathcal{F}$ *be a lisse* A*-sheaf on* $\mathbf{G}_m \otimes k$ *which is tame at zero, and totally wild at* ∞, *with* $\mathrm{Swan}_\infty(\mathcal{F}) = 1$. *Suppose that the geometric local monodromy of* $\mathcal{F}$ *at zero is quasi-unipotent (this is automatic if* $A = \mathbf{F}_\lambda$ *or if* k *is finite), and that all its eigenvalues are* $q-1$*'st roots of unity which* *lie in A. Then there exist*

$$\begin{cases} n = \operatorname{rank}(\mathcal{F}) \text{ multiplicative characters } \chi_i : \mathbf{F}_q^\times \to A^\times \\ \\ \text{a point } b \in k^\times = \mathbf{G}_m(k) \\ \\ \text{lisse, rank one, geometrically constant } A\text{-sheaves} \\ T, \tilde{T} \text{ (for "twist") on } \mathbf{G}_m \otimes k \end{cases}$$

and isomorphisms of lisse sheaves on $\mathbf{G}_m \otimes k$

$$\begin{cases} \mathcal{F} \underset{A}{\otimes} T \simeq (\mathcal{L}_{\psi(bx)} \otimes \mathcal{L}_{\chi_1}) * (\mathcal{L}_\psi \otimes \mathcal{L}_{\chi_2}) * \cdots * (\mathcal{L}_\psi \otimes \mathcal{L}_{\chi_n}). \\ \mathcal{F} \underset{A}{\otimes} \tilde{T} \simeq (\mathrm{Trans}_b)^*(\mathrm{Kl}(\psi; \chi_1, \ldots, \chi_n; 1, \ldots, 1)). \end{cases}$$

Proof. We first explain why the second isomorphism is just a rewriting of the first. For any character $\chi : \mathbf{F}_q^\times \to A^\times$, and any $b \in k^\times$, we claim that the sheaf $\mathrm{Trans}_b^*(\mathcal{L}_\chi)$ on $\mathbf{G}_m \otimes k^{\mathrm{sep}}$ is isomorphic to $\mathcal{L}_\chi$. Indeed we know that under the multiplication morphism

$$\begin{aligned} (\mathbf{G}_m \otimes \mathbf{F}_q) \times (\mathbf{G}_m \otimes \mathbf{F}_q) &\xrightarrow{\pi} \mathbf{G}_m \otimes \mathbf{F}_q \\ (x, y) &\longmapsto xy \end{aligned}$$

we have a canonical isomorphism on $(\mathbf{G}_m \otimes \mathbf{F}_q) \times (\mathbf{G}_m \otimes \mathbf{F}_q)$

$$\pi * (\mathcal{L}_\chi) \simeq \mathcal{L}_\chi \boxtimes \mathcal{L}_\chi.$$

Extending scalars to k, and passing to the closed subscheme $\{b\} \times (\mathbf{G}_m \otimes k)$, this isomorphism yields an isomorphism on $\mathbf{G}_m \otimes k$

$$\mathrm{Trans}_b^*(\mathcal{L}_\chi) \simeq b^*(\mathcal{L}_\chi) \otimes \mathcal{L}_\chi,$$

as required.

Applying this to χ_1, we may rewrite the first isomorphism of the theorem in terms of the twist sheaf $\tilde{T} = T \otimes b^*(\mathcal{L}_\chi)$ as

$$\begin{aligned} \mathcal{F} \underset{A}{\otimes} \tilde{T} &\simeq ((\mathrm{Trans}_b)^*(\mathcal{L}_\psi \otimes \mathcal{L}_{\chi_1})) * (\mathcal{L}_\psi \otimes \mathcal{L}_{\chi_2}) * \cdots * (\mathcal{L}_\psi \otimes \mathcal{L}_{\chi_n}) \\ &\simeq (\mathrm{Trans}_b)^*(\mathrm{Kl}(\psi; \chi_1, \ldots, \chi_n; 1, \ldots, 1)), \end{aligned}$$

the last isomorphism by property (11) of convolution (cf. 5.1).

To establish the first isomorphism, we proceed by induction on $n = \operatorname{rank}(\mathcal{F})$. For $n = 1$, $\mathcal{F}$ has break 1 at ∞, so by the Break-Depression Lemma, there exists a unique $b \neq 0$ in k (in k because unique in k^{sep}) such that $\mathcal{F} \otimes \mathcal{L}_{\psi(-bx)}$ has break < 1 at ∞. But in rank 1, break < 1 at ∞ implies break $= 0$ at ∞ (integrality of Swan_∞), so $\mathcal{F} \otimes \mathcal{L}_{\psi(-bx)}$ is lisse of rank one on $\mathbf{G}_m \otimes k$, tame at both zero and ∞, with local monodromy at zero of order dividing $q-1$. Therefore for some $\chi : \mathbf{F}_q^\times \to A^\times$, $\mathcal{F}\otimes\mathcal{L}_{\psi(-bx)}\otimes\mathcal{L}_{\chi^{-1}}$ is geometrically constant. Calling it T^{-1}, we obtain

$$\mathcal{F}\otimes T \simeq \mathcal{L}_{\psi(bx)}\otimes\mathcal{L}_{\chi},$$

as required in rank one.

In general, the hypothesis on the local monodromy of $\mathcal{F}$ at zero shows that for some character $\chi_n : \mathbf{F}_q^\times \to A^\times$, $\mathcal{F}\otimes\mathcal{L}_{\overline{\chi}_n}$ has

$$\dim(\mathcal{F}\otimes\mathcal{L}_{\overline{\chi}_n})^{I_0} \geq 1.$$

But as $\mathcal{F}\in\mathcal{C}$, with $\mathrm{Swan}_\infty(\mathcal{F})=1$, we always have (cf. 7.5.1.3)

$$\dim(\mathcal{F}\otimes\mathcal{L}_{\overline{\chi}_n})^{I_0} \leq \mathrm{Swan}_\infty(\mathcal{F}) = 1.$$

Because $n\geq 2$, all the breaks of $\mathcal{F}\otimes\mathcal{L}_{\overline{\chi}_n}$ at ∞ are $(1/n)<1$. Therefore there exists $\mathcal{G}\in\mathcal{C}$ of rank $n-1$, $\mathrm{Swan}_\infty(\mathcal{G})=1$, with

$$\mathcal{L}_\psi * \mathcal{G} \simeq \mathcal{F}\otimes\mathcal{L}_{\overline{\chi}_n},$$

thanks to the previous theorem. Tensoring with $\mathcal{L}_{\chi_n}$ gives

$$\mathcal{F}\simeq(\mathcal{L}_\psi\otimes\mathcal{L}_{\chi_n}) * (\mathcal{G}\otimes\mathcal{L}_{\chi_n}).$$

Now begin again with $\mathcal{G}\otimes\mathcal{L}_{\chi_n}\ldots$. ∎

8.7.2. Corollary. *If k is a finite field, and $A=E_\lambda$, then any $\mathcal{F}\in\mathcal{C}$ with* $\mathrm{Swan}_\infty(\mathcal{F})=1$ *and* $\det(\mathcal{F})$ *of finite order is pure of weight zero.*

Proof. Because k is finite, the local monodromy of $\mathcal{F}$ at zero is quasi-unipotent; being tame, its eigenvalues are roots of unity of order prime to p, so of order dividing $q-1$ for some power q of p. Enlarging E_λ and k, we find ourselves in the hypotheses of the previous theorem: up to a twist, $\mathcal{F}$ is a translate of a Kloosterman sheaf, so pure of some weight, and this weight can be none other than $(1/\mathrm{rank}(\mathcal{F}))\times$ (weight of $\det(\mathcal{F})$). ∎

8.8. Fine Structure over $\mathbf{F}_q$: Canonical Descents of Kloosterman Sheaves

8.8.1. In this section, we work with $k=\mathbf{F}_q$, $l\neq p$, $A=E_\lambda$ or $\mathbf{F}_\lambda$, and $\psi : (\mathbf{F}_q,+)\to A^\times$ a non-trivial additive character. We suppose that A contains the $q-1$'st roots of unity.

8.8.2. Let $\mathcal{F}\in\mathcal{C}$ have $\mathrm{Swan}_\infty(\mathcal{F})=1$. We have seen that if the characters χ_i of I_0^{tame} which occur in the geometric local monodromy of $\mathcal{F}$ at zero all have order dividing $q-1$, then $\mathcal{F}$ is isomorphic to a twist of an $\mathbf{F}_q^\times$-translate of the Kloosterman sheaf

$$\mathrm{Kl}(\psi; \chi_1, \ldots, \chi_n; 1, \ldots, 1).$$

8.8.3. What can we say about $\mathcal{F}$ if the characters χ_i of I_0^{tame} occurring in $\mathcal{F}$ are *not* all of order dividing $q-1$? To analyze this question, we pass to a finite extension field of $A = E_\lambda$ or $\mathbf{F}_\lambda$ over which these characters exist, i.e., we adjoin to A the eigenvalues of $\rho(\gamma_0)$ for γ_0 a topological generator of I_0^{tame}. Because $\mathcal{F}$ is "defined over $\mathbf{F}_q$" the characters (with multiplicity) $\chi_1, \ldots, \chi_n$ of I_0^{tame} occurring in $\mathcal{F}$ are *permuted* by $\chi \mapsto \chi^q$ (this is the "trick" underlying Grothendieck's proof of the local monodromy theorem, cf. [Se-Ta]).

Break $\{\chi_1, \ldots, \chi_n\}$ into orbits under this action. A typical orbit has the form

$$\chi, \chi^q, \ldots, \chi^{q^{\nu-1}};$$

we may form the "corresponding" Kloosterman sheaf of rank ν,

$$\mathrm{Kl}(\text{orbit}) \overset{\text{dfn}}{=} \mathrm{Kl}(\psi \circ \mathrm{trace}_{\mathbf{F}_{q^\nu}/\mathbf{F}_q}; \chi, \chi^q, \ldots, \chi^{q^{\nu-1}}; 1, \ldots, 1).$$

If $d \geq 1$ is a common multiple of all the orbit lengths ν which occur, then $\mathbf{F}_{q^d}$ contains each $\mathbf{F}_{q^\nu}$, and, on $\mathbf{G}_m \otimes \mathbf{F}_{q^d}$ we have an isomorphism

$$\mathrm{Kl}(\psi \circ \mathrm{trace}_{\mathbf{F}_{q^\nu}/\mathbf{F}_q}; \chi_1, \ldots, \chi_n; 1, \ldots, 1) \simeq \underset{\text{orbits}}{*} \mathrm{Kl}(\text{orbit}).$$

Applying the theorem 8.7.1 over $\mathbf{F}_{q^d}$, we conclude that when $\mathcal{F}$ is pulled back to $\mathbf{G}_m \otimes \mathbf{F}_{q^d}$, it becomes a twist of a translate of a convolution of "orbit" sheaves Kl(orbit).

In fact, each orbit sheaf Kl(orbit), *à priori* defined on $\mathbf{G}_m \otimes \mathbf{F}_{q^\nu}$, has a natural descent to $\mathbf{G}_m \otimes \mathbf{F}_q$. To explain this, it is convenient to consider a slightly more general situation (cf. [De-3], second part of Remarque 7.18).

8.8.4. Let B be a finite etale $\mathbf{F}_q$-algebra of rank $\nu \geq 1$. Any additive character of B,

$$\tilde{\psi} : (B, +) \to A^\times,$$

is uniquely of the form

$$\tilde{\psi}(b) = \psi(\mathrm{trace}_{B/\mathbf{F}_q}(b_1 b))$$

for some element $b_1 \in B$. We say that $\tilde{\psi}$ is *non-degenerate* if b_1 is invertible in B, or equivalently if $(b, b') \mapsto \tilde{\psi}(bb')$ makes $(B, +)$ into its own Pontrajagin dual.

For any $\mathbf{F}_q$-algebra R, we define

$$\begin{cases} B(R) = B \underset{\mathbf{F}_q}{\otimes} R, \text{ a finite free } R\text{-algebra of rank } \nu \\ B^{\times}(R) = (B \underset{\mathbf{F}_q}{\otimes} R)^{\times}. \end{cases}$$

The norm map for $B(R)/R$ maps $B^{\times}(R)$ to $R^{\times}$.

If R is an $\mathbf{F}_q$-algebra, and S is a finite free R-algebra of rank $r \geq 1$, then $B(S)$ is a finite free $B(R)$-algebra of the same rank r. The corresponding norm and trace maps are denoted simply $\mathbf{N}_{S/R}$, $\text{trace}_{S/R}$.

The functors on $(\text{Sch}/\mathbf{F}_q)$ defined by

$$R \mapsto B(R), B^{\times}(R)$$

are represented by smooth commutative $\mathbf{F}_q$-group schemes, denoted $\mathbf{B}$ and $\mathbf{B}^{\times}$ respectively, with $\mathbf{B}^{\times}$ an open subscheme of $\mathbf{B}$. For any additive character $\tilde{\psi} : \mathbf{B}(\mathbf{F}_q) \to A^{\times}$ (resp. any multiplicative character $\chi : \mathbf{B}^{\times}(\mathbf{F}_q) \to A^{\times}$) we denote by $\mathcal{L}_{\tilde{\psi}}$ (resp. $\mathcal{L}_{\chi}$) the corresponding lisse rank one A-sheaf on $\mathbf{B}$ (resp. $\mathbf{B}^{\times}$), obtained by pushout from the Lang torsor (cf. 4.3).

8.8.5. Theorem. *Let $\mathbf{F}_q$ be a finite field, B a finite etale $\mathbf{F}_q$-algebra of rank $\nu \geq 1$, $l \neq p$, A an l-adic coefficient ring whose residue field contains the p'th roots of unity,*

$$\tilde{\psi} : (B, +) \longrightarrow A^{\times} \text{ a non-degenerate additive character.}$$
$$\chi : B^{\times} \longrightarrow A^{\times} \text{ a multiplicative character.}$$

Then denoting by $\mathbf{N} : \mathbf{B}^{\times} \to \mathbf{G}_m \otimes \mathbf{F}_q$ the norm map, we have

(1)

$$R^i\mathbf{N}_!(\mathcal{L}_{\tilde{\psi}} \otimes \mathcal{L}_{\chi}) = \begin{cases} 0 & \text{if } i \neq \nu - 1 \\ \text{lisse, } A\text{-free of rank } \nu & \text{if } i = \nu - 1 \end{cases}$$

and the "forget supports" map is an isomorphism

$$R\mathbf{N}_!(\mathcal{L}_{\tilde{\psi}} \otimes \mathcal{L}_{\chi}) \xrightarrow{\sim} R\mathbf{N}_*(\mathcal{L}_{\tilde{\psi}} \otimes \mathcal{L}_{\chi})$$

whose formation commutes with passage to fibres.

(2) *For any finite extension k of $\mathbf{F}_q$, and any point $a \in k^\times = \mathbf{G}_m(k)$, we have*

$$\operatorname{trace}\big(F_{a,k} \mid R^{\nu-1}\mathbf{N}_!(\mathcal{L}_{\tilde{\psi}} \otimes \mathcal{L}_\chi)_a\big)$$

$$= (-1)^{\nu-1} \sum_{\substack{x \in B(k) \\ \mathbf{N}(x)=a}} \tilde{\psi}(\operatorname{trace}_{k/\mathbf{F}_q}(x))\chi(\mathbf{N}_{k/\mathbf{F}_q}(x)).$$

(3) *If $A = E_\lambda$ or $\mathcal{O}_\lambda$, $R^{\nu-1}\mathbf{N}_!(\mathcal{L}_{\tilde{\psi}} \otimes \mathcal{L}_\chi)$ is pure of weight $\nu - 1$.*

Proof. Assertion (2) follows from (1) via the Lefschetz trace formula and the formalism of the Lang torsor. Assertions (1) and (3) are invariant under finite extension of the ground field $\mathbf{F}_q$, so may be checked over a finite extension of $\mathbf{F}_q$ over which B as $\mathbf{F}_q$-algebra becomes isomorphic to the ν-fold direct product $\mathbf{F}_q \times \cdots \times \mathbf{F}_q$. In terms of the corresponding coordinates $x_1, \ldots, x_\nu$, we have

$$\begin{cases} \tilde{\psi}(x) = \prod \psi(a_i x_i) & \text{all } a_i \in \mathbf{F}_q^\times \\ \\ \chi(x) = \prod \chi_i(x_i) \end{cases}$$

and canonical isomorphisms $\mathbf{B} = \mathbf{A}^\nu$, $\mathbf{B}^\times = (\mathbf{G}_m)^\nu$, with $\mathbf{N} : \mathbf{B}^\times \to \mathbf{G}_m$ the ν-fold product map $x_1 \cdots x_\nu$. Therefore we have (cf. 5.4) a "multiple convolution," whence

$$R^i\mathbf{N}_!(\mathcal{L}_{\tilde{\psi}} \otimes \mathcal{L}_\chi) = \begin{cases} 0 & i \neq \nu - 1 \\ \\ \mathop{*}\limits_{i=1}^{\nu}(\mathcal{L}_{\psi(a_i x)} \otimes \mathcal{L}_{\chi_i(x)}), & i = \nu - 1. \ \blacksquare \end{cases}$$

8.8.6. Definition. Given a finite etale $\mathbf{F}_q$-algebra B of rank $\nu \geq 1$, a non-degenerate additive character $\tilde{\psi} : (B, +) \to A^\times$, and a multiplicative character $\chi : B^\times \to A^\times$, we denote by $\mathrm{Kl}(B; \tilde{\psi}, \chi)$ the lisse A-sheaf of rank ν on $\mathbf{G}_m \otimes \mathbf{F}_q$ given by

$$\mathrm{Kl}(B; \tilde{\psi}, \chi) \stackrel{\mathrm{dfn}}{=} R^{\nu-1}\mathbf{N}_!(\mathcal{L}_{\tilde{\psi}} \otimes \mathcal{L}_\chi).$$

As the *proof* of the last theorem shows, we have

$$\mathrm{Kl}(B; \tilde{\psi}; \chi) \in \mathcal{C}, \quad \mathrm{Swan}_\infty(\mathrm{Kl}(B; \tilde{\psi}, \chi)) = 1.$$

8.8.7. Lemma. *Let* $\nu \geq 1$, $\chi : \mathbf{F}_{q^\nu}^\times \to A^\times$ *a multiplicative character. For* $B = \mathbf{F}_{q^\nu}$, $\tilde{\psi} = \psi \circ \mathrm{trace}_{\mathbf{F}_{q^\nu}/\mathbf{F}_q}$, χ, *we have a natural isomorphism on* $\mathbf{G}_m \otimes \mathbf{F}_{q^\nu}$

$$\mathrm{Kl}(\mathbf{F}_{q^\nu}; \psi \circ \mathrm{trace}, \chi) \simeq \mathrm{Kl}(\psi \circ \mathrm{trace}; \chi, \chi^q, \ldots, \chi^{q^{\nu-1}}; 1, \ldots, 1).$$

Proof. Indeed $B = \mathbf{F}_{q^\nu}$ is split as an $\mathbf{F}_q$-algebra by the finite extension $\mathbf{F}_{q^\nu}$ of $\mathbf{F}_q$, and the resulting isomorphism is given explicitly by $x \in B \mapsto (x, x^q, \ldots, x^{q^{\nu-1}})$ in $\mathbf{F}_{q^\nu} \times \cdots \times \mathbf{F}_{q^\nu}$. ∎

8.8.8. Given a finite collection of non-zero finite etale $\mathbf{F}_q$-algebras B_i, each equipped with a non-degenerate additive character $\tilde{\psi}_i : (B_i, +) \to A^\times$ and a multiplicative character $\chi_i : B^\times \to A^\times$, we denote by ΠB_i the product finite etale $\mathbf{F}_q$-algebra, and by

$$\begin{cases} \prod \tilde{\psi} : (\Pi B_i, +) \longrightarrow A^\times \\ \qquad (x_i)_i \longmapsto \Pi \tilde{\psi}_i(x_i) \\ \\ \prod \chi_i : (\Pi B_i)^\times \longrightarrow A^\times \\ \qquad (x_i)_i \longmapsto \Pi \chi_i(x_i) \end{cases}$$

the product characters, additive and multiplicative respectively, of this product algebra. One checks easily that $\Pi \tilde{\psi}_i$ is non-degenerate.

8.8.9. Lemma. *We have a natural isomorphism*

$$\mathrm{Kl}(\Pi B_i, \Pi \tilde{\psi}_i, \Pi \chi_i) = \mathop{*}_i \mathrm{Kl}(B_i; \tilde{\psi}_i, \chi_i).$$

Proof. By induction it suffices to treat the case of two factors, in which case it results from the commutative diagram

$$(\mathbf{B}_1 \times \mathbf{B}_2)^\times = (\mathbf{B}_1)^\times \times (\mathbf{B}_2)^\times \xrightarrow{\mathbf{N} \times \mathbf{N}} \mathbf{G}_m \times \mathbf{G}_m \xrightarrow{\pi} \mathbf{G}_m$$

(with the composite map $(\mathbf{B}_1 \times \mathbf{B}_2)^\times \to \mathbf{G}_m$ labelled $\mathbf{N}$)

and the Kunneth formula. ∎

8.9. Embedding in a compatible system

8.9.1. Structure Theorem. *Let* $k = \mathbf{F}_q$, $l \neq p$, $A = E_\lambda$ *or* $\mathbf{F}_\lambda$, *and* $\mathcal{F} \in \mathcal{C}$ *with* $\mathrm{Swan}_\infty(\mathcal{F}) = 1$. *We suppose that* A *contains the p'th roots of unity* *and the* $q^\nu - 1$*'st roots of unity for all* ν *in the range* $1 \leq \nu \leq \mathrm{rank}(\mathcal{F})$. *Then there exist*

$$\begin{cases} \text{a finite etale } \mathbf{F}_q\text{-algebra } B \text{ of } \mathrm{rank} = \mathrm{rank}(\mathcal{F}) \\ \text{a non-degenerate additive character } \tilde{\psi} : (B, +) \to A^\times \\ \text{a multiplicative character } \chi : B^\times \to A^\times \\ \text{a lisse geometrically constant } A\text{-sheaf of rank 1, } T, \text{ on } \mathbf{G}_m \otimes \mathbf{F}_q \end{cases}$$

and an isomorphism of lisse sheaves on $\mathbf{G}_m \otimes \mathbf{F}_q$

$$\mathcal{F} \underset{A}{\otimes} T \simeq \mathrm{Kl}(B; \tilde{\psi}, \chi).$$

Proof. The preceding discussion shows that for $\mathcal{F}$ as above, there exists data $(B, \tilde{\psi}, \chi)$ of the asserted type for which the geometric local monodromy at zero of $\mathrm{Kl}(B; \tilde{\psi}, \chi)$ is isomorphic to that of $\mathcal{F}$.

Because both $\mathcal{F}$ and $\mathrm{Kl}(B; \tilde{\psi}, \chi)$ lie in $\mathcal{C}$ and have $\mathrm{Swan}_\infty = 1$, it follows from the theorem 8.7.1, applied over $\overline{\mathbf{F}}_q$, that there exists $b \neq 0$ in $(\overline{\mathbf{F}}_q)^\times$ and an isomorphism on $\mathbf{G}_m \otimes \overline{\mathbf{F}}_q$

$$\mathcal{F} \simeq (\mathrm{Trans}_b)^*(\mathrm{Kl}(B; \tilde{\psi}, \chi)).$$

Because $\mathcal{F}$ lies in $\mathcal{C}$ and has $\mathrm{Swan}_\infty(\mathcal{F}) = 1$, it follows from 4.1.6 that the quantity $b \in \overline{\mathbf{F}}_q$ is *unique*. Because both $\mathcal{F}$ and $\mathrm{Kl}(B; \tilde{\psi}, \chi)$ are "defined over $\mathbf{F}_q$," it follows from the *uniqueness* of b that b lies in $\mathbf{F}_q^\times$.

The two sheaves $\mathcal{F}$ and $(\mathrm{Trans}_b)^*(\mathrm{Kl}(B; \tilde{\psi}, \chi))$ on $\mathbf{G}_m \otimes \mathbf{F}_q$ are now two lisse sheaves on $\mathbf{G}_m \otimes \mathbf{F}_q$ which on $\mathbf{G}_m \otimes \overline{\mathbf{F}}_q$ become isomorphic, and both are absolutely irreducible as representations of $\pi_1(\mathbf{G}_m \otimes \overline{\mathbf{F}}_q, \bar{\eta})$. Therefore

$$T_0 = H^0(\mathbf{G}_m \otimes \overline{\mathbf{F}}_q, \mathcal{F}^\vee \underset{A}{\otimes} (\mathrm{Trans}_b)^*(\mathrm{Kl}(B; \tilde{\psi}, \chi)))$$

is a 1-dimensional A-space, on which $\mathrm{Gal}(\overline{\mathbf{F}}_q / \mathbf{F}_q)$ acts continuously, and the natural map of lisse sheaves on $\mathbf{F}_q$

$$\mathcal{F} \underset{A}{\otimes} T_0 \longrightarrow (\mathrm{Trans}_b)^*(\mathrm{Kl}(B; \tilde{\psi}, \chi))$$

is an isomorphism.

Because B is a non-zero finite etale $\mathbf{F}_q$-algebra, the norm map $\mathbf{N} : B^\times \to \mathbf{F}_q^\times$ is surjective; indeed writing B as a product of finite field extensions $B \simeq \Pi \mathbf{F}_{q^\nu}$, each factor $\mathbf{F}_{q^\nu}^\times$ maps onto $\mathbf{F}_q^\times$ by the norm. Therefore there exists $\tilde{b} \in B^\times$ with $\mathbf{N}(\tilde{b}) = b$. Then for $n = \mathrm{rank}(B)$, we have

$$\begin{aligned}(\mathrm{Trans}_b)^* \,\mathrm{Kl}(B : \tilde{\psi}, \chi) &= (\mathrm{Trans}_b)^* R^{n-1}\mathbf{N}_!(\mathcal{L}_{\tilde{\psi}} \otimes \mathcal{L}_\chi) \\ &= (R^{n-1}\mathbf{N}_!)(\mathrm{Trans}_{\tilde{b}})^*(\mathcal{L}_{\tilde{\psi}} \otimes \mathcal{L}_\chi).\end{aligned}$$

Just as explained in the proof of 8.7.1, the "primitivity" of $\mathcal{L}_\chi$ on $\mathbf{B}^\times$ assures that for any $\tilde{b} \in B^\times$, we have

$$(\mathrm{Trans}_{\tilde{b}})^*(\mathcal{L}_\chi) = \mathcal{L}_\chi \otimes T_1$$

where T_1 is the inverse image on $\mathbf{B}^\times$ of the lisse rank one A-sheaf $F^n \mapsto \chi(\tilde{b})^n$ on $\mathrm{Spec}(\mathbf{F}_q)$. On the $\tilde{\psi}$ side, we have

$$(\mathrm{Trans}_{\tilde{b}})^*(\mathcal{L}_{\tilde{\psi}}) = \mathcal{L}_{\approx \atop \psi}$$

for the (still non-degenerate) additive character of B given by

$$\overset{\approx}{\psi}(x) = \tilde{\psi}(\tilde{b}x).$$

Thus we find

$$\begin{aligned}(\mathrm{Trans}_b)^*(\mathrm{Kl}(B;\tilde{\psi},\chi)) &= R^{n-1}\mathbf{N}_!(\mathcal{L}_{\approx \atop \psi} \otimes \mathcal{L}_\chi \otimes T_1)\\ &= \mathrm{Kl}(B;\overset{\approx}{\psi},\chi) \otimes T_1,\end{aligned}$$

whence the required isomorphism

$$\mathcal{F} \underset{A}{\otimes} (T_0 \otimes T_1) \xrightarrow{\sim} \mathrm{Kl}(B;\overset{\approx}{\psi},\chi). \quad ∎$$

8.9.2. Corollary. *Let $k = \mathbf{F}_q$, $l \neq p$, and $\mathcal{F}$ a lisse $\overline{\mathbf{Q}}_l$-sheaf on $\mathbf{G}_m \otimes \mathbf{F}_q$ which is tame at zero and totally wild at ∞ with* $\mathrm{Swan}_\infty(\mathcal{F}) = 1$. *Then if* $\det(\mathcal{F})$ *takes algebraic values on Frobenius elements, there exists a subfield E of $\overline{\mathbf{Q}}_l$ which is a finite extension of $\mathbf{Q}$, such that*

(1) *all the local reversed characteristic polynomials,*

$$P_{a,k} = \det(1 - TF_{a,k}|\mathcal{F}_a),$$

for k an arbitrary finite extension of $\mathbf{F}_q$ and any $a \in \mathbf{F}_q^\times$, have coefficients in E.

(2) *for every finite place λ of E or residue characteristic different from p, there exists a lisse E_λ-sheaf $\mathcal{F}_\lambda$ on $\mathbf{G}_m \otimes \mathbf{F}_q$ whose local reversed characteristic polynomials $P_{a,k}^{(\lambda)}$ have coefficients in E, equal to those of $P_{a,k}$.*

Proof. That $\mathrm{Kl}(B;\tilde{\psi},\chi)$ is part of a compatible system is obvious. The hypothesis on the determinant of $\mathcal{F}$ insures that the geometrically constant twisting sheaf T is of the form $\mathrm{trace}(F_x|T) = \alpha^{\deg(x)}$ for some *algebraic* number α. ∎

CHAPTER 9

Equidistribution in $(S^1)^r$ of r-tuples of Angles of Gauss Sums

9.0. In order to motivate the equidistribution problem which we will study in this chapter, we first recall a beautiful result of Davenport which dates from 1931 (cf. [Dav.]).

Let $r \geq 1$ be an integer, and

$$a_1 < a_2 < \cdots < a_r$$

a sequence of r necessarily distinct integers. For p large (e.g., $p > a_r - a_1$), the reductions mod p of $a_1, \ldots, a_r$ will all be distinct in $\mathbf{F}_p$. For such a p, and for any element

$$x \in \mathbf{F}_p - \{-a_1, \ldots, -a_r\},$$

the sums

$$x + a_1, x + a_2, \ldots, x + a_r$$

all lie in $\mathbf{F}_p^\times$. Davenport asks the following question:

How many of the $p - r$ elements $x \in \mathbf{F}_p - \{-a_1, \ldots, -a_r\}$ have the property that $x + a, x + a_2, \ldots, x + a_r$ are all *squares* in $\mathbf{F}_p^\times$?

For example, if

$$(a_1, \ldots, a_r) = (0, 1, 2, \ldots, r - 1),$$

we are asking "how many" sequences

$$(x, x + 1, x + 2, \ldots, x + r - 1)$$

of r successive integers are all squares mod p.

More generally, for any odd p, let us denote by

$$\left(\frac{x}{p}\right) : \mathbf{F}_p^\times \to \pm 1$$

the quadratic character. Then we may consider the map

$$\mathbf{F}_p - \{-a_1, \ldots, -a_r\} \to (\pm 1)^r$$

$$x \mapsto \left(\left(\frac{x+a_1}{p}\right), \ldots, \left(\frac{x+a_r}{p}\right)\right).$$

For any of the 2^r elements

$$\varepsilon = (\varepsilon_1, \ldots, \varepsilon_r) \in (\pm 1)^r,$$

let us denote by $N(\varepsilon; p)$ the integer

$$N(\varepsilon;p) = \text{the number of elements } x \in \mathbf{F}_p - \{-a_q, \ldots, -a_r\}$$
$$\text{such that } \left(\frac{x+a_i}{p}\right) = \varepsilon_i \text{ for all } i = 1, 2, \ldots, r.$$

9.0.1. Theorem (Davenport). *For any element ε in $(\pm 1)^r$, we have*

$$\lim_{p \uparrow \infty} \frac{N(\varepsilon;p)}{p-r} = \frac{1}{2^r}.$$

Davenport's proof is based on the Riemann Hypothesis for the hyperelliptic curves

$$y^2 = \prod_{\text{some } i} (x + a_i) \qquad \text{over } \mathbf{F}_p.$$

Here is a minor reformulation of the theorem. For odd $p > a_r - a_1$, let us denote by μ_p the positive measure of total mass 1 on the group $(\pm 1)^r$ defined by

$$\mu_p = \frac{1}{p-r} \sum_{x \in \mathbf{F}_p - \{-a_i, \ldots, -a_r\}} \begin{pmatrix} \text{the Dirac } \delta\text{-measure supported} \\ \text{at } ((\frac{x-a_1}{p}), \ldots, (\frac{x-a_r}{p})) \end{pmatrix}.$$

Then the theorem asserts that as $p \uparrow \infty$, the measures μ_p converge "weak *" to Haar measure μ_{Haar} on $(\pm 1)^r$, normalized to have total mass one. Concretely, this last statement just means that for any continuous $\mathbf{C}$-valued function

$$f : (\pm 1)^r \to \mathbf{C}$$

we have

$$\lim_{p \uparrow \infty} \int f \, d\mu_p = \int f \, d\mu_{\text{Haar}}.$$

The relation of this formulation to the original one is this: it suffices to check the above limit formula for f the characteristic function of an arbitrary element ε of $(\pm 1)^r$, and for such f the above limit formula in question is

$$\lim_{p\uparrow\infty} \frac{N(\varepsilon;p)}{p-r} = \frac{1}{2^r}.$$

The other "obvious" choice of functions f on which to check such a limit formula are the *characters* of the group $(\pm 1)^r$. The case of the *trivial* character offers no difficulty, because each μ_p already has total mass one. For a non-trivial character, necessarily of the form

$$\varepsilon = (\varepsilon_1, \ldots, \varepsilon_r) \mapsto \prod_{i\in S} \varepsilon_i$$

for some non-void subset S of $\{1, \ldots, r\}$, its integral against μ_{Haar} vanishes, and what must be proven is that

$$\int \left(\prod_{i\in S} \varepsilon_i\right) d\mu_p \overset{\text{dfn}}{=} \frac{1}{p-r} \sum_{x\in F_p - \{-a_1,\ldots,-a_r\}} \left(\frac{\prod_{i\in S}(x+a_i)}{p}\right)$$

tends to zero as $p \uparrow \infty$. By the Riemann Hypothesis for the corresponding hyperelliptic curve

$$y^2 = \prod_{i\in S}(x+a_i)$$

over $\mathbf{F}_p$, one finds that the above integral is $O(1/\sqrt{p})$. (Of course when Davenport was writing, in 1931, the full strength of the Riemann Hypothesis for curves was not available, but enough was known about hyperelliptic curves to give Davenport an estimate $O(1/p^{\varepsilon})$ for the integral....)

9.1. We now turn to the subject proper of this chapter. Once again we fix an integer $r \geq 1$, and a sequence

$$a_1 < a_2 < \cdots < a_r$$

of r necessarily distinct integers.

Let $\mathbf{F}_q$ be a finite field,

$$\psi : (\mathbf{F}_q, +) \to \mathbf{C}^\times$$

a non-trivial additive character, and

$$\chi : \mathbf{F}_q^\times \to \mathbf{C}^\times$$

a multiplicative character of exact order $q-1$ (e.i., χ is a *generator* of the cyclic character group of $\mathbf{F}_q^\times$). If

$$q - 1 > a_r - a_1,$$

which we will henceforth assume, the characters

$$\chi^{a_1}, \ldots, \chi^{a_r}$$

are all distinct. Therefore for any element

$$x \in \mathbf{Z}/(q-1)\mathbf{Z} - \{-a_1, \ldots, -a_r\},$$

the characters

$$\chi^{x+a_1}, \ldots, \chi^{x+a_r}$$

are all non-trivial. The corresponding gauss sums

$$g(\psi, \chi^{x+a_i}), \quad i = 1, \ldots, r$$

are therefore complex numbers of absolute value $\sqrt{q}$, so we may view

$$\theta(x; \mathbf{F}_q, \psi, \chi) \overset{\text{dfn}}{=} \left(\frac{g(\psi, \chi^{x+a_1})}{\sqrt{q}}, \ldots, \frac{g(\psi, \chi^{x+a_r})}{\sqrt{q}} \right)$$

as being a point of $(S^1)^r$, where S^1 denotes the unit circle. As x varies over

$$\mathbf{Z}/(q-1)\mathbf{Z} - \{-a_1, \ldots, -a_r\},$$

we obtain in this way a collection of $q-1-r$ points in $(S^1)^r$. We will prove that this set of $q-1-r$ points in $(S^1)^r$ becomes equidistributed with respect to normalized Haar measure on $(S^1)^r$, as q tends to infinity.

9.2. More precisely, for each datum $(\mathbf{F}_q, \psi, \chi)$ as above, let us denote by $\mu_{(\mathbf{F}_q,\psi,\chi)}$ the positive measure of total mass one on $(S^1)^r$ defined by

$$\mu_{(\mathbf{F}_q,\psi,\chi)} = \frac{1}{q-1-r} \sum_{x \in \mathbf{Z}/(q-1)\mathbf{Z} - \{-a_1,\ldots,-a_r\}} \begin{pmatrix} \text{the Dirac delta} \\ \text{measure supported at} \\ \theta(x; \mathbf{F}_q, \psi, \chi) \end{pmatrix}.$$

9.3. Theorem *In any sequence of data $(\mathbf{F}_q, \psi, \chi)$ as above in which q tends to ∞, the measures $\mu_{(\mathbf{F}_q,\psi,\chi)}$ on $(S^1)^r$ converge (weak $*$) to normalized (total mass one) Haar measure on $(S^1)^r$, i.e., for any continuous $\mathbf{C}$-valued function f on $(S^1)^r$, we have*

$$\lim_{q \uparrow \infty} \int f \, d\mu_{(\mathbf{F}_q,\psi,\chi)} = \int f \, d\mu_{\text{Haar}},$$

the limit taken over any sequence of data $(\mathbf{F}_q, \psi, \chi)$ in which q tends to ∞.

9.4. Before proving the theorem, it will be convenient to give a version of it which is both slightly stronger and slightly more intrinsic. To motivate it, let us fix a datum $(\mathbf{F}_q, \psi, \chi)$ as above, with $q-1 > a_r - a_1$. Then the characters

$$\chi_1 = \chi^{a_1}, \ldots, \chi_r = \chi^{a_r}$$

are all distinct, and the sequences of characters

$$\chi^{x+a_1}, \ldots, \chi^{x+a_r}$$

with

$$x \in \mathbf{Z}/(q-1)\mathbf{Z} - \{-a_1, \ldots, -a_r\}$$

may be described intrinsically as the sequences $(\rho = \chi^x)$

$$(\rho\chi_1, \rho\chi_2, \ldots, \rho\chi_r)$$

as ρ runs over

$$\rho \in (\text{char. group of } \mathbf{F}_q^\times) - \{\overline{\chi}_1, \ldots, \overline{\chi}_r\}.$$

This being so, we consider data of the following type:

$$(\mathbf{F}_q; \psi; \chi_1, \ldots, \chi_r)$$

where $\mathbf{F}_q$ is a finite field, ψ is a non-trivial additive character

$$\psi : (\mathbf{F}_q, +) \to \mathbf{C}^\times,$$

and $\chi_1, \ldots, \chi_r$ are r distinct multiplicative characters $\chi_i : \mathbf{F}_q^\times \to \mathbf{C}^\times$ (one of which may be trivial). Having fixed such a datum, we consider for variable

$$\chi \in (\text{char. group of } \mathbf{F}_q^\times) - \{\overline{\chi}_1, \ldots, \overline{\chi}_n\}$$

the element

$$\theta(\chi; \mathbf{F}_q, \psi, \chi_1, \ldots, \chi_r) \in (S_1)^r$$

defined by

$$\theta(\chi; \mathbf{F}_q, \psi, \chi_1, \ldots, \chi_r) = \left(\ldots, \frac{g(\psi; \chi_i\chi)}{\sqrt{q}}, \ldots\right),$$

and we denote by

$$\mu(\mathbf{F}_q, \psi; \chi_1, \ldots, \chi_r)$$

the positive measure of total mass one on $(S^1)^r$ defined by

$$\mu(\mathbf{F}_q, \psi; \chi_1 \ldots \chi_r) = \frac{1}{q-1-r} \sum_{\substack{\chi \in (\text{char.group of } \mathbf{F}_q^\times) \\ \chi \notin \{\overline{\chi}_1, \ldots, \overline{\chi}_r\}}} \begin{pmatrix} \text{the Dirac delta measure} \\ \text{supported at} \\ \theta(\chi; \mathbf{F}_q, \psi, \chi_1, \ldots, \chi_r) \end{pmatrix}.$$

9.5. Theorem. *Fix $r \geq 1$. In any sequence of data $(\mathbf{F}_q, \psi; \chi_1, \ldots, \chi_r)$ as above (ψ non-trivial additive character of $\mathbf{F}_q$, $\chi_1, \ldots, \chi_r$ r distinct multiplicative characters of $\mathbf{F}_q^\times$) in which q tends to ∞, the measures*

$$\mu(\mathbf{F}_q, \psi; \chi_1, \ldots, \chi_r)$$

tend (weak $*$) to normalized Haar measure on $(S^1)^r$.

Proof. By the Weyl criterion for equidistribution, it suffices to show that

$$\lim_{q \uparrow \infty} \int f \, d\mu_{(\mathbf{F}_q, \psi; \chi_1, \ldots, \chi_r)} = \int f \, d\mu_{\text{Haar}}$$

for f any unitary character of the compact abelian group $(S^1)^r$. For f the trivial character, there is nothing to prove, for each individual

$$\mu(\mathbf{F}_q, \psi; \chi_1, \ldots, \chi_r)$$

already has total mass one.

The case of non-trivial unitary characters is more involved. If we view a point of $(S^1)^r$ as being an r-tuple

$$(z_1, \ldots, z_r)$$

of complex numbers of absolute value one, then the unitary characters of $(S^1)^r$ are just the monomials in the z_i

$$\chi(z_i, \ldots, z_r) = \prod_{i=1}^{r} z_i^{n_i}.$$

Given a character χ as above, we break the index set $\{1, \ldots, r\}$ into three disjoint subsets

$$A = \{i \text{ for which } n_i > 0\}$$
$$B = \{i \text{ for which } n_i < 0\}$$
$$C = \{i \text{ for which } n_i = 0\}.$$

Writing

$$m_i = -n_i \text{ for } i \in B,$$

we thus obtain an express for our unitary character χ as

$$\chi(z) = \left(\prod_{i \in A} z_i^{n_i}\right) \left(\prod_{j \in B} \overline{z_j}^{m_j}\right), \text{ with } n_i, m_j > 0$$

just using the relation $z_i \bar{z}_i = 1$. For χ non-trivial, at least one of the sets A or B is non-empty.

For non-trivial χ, we have the orthogonality relation

$$\int \chi \, d\mu_{\text{Haar}} = 0,$$

so what must be proven is that for χ non-trivial we have

$$\lim_{q \uparrow \infty} \int \chi \, d\mu_{(\mathbf{F}_q, \psi; \chi_1, \ldots, \chi_r)} = 0.$$

In fact, we will prove

9.6. Theorem *For χ non-trivial written as above, we have the estimate*

$$\left| \int \chi \, d\mu_{(\mathbf{F}_q, \psi; \chi_1, \ldots, \chi_r)} \right| \leq \frac{\sup\left(\sum_{i \in A} n_i, \sum_{i \in B} m_i \right)}{\sqrt{q}} + \frac{2r}{q-1},$$

for any data $(\mathbf{F}_q, \psi; \chi_1, \ldots, \chi_r)$ with ψ a non-trivial additive character and $\chi_1, \ldots, \chi_r$ $r \geq 1$ distinct multiplicative characters of $\mathbf{F}_q$.

We now turn to the proof of this last theorem. By definition, we have

$$\int \chi \, d\mu_{(\mathbf{F}_q, \psi; \chi_1, \ldots, \chi_r)}$$

$$= \frac{1}{q-1-r} \sum_{\substack{\chi \text{ char of } \mathbf{F}_q^\times \\ \chi \notin \{\overline{\chi}_1, \ldots, \overline{\chi}_r\}}} \prod_{i \in A} \left(\frac{g(\psi, \chi_i \chi)}{\sqrt{q}} \right)^{n_i} \prod_{i \in B} \left(\frac{\overline{g(\psi, \chi_i \chi)}}{\sqrt{q}} \right)^{m_i}$$

Each term after the $\sum$ is a product of numbers of absolute value one. The "excluded" terms, i.e., those where χ is one of the $\overline{\chi}_i$, are products of terms each of absolute value ≤ 1 (because $|g(\psi, \chi_i \chi)| = 1$ if $\chi = \overline{\chi}_i$). So the above integral is fairly well approximated by the quantity "S" defined as

$$S \overset{\text{dfn}}{=} \frac{1}{q-1} \sum_{\substack{\text{all chars } \chi \\ \text{of } \mathbf{F}_q^\times}} \prod_{i \in A} \left(\frac{g(\psi, \chi_i \chi)}{\sqrt{q}} \right)^{n_i} \prod_{i \in B} \left(\frac{\overline{g(\psi, \chi_i \chi)}}{\sqrt{q}} \right)^{m_i};$$

in fact

$$\int - S = \frac{1}{q-1-r} \sum_{q-1-r\,\chi's} - \frac{1}{q-1} \sum_{q-1\,\chi's}$$

$$= \left(\frac{1}{q-1-r} - \frac{1}{q-1}\right) \sum_{q-1-r\,\chi's} - \frac{1}{q-1} \sum_{r\,\chi's}$$

$$= \frac{r}{(q-1)(q-1-r)} \sum_{q-1-r\,\chi's} - \frac{1}{q-1} \sum_{r\,\chi's},$$

and since each monomial after the $\sum$ has $\|\ \| \leq 1$, we find

$$\left| \int - S \right| \leq \frac{2r}{(q-1)}.$$

Therefore we are reduced to proving the estimate

$$|S| \leq \frac{\sup\left(\sum_{i \in A} n_i, \sum_{i \in B} m_i\right)}{\sqrt{q}}.$$

To go further, we observe that the sum S is of the form

$$\frac{1}{q-1} \sum_{\substack{\chi \text{ chars} \\ \text{of } \mathbf{F}_q^\times}} \hat{f}(\chi)\overline{\hat{g}(\chi)}$$

for the unique functions f, g on $\mathbf{F}_q^\times$ whose multiplicative Fourier transforms are

$$\hat{f}(\chi) = \left(\frac{1}{\sqrt{q}}\right)^{\sum_{i \in A} n_i} \prod_{i \in A} g(\psi, \chi_i \chi)^{n_i}$$

$$\hat{g}(\chi) = \left(\frac{1}{\sqrt{q}}\right)^{\sum_{i \in B} m_i} \prod_{i \in B} g(\psi, \chi_i \chi)^{m_i}.$$

Case 1. If A is *empty*, then $\hat{f}(\chi)$ is identically 1, whence

$$f = \text{ the Dirac } \delta\text{-function on } \mathbf{F}_q^\times \text{ supported at } 1.$$

Case 2. If A is non-empty, then f is the inverse Fourier transform of a monomial in gauss sums, so given by the trace function of a Kloosterman sheaf on $\mathbf{G}_m \otimes \mathbf{F}_q$. More precisely, for any prime $l \neq p = \mathrm{char}(\mathbf{F}_q)$, and any l-adic place λ of the field $E = Q(\zeta_p, \zeta_{q-1})$, viewed as a subfield of C where ψ and the χ_i take values, we have

$$-f(a) = \left(\frac{-1}{\sqrt{q}}\right)^{\sum_{i\in A} n_i} \operatorname{trace}(F_a \mid (\mathrm{Kl}_A)_{\bar{a}})$$

where Kl_A denotes the lisse E_λ-sheaf on $\mathbf{G}_m \otimes \mathbf{F}_q$ given by

$$\mathrm{Kl}_A \overset{\mathrm{dfn}}{=} \mathrm{Kl}(\psi;\ \text{the } \chi_i \text{ for } i \in A, \text{ each repeated } n_i \text{ times; } 1,1,\ldots,1).$$

Symmetrically in B, we have
Case 1. If B is empty, then

$$g = \text{ the Dirac } \delta\text{-function on } \mathbf{F}_q^\times \text{ supported at 1.}$$

Case 2. If B is non-empty, then

$$-g(a) = \left(\frac{-1}{\sqrt{q}}\right)^{\sum_{i\in B} m_i} \operatorname{trace}(F_a \mid (\mathrm{Kl}_B)_{\bar{a}}).$$

Let us now make use of the Parsefal identity (cf. 4.0)

$$\sum_{a\in\mathbf{F}_q^\times} f(a)\overline{g(a)} = \frac{1}{q-1}\sum_{\chi} \hat{f}(\chi)\overline{\hat{g}(\chi)},$$

to evaluate the second term, which is S.
Case 1. A empty, B non-empty. Then

$$-S = \left(\frac{-1}{\sqrt{q}}\right)^{\sum_{i\in B} m_i} \overline{\operatorname{trace}(F_1 \mid (\mathrm{Kl}_B)_1)}.$$

As Kl_B is lisse of rank $\sum_{i\in B} m_i$ and pure of weight $\left(\sum_{i\in B} m_i\right) - 1$, we have

$$|\operatorname{trace}(F_1(\mathrm{Kl}_B)_1| \leq \left(\sum_{i\in B} m_i\right)(\sqrt{q})^{\left(\sum_{i\in B} m_i\right)-1}$$

whence

$$|S| \leq \frac{\sum_{i\in B} m_i}{\sqrt{q}}, \text{ as required.}$$

Case 2. A non-empty, B empty. Then

$$-S = \left(\frac{-1}{\sqrt{q}}\right)^{\sum_{i\in A} n_i} \operatorname{trace}(F_1 \mid \mathrm{Kl}_A)_1),$$

and just as above we find

$$|S| \le \frac{\sum_{i\in A} n_i}{\sqrt{q}}, \text{ as required.}$$

Case 3. A, B both non-empty. This is the only essentially new case. Here we have

$$S = \left(\frac{-1}{\sqrt{q}}\right)^{\sum_{i\in A} n_i + \sum_{j\in B} m_j} \sum_{a\in \mathbf{F}_q^\times} \operatorname{trace}(F_a \mid (\mathrm{Kl}_A)_{\bar{a}})\overline{\operatorname{trace}(F_a \mid (\mathrm{Kl}_B)_{\bar{a}})}.$$

Let us denote by $\overline{\mathrm{Kl}}_B$ the Kloosterman sheaf which is the "complex conjugate" of Kl_B, i.e.,

$$\overline{\mathrm{Kl}}_B =$$

$$\mathrm{Kl}(\overline{\psi}; \text{ those characters } \overline{\chi}_i \text{ for } i \in B, \text{ each repeated } m_i \text{ times; } 1,\ldots,1).$$

Then we have

$$\overline{\operatorname{trace}(F_a \mid (\mathrm{Kl}_B)_{\bar{a}})} = \operatorname{trace}(F_a \mid (\overline{\mathrm{Kl}}_B)_{\bar{a}}),$$

so we may write

$$S = \left(\frac{-1}{\sqrt{q}}\right)^{\sum n_i + \sum m_j} \sum_{a\in \mathbf{F}_q^\times} \operatorname{trace}(F_a \mid (\mathrm{Kl}_A \otimes \overline{\mathrm{Kl}}_B)_{\bar{a}}).$$

Applying the Lefschetz trace formula to the lisse sheaf $\mathrm{Kl}_A \otimes \overline{\mathrm{Kl}}_B$ on $\mathbf{G}_m \otimes \mathbf{F}_q$, we have

$$\sum_{a\in \mathbf{F}_q^\times} \operatorname{trace}(F_a \mid (\mathrm{Kl}_A \otimes \overline{\mathrm{Kl}}_B)_{\bar{a}})$$

$$= \sum_{i=1}^{2} (-1)^i \operatorname{trace}(F \mid H_c^i(\mathbf{G}_m \otimes \overline{\mathbf{F}}_q, \mathrm{Kl}_A \otimes \overline{\mathrm{Kl}}_B)).$$

Because $\mathrm{Kl}_A \otimes \overline{\mathrm{Kl}}_B$ is pure of weight

$$\left(\sum_{i\in A} n_i\right) - 1 + \left(\sum_{j\in B} m_j\right) - 1,$$

its H_c^1 is mixed of weight

$$w(H_c^1) \le \left(\sum_{i\in A} n_i\right) + \left(\sum_{j\in B} m_j\right) - 1,$$

while its H_c^2 is pure of weight

$$\left(\sum_{i\in A} n_i\right) + \left(\sum_{j\in B} m_j\right),$$

all this by Deligne's results in Weil II. Therefore to prove the required estimate

$$|S| \le \frac{\sup\left(\sum_{i\in A} n_i, \sum_{j\in B} m_j\right)}{\sqrt{q}},$$

it suffices to prove the following

9.7. Proposition. *In the above situation, we have*

$$H_c^2(\mathbf{G}_m \otimes \overline{\mathbf{F}}_q, \mathrm{Kl}_A \otimes \mathrm{Kl}_B) = 0,$$

$$\dim H_c^1(\mathbf{G}_m \otimes \overline{\mathbf{F}_q}, \mathrm{Kl}_A \otimes \overline{\mathrm{Kl}}_B) \le \sup\left(\sum_{i\in A} n_i, \sum_{j\in B} m_j\right).$$

Proof. To prove that H_c^2 vanishes for a lisse sheaf $\mathcal{F}$ on $\mathbf{G}_m \otimes \overline{\mathbf{F}}_q$, it certainly suffices to show that as representation of I_0, $\mathcal{F}$ has no non-zero coinvariants (because up to a Tate twist the H^2_{comp} is a quotient of the coinvariants, namely the coinvariants under all of π_1^{geom}). But

$$\begin{aligned} \mathrm{Kl}_A &= \mathrm{Kl}(\psi; \text{various of the } \chi_i,\ i \in A, \text{ repeated}; 1,\dots,1) \\ \overline{\mathrm{Kl}}_B &= \mathrm{Kl}(\overline{\psi}; \text{various of the } \overline{\chi}_j,\ j \in B, \text{ repeated}; 1,\dots,1). \end{aligned}$$

So by the explicit determination of the local monodromy at zero of Kloosterman sheaves, we have

As I_0-representation, Kl_A is a successive extension of various $\mathcal{L}_{\chi_i}$ with $i \in A$,

while Kl_B is a successive extension of various $\mathcal{L}_{\overline{\chi}_j}$ with $j \in B$.

Therefore as I_0-representation, $\mathrm{Kl}_A \otimes \overline{Kl}_B$ is a successive extension of the characters

$$\mathcal{L}_{\chi_i\overline{\chi}_j} \text{ with } i \in A \text{ and } j \in B.$$

By the original hypothesis, the characters $\chi_1, \dots, \chi_r$ are all *distinct*, so the fact that A and B are *disjoint* guarantees that

$$\chi_i\overline{\chi}_j \ne \mathbf{1} \text{ if } i \in A \text{ and } j \in B.$$

Therefore $\mathrm{Kl}_A \otimes \overline{\mathrm{Kl}}_B$ as I_0-representation is a successive extension of non-trivial characters of I_0, so it certainly has no non-zero I_0-coinvariants. This proves the asserted vanishing of H_c^2.

Once we know $H_c^2 = 0$, then $\dim H_c^1$ is just minus the Euler characteristic. As both Kl_A and Kl_B are tame at zero, so is $\mathrm{Kl}_A \otimes \overline{\mathrm{Kl}}_B$, so by the Euler–Poincaré formula we have

$$\dim H_c^1(\mathbf{G}_m \otimes \overline{\mathbf{F}}_q, \mathrm{Kl}_A \otimes \overline{\mathrm{Kl}}_B) = \mathrm{Swan}_\infty(\mathrm{Kl}_A \otimes \overline{\mathrm{Kl}}_B).$$

By the structure at ∞ of Kloosterman sheaves, we know that

$$\begin{cases} \mathrm{Kl}_A \text{ has } \mathrm{Swan}_\infty = 1, \text{ all breaks} = 1/\sum\limits_{i\in A} n_i \\ \overline{\mathrm{Kl}}_B \text{ has } \mathrm{Swan}_\infty = 1, \text{ all breaks} = 1/\sum\limits_{m\in B} m_j, \end{cases}$$

Therefore $\mathrm{Kl}_A \otimes \overline{\mathrm{Kl}}_B$ has all breaks at ∞

$$\leq \max\left(1/\sum_{i\in A} n_i,\ 1/\sum_{j\in B} m_j\right),$$

while its rank is

$$\left(\sum_{i\in A} n_i\right)\left(\sum_{j\in B} m_j\right);$$

because its Swan_∞ is $\leq$ (rk)$\times$ (biggest break), we find

$$\mathrm{Swan}_\infty(\mathrm{Kl}_A \otimes \overline{\mathrm{Kl}}_B) \leq \max\left(\sum_{j\in B} m_j,\ \sum_{i\in A} n_i\right). \quad \blacksquare$$

9.8. Remark. If $\sum_{i\in A} \neq \sum_{j\in B}$, i.e., if Kl_A and Kl_B have different ranks, then their breaks at ∞ are disjoint, so that *every* break at ∞ of $\mathrm{Kl}_A \otimes \overline{\mathrm{Kl}}_B$ is *equal* to

$$\max(1/\sum n_i, 1/\sum m_j).$$

This shows that $\mathrm{Kl}_A \otimes \overline{\mathrm{Kl}}_B$ is *totally wild* at ∞ in this case, and so gives *another* proof of the vanishing of H_c^2 in the case when Kl_A and Kl_B have different ranks. Continuing with this case, our exact knowledge of the breaks at ∞ gives

$$\mathrm{Swan}_\infty(\mathrm{Kl}_A \otimes \overline{\mathrm{Kl}}_B) = \max\left(\sum_{i\in A} n_i,\ \sum_{j\in B} m_j\right)$$

provided $\sum n_i \neq \sum m_j$, i.e., the inequality for $\dim H_c^1$ in the proposition is an equality in this case.

Now, quite generally, the fact that all Kloosterman sheaves are geometrically irreducible (cf. 4.1.2) shows that for any two Kloosterman sheaves Kl and Kl′, we have

$$H^2_c(\mathbf{G}_m \otimes \overline{\mathbf{F}}_q, \mathrm{Kl} \otimes \mathrm{Kl}') \neq 0 \Longleftrightarrow \begin{array}{l}\text{there exists a geometric}\\ \text{isomorphism } \mathrm{Kl}' \simeq (\mathrm{Kl})^\vee \text{ on}\\ \mathbf{G}_m \otimes \overline{\mathbf{F}}_q\end{array}$$

and so in particular (cf. 4.1.3)

$$H^2_c(\mathbf{G}_m \otimes \overline{\mathbf{F}}_q, \mathrm{Kl} \otimes \overline{\mathrm{Kl}}_B) = 0 \Longleftrightarrow \begin{array}{l}\text{there exists a geometric}\\ \text{isomorphism } \mathrm{Kl}_A \simeq \mathrm{Kl}_B \text{ on}\\ \mathbf{G}_m \otimes \overline{\mathbf{F}}_q.\end{array}$$

In the case when Kl_A and Kl_B have different ranks, they are certainly not geometrically isomorphic, so we have "another" proof of the vanishing of H^2_c in this case. It is in the case of equal rank that we must (jusqu'*à* nouvel ordre!) make use of the precise description of local monodromy at zero to show that Kl_A and Kl_B are not geometrically isomorphic because they are not I_0-isomorphic.

CHAPTER 10

Local Monodromy at ∞ of Kloosterman Sheaves

10.0. We begin by recalling some "independence of l" results. Thus we fix

a finite extension E of $\mathbf{Q}$

a finite field $\mathbf{F}_q$, whose characteristic is denoted p

a non-trivial additive character $\psi : (\mathbf{F}_q, +) \to \mathcal{O}_E^\times$

an integer $n \geq 1$

multiplicative characters $\chi_1, \ldots, \chi_n : \mathbf{F}_q^\times \to \mathcal{O}_E^\times$

integers $b_1, \ldots, b_n$, all ≥ 1.

For every prime number $l \neq p$, and every l-adic place λ of E, we denote by

$$\mathrm{Kl}_\lambda = \mathrm{Kl}_\lambda(\psi; \chi_1, \ldots, \chi_n; b_1, \ldots, b_n)$$

the lisse E_λ-sheaf on $\mathbf{G}_m \otimes \mathbf{F}_q$ whose existence is guaranteed by (4.1.1). For variable λ (i.e., λ running over the finite places of E of residue characteristic different from p), the lisse E_λ-sheaves Kl_λ on $\mathbf{G}_m \otimes \mathbf{F}_q$ form a "compatible system," in the sense that for every closed point x of $\mathbf{G}_m \otimes \mathbf{F}_q$, the trace of Frobenius at x,

$$\mathrm{trace}(F_x \mid (\mathrm{Kl}_\lambda)_{\bar{x}}),$$

which *à* priori lies in E_λ, in fact lies in E and is independent of λ. By a result of Deligne ([De-1], 1.9), the fact that the Kl_λ are compatible in the above sense implies that *as representations of* I_∞ (or of I_0, but this case we have already analyzed), the Kl_λ have character with values in E, independent of λ.

10.1. Proposition. *As representation of* P_∞, *the lisse* E_λ*-sheaf on* $\mathbf{G}_m \otimes \mathbf{F}_q$

$$\mathrm{Kl}_\lambda(\psi; \chi_1, \ldots, \chi_n; b_1, \ldots, b_n)$$

has character with values in E *which is independent of both* λ *and of the particular choice of the multiplicative characters* $\chi_1, \ldots, \chi_n$.

Proof. The E-valuedness and independence of λ, for fixed choice of χ_i's, has already been seen, just above. To compare $(\chi_1, \ldots, \chi_n)$ to $(\mathbf{1}, \ldots, \mathbf{1})$, factor each χ_i as a product of characters of prime power orders. Because χ_i has order dividing $q-1$, the prime power orders which occur are necessarily prime to p. Proceeding one $l \neq p$ at a time, it suffices to show that if $\rho_1, \ldots, \rho_n$ are characters of l-power order of $\mathbf{F}_q^\times$, then for λ any l-adic place of E, the E_λ-sheaves

$$\mathrm{Kl}_\lambda(\psi; \chi_1, \ldots, \chi_n; b_1, \ldots, b_n)$$

and

$$\mathrm{Kl}_\lambda(\psi; \chi_1\rho_1, \ldots, \chi_n\rho_n; b_1, \ldots, b_n)$$

have E-valued and equal characters on P_∞. As we already have seen the E-valuedness, it suffices to show the characters on P_∞ agree in E_λ. Because in both representations P_∞ acts through a finite p-group quotient, and $l \neq p$, it suffices to show that the two E_λ-representations of P_∞ are isomorphic when reduced mod λ. But if we pick $\mathcal{O}_\lambda$-forms of the two sheaves in question, their $\otimes \mathbf{F}_\lambda$'s are in fact isomorphic as representations of the *entire* (arithmetic) fundamental group of $\mathbf{G}_m \otimes \mathbf{F}_q$, because they are absolutely irreducible (cf. 4.1.2) and because for every closed point x, the traces of F_x on the two agree in $\mathbf{F}_\lambda$ (because the characters ρ_i, being of l-power order, become *trivial* after $\otimes \mathbf{F}_\lambda$). ∎

10.2. Corollary. *The set of breaks with multiplicities at ∞ of a tensor product*

$$\mathrm{Kl}_\lambda(\psi; \chi_1, \ldots, \chi_n; b_1, \ldots, b_n) \otimes \mathrm{Kl}_\lambda(\psi'; \chi'_1, \ldots, \chi'_m; b'_1, \ldots, b'_m)$$

of Kloosterman sheaves is independent of the particular choice of λ and of the multiplicative characters $\chi_1, \ldots, \chi_n$ and $\chi'_1, \ldots, \chi'_m$.

Proof. The breaks with multiplicities are determined by the upper numbering filtration of P_∞ and by the values of the character of the representation on each $P_\infty^{(x)}$ subgroup. By the proposition this data is independent of both λ and of the choice of the χ_i and the χ'_i. ∎

10.3. Corollary. *If $n = m$ and if $(b_1, \ldots, b_n) = (b'_1, \ldots, b'_n)$, then*

$$\mathrm{Kl}_\lambda(\psi; \chi_1, \ldots, \chi_n; b_1, \ldots, b_n) \otimes \mathrm{Kl}_\lambda(\overline{\psi}; \chi'_1, \ldots, \chi'_n; b_1, \ldots, b_n)$$

is not totally wild at ∞, i.e., at least one of its breaks at ∞ vanishes.

Proof. The question being independent of the choice of χ_i', we take $\chi_i' = \overline{\chi}_i$. Then the sheaf in question is a Tate twist of $\underline{\mathrm{End}}(\mathrm{Kl}_\lambda(\psi; \chi_i\text{'s}, b_i\text{'s}))$, which has a global (i.e. under π_1^{geom}) invariant, namely the identity endomorphism, so it certainly has non-zero invariants under P_∞. ∎

10.4. We now consider in some detail the case when all the $b_i = 1$, and we write simply

$$\mathrm{Kl}(\psi; \chi_1, \ldots, \chi_n) \stackrel{\mathrm{dfn}}{=} \mathrm{Kl}(\psi; \chi_1, \ldots, \chi_n; 1, \ldots, 1).$$

If $m \neq n$, then we have already observed that

$$\mathrm{Kl}(\psi; \chi_1, \ldots, \chi_n) \otimes \mathrm{Kl}(\psi'; \chi_1', \ldots, \chi_m')$$

is totally wild at ∞, with all its mn breaks equal to $\max(1/n, 1/m)$.

10.4.1. Proposition. *Let ψ_1 and ψ_2 be non-trivial E-valued additive characters of $\mathbf{F}_q$, and denote by $\alpha \in \mathbf{F}_q^\times$ their ratio, i.e.*

$$\psi_2(x) = \psi_1(\alpha x) \quad \textit{for} \quad x \in \mathbf{F}_q.$$

Let $n \geq 1$ be an integer, and $\chi_1, \ldots, \chi_n, \chi_1', \ldots, \chi_n'$ E-valued multiplicative characters of $\mathbf{F}_q^\times$. Consider the tensor product

$$\mathrm{Kl}_1 \otimes \mathrm{Kl}_2 = \mathrm{Kl}(\psi_1; \chi_1, \ldots, \chi_n) \otimes \mathrm{Kl}(\psi_2; \chi_1', \ldots, \chi_n').$$

Then we have

$$\mathrm{Swan}_\infty(\mathrm{Kl}_1 \otimes \mathrm{Kl}_2) = \begin{cases} n \textit{ if } (-\alpha)^n \neq 1 \textit{ in } \mathbf{F}_q \\ \\ n-1 \textit{ if } (-\alpha)^n = 1 \textit{ in } \mathbf{F}_q \end{cases}$$

Proof. As we have seen above (cf. 10.2), the breaks with their multiplicities of this tensor product are independent of the choice of the χ_i and the χ_i'. We will choose all $\chi_i = \mathbf{1}$ and all $\chi_i' = \mathbf{1}$. We will denote simply by $\mathrm{Kl}_n(\psi)$ the lisse sheaf

$$\mathrm{Kl}_n(\psi) \stackrel{\mathrm{dfn}}{=} \mathrm{Kl}(\psi; \underbrace{\mathbf{1}, \ldots, \mathbf{1}}_{n \text{ times}}; \underbrace{1, \ldots, 1}_{n \text{ times}}).$$

Thus our problem is to compute Swan_∞ for

$$\mathrm{Kl}_n(\psi_1) \otimes \mathrm{Kl}_n(\psi_2).$$

As this sheaf is lisse on $\mathbf{G}_m \otimes \mathbf{F}_q$, and tame at zero, we have

$$\mathrm{Swan}_\infty = -\chi_c(\mathbf{G}_m \otimes \overline{\mathbf{F}}_q, \mathrm{Kl}_n(\psi_1) \otimes \mathrm{Kl}_n(\psi_2)),$$

so by the Lefschetz trace formula we have

$$\begin{aligned}\mathrm{Swan}_\infty = &\textit{degree} \text{ as rational function of the } L\text{-function of}\\ &\mathbf{G}_m \otimes \mathbf{F}_q \text{ with coefficients in } \mathrm{Kl}_n(\psi_1)\otimes\mathrm{Kl}_n(\psi_2).\end{aligned}$$

Explicitly, we have

$$L(\mathbf{G}_m \otimes \mathbf{F}_q, \mathrm{Kl}_n(\psi_1)\otimes\mathrm{Kl}_n(\psi_2), T) = \exp\left(\sum_{r\geq 1} S_r \frac{T^r}{r}\right)$$

where for each $r \geq 1$ we have

$$\begin{aligned}S_r &= \sum_{a\in\mathbf{F}_{q^r}^\times} \mathrm{trace}(F_{a;\mathbf{F}_{q^r}} | (\mathrm{Kl}_n(\psi_1)\otimes\mathrm{Kl}_n(\psi_2))_{\bar{a}})\\ &= \sum_{a\in\mathbf{F}_{q^r}^\times} (-1)^{2n-2} \sum_{\substack{x_1\ldots x_n = a\\ \text{all } x_i\in\mathbf{F}_{q^r}}} (\psi_1 \circ \mathrm{trace})(\Sigma x_i) \sum_{\substack{y_1\ldots y_n = a\\ \text{all } y_i\in\mathbf{F}_{q^r}}} (\psi_2 \circ \mathrm{trace})(\Sigma y_i)\end{aligned}$$

where "trace" is that of $\mathbf{F}_{q^r}/\mathbf{F}_q$. Writing $\psi_2(x) = \psi_1(\alpha x)$, and remembering that $\psi(x)\psi(y) = \psi(x+y)$, we find

$$S_r = \sum_{a\in\mathbf{F}_{q^r}^\times} \sum_{\substack{x_1\ldots x_n = a\\ \text{all } x\in\mathbf{F}_{q^r}}} \sum_{\substack{y_1\ldots y_n = a\\ \text{all } y_i\in\mathbf{F}_{q^r}}} (\psi_1 \circ \mathrm{trace})(\Sigma x_i + \alpha\Sigma y_i)$$

whence

$$S_r = \sum_{\substack{x_1\ldots x_n = y_1\ldots y_n\\ \text{all } x_i, y_j\in\mathbf{F}_{q^r}^\times}} (\psi_1 \circ \mathrm{trace})(\Sigma x_i + \alpha\Sigma y_i).$$

Now let us explicitly keep track of the dependence on $n \geq 1$ and on $\alpha \in \mathbf{F}_q^\times$. For this it is convenient to introduce, for $n \geq 1$, $\alpha \in \mathbf{F}_q^\times$ and $\beta \in \mathbf{F}_q^\times$, the sums

$$S_r(n,\alpha,\beta) = \sum_{\substack{\beta x_1\ldots x_n = y_1\ldots y_n\\ \text{all } x_i, y_i \text{ in } \mathbf{F}_{q^r}^\times}} (\psi_1 \circ \mathrm{trace})\left(\sum_1^n x_i + \alpha\sum_1^n y_i\right).$$

For $n = 1$, we readily calculate

$$S_r(1, \alpha, \beta) = \sum_{\substack{y=\beta x, \\ x,y \text{ in } \mathbf{F}_{q^r}^{\times}}} (\psi_1 \circ \text{trace})(x + \alpha y)$$

$$= \sum_{x \in \mathbf{F}_{q^r}^{\times}} (\psi_1 \circ \text{trace})((1 + \alpha\beta)x)$$

$$= \begin{cases} -1 & \text{if} \quad \alpha\beta \neq -1 \\ \\ q^r - 1 & \text{if} \quad \alpha\beta = -1. \end{cases}$$

10.4.2. Lemma. *For $n \geq 1$, $S_r(n+1, \alpha, \beta) = -1 + q^r S_r(n, \alpha, -\alpha\beta)$.*

Proof. We have

$$S_r(n+1, \alpha, \beta) = \sum_{\substack{\beta x_1 \ldots x_{n+1} = y_1 \ldots y_{n+1} \\ x_i, y_i \text{ all } \neq 0 \text{ in } \mathbf{F}_{q^r}}} (\psi_1 \circ \text{trace}) \left(\sum_1^{n+1} x_i + \alpha \sum_1^{n+1} y_i \right).$$

Using $x_1, \ldots, x_n, y_1, \ldots, y_{n+1}$ as free variables, we have

$$x_{n+1} = \frac{y_1 \cdots y_{n+1}}{\beta x_1 \ldots x_n},$$

so we may write

$$S_r(n+1,\alpha,\beta) =$$

$$\sum_{\substack{x_1,\ldots,x_n,\\ y_1,\ldots,y_{n+1},\\ \text{all } \neq 0 \text{ in } \mathbf{F}_{q^r}}} (\psi_1 \circ \text{trace})\left(\frac{y_1\cdots y_{n+1}}{\beta x_1\ldots x_n} + \sum_1^n x_i + \alpha\sum_1^n y_i + \alpha y_{n+1}\right)$$

$$= \sum_{\substack{x_1,\ldots,x_n\\ y_1,\ldots,y_n\\ \text{all } \neq 0 \text{ in } \mathbf{F}_{q^r}}} (\psi_1 \circ \text{trace})\left(\sum_1^n x_i + \alpha\sum_1^n y_i\right) \sum_{\substack{y_{n+1}\neq 0\\ \text{in } \mathbf{F}_{q^r}}} (\psi_1 \circ \text{trace})\left(y_{n+1}\left(\alpha + \frac{y_1\cdots y_n}{\beta x_1\ldots x_n}\right)\right)$$

$$= - \sum_{\substack{x_1,\ldots,x_n\\ y_1,\ldots,y_n\\ \text{all } \neq 0 \text{ in } \mathbf{F}_{q^r}}} (\psi_1 \circ \text{trace})\left(\sum_1^n x_i + \alpha\sum_1^n y_i\right)$$

$$+ \sum_{\substack{x_1,\ldots,x_n\\ y_1,\ldots,y_n\\ \text{all } \neq 0 \text{ in } \mathbf{F}_{q^r}}} (\psi_1 \circ \text{trace})\left(\sum_1^n x_i + \alpha\sum_1^n y_i\right) \sum_{\substack{\text{all } y_{n+1}\\ \text{in } \mathbf{F}_{q^r}}} (\psi_1 \circ \text{trace})\left(y_{n+1}\left(\alpha + \frac{y_1\cdots y_n}{\beta x_1\ldots x_n}\right)\right)$$

$$= -1 + q^r \sum_{\substack{x_1,\ldots,x_n\\ y_1,\ldots,y_n\\ \text{all } \neq 0 \text{ in } \mathbf{F}_{q^r},\\ \alpha + \frac{y_1\cdots y_n}{\beta x_1\ldots x_n} = 0}} (\psi_1 \circ \text{trace})\left(\sum_1^n x_i + \alpha\sum_1^n y_i\right)$$

$$= -1 + q^r S_r(n,\alpha,-\alpha\beta). \blacksquare$$

Using this, we find

$$\begin{aligned} S_r(n+1,\alpha,\beta) &= -1 - q^r + q^{2r} S_r(n-1,\alpha,(-\alpha)^2\beta) \\ &= \cdots \\ &= -1 - q^r - q^{2r} - \cdots - q^{(n-1)r} + q^{nr} S_r(1,\alpha,(-\alpha)^n\beta) \\ &= \begin{cases} -1 - q^r - q^{2r} - \cdots - q^{nr} & \text{if } \alpha(-\alpha)^n\beta \neq -1 \\ -1 - q^r - q^{2r} - \cdots - q^{nr} + q^{(n+1)r} & \text{if } \alpha(-\alpha)^n\beta = 1. \end{cases} \end{aligned}$$

Taking $\beta = 1$, and replacing $n+1$ by n, we see that the L function of $\mathrm{Kl}_n(\psi_1) \otimes \mathrm{Kl}_n(\psi_2)$ is given by

$$L = \begin{cases} \prod_{i=0}^{n-1}(1 - q^i T) & \text{if } (-\alpha)^n \neq 1 \\ \prod_{i=0}^{n-1}(1 - q^i T)/(1 - q^n T) & \text{if } (-\alpha)^n = 1. \end{cases}$$

10.4.3. Corollary. *Notations as in the proposition 10.4.1, if* $(-\alpha)^n \neq 1$ *in* $\mathbf{F}_q$ *then*

$$\mathrm{Kl}(\psi_1; \chi_1, \ldots, \chi_n) \otimes \mathrm{Kl}(\psi_2; \chi'_1, \ldots, \chi'_n)$$

with $\psi_2(x) = \psi_1(\alpha x)$, *has all its breaks at* ∞ *equal to* $1/n$.

Proof. All its breaks at ∞ are $\leq 1/n$, and its rank is n^2, so "all breaks $= 1/n$" is equivalent to "$\mathrm{Swan}_\infty = n$." ∎

10.4.4. Corollary. *If* $(-1)^n \neq 1$ *in* $\mathbf{F}_q$, *then for any non-trivial additive character* ψ, *and any multiplicative characters* $\chi_1, \ldots, \chi_n, \chi'_1, \ldots, \chi'_n$

$$\mathrm{Kl}(\psi; \chi_1, \ldots, \chi_n) \otimes \mathrm{Kl}(\psi; \chi'_1, \ldots, \chi'_n)$$

has all its breaks at ∞ *equal to* $1/n$. *In particular, when* $(-1)^n \neq 1$ *in* $\mathbf{F}_q$, *the sheaf* $\mathrm{Kl}(\psi; \chi_1, \ldots, \chi_n)$ *admits no non-zero* P_∞*-invariant bilinear form* $\mathrm{Kl} \times \mathrm{Kl} \to E_\lambda$.

Proof. This is the case $\alpha = 1$ of the above corollary. ∎

10.4.5. Remark. If n is *prime* to p, this corollary also results from 5.6, according to which if we take $\rho_1, \ldots, \rho_n$ to be the n characters of order dividing n of some $\mathbf{F}_q^\times$ with $q \equiv 1 \mod n$, then $\mathrm{Kl}(\psi; \rho_1, \ldots, \rho_n)$ is geometrically isomorphic to $[n]_* \mathcal{L}_{\psi_{1/n}}$. Therefore we have

$$[n]^*(\mathrm{Kl}(\psi; \chi_1, \ldots, \chi_n)) \simeq \bigoplus_{\zeta^n = 1 \text{ in } \mathbf{F}_q} \mathcal{L}_{\psi_{\zeta/n}}$$

as representations of P_∞, for *any* $\chi_1, \ldots, \chi_n$, thanks to 10.1. Thus we have, as P_∞-representations,

$$[n]^*(\mathrm{Kl}(\psi; \chi_1, \ldots, \chi_n) \otimes \mathrm{Kl}(\psi; \chi'_1, \ldots, \chi'_n)) \simeq \bigoplus_{\substack{\zeta_1^n = 1, \zeta_2^n = 1 \\ \zeta_i \in \mathbf{F}_q}} \mathcal{L}_{\psi_{\frac{\zeta_1 + \zeta_2}{n}}}.$$

Now if $(-1)^n \neq 1$ in $\mathbf{F}_q$, then $(\zeta_1 + \zeta_2)/n$ is always non-zero in $\mathbf{F}_q$, so each $\mathcal{L}_{\psi_{\frac{\zeta_1 + \zeta_2}{n}}}$ has break 1, so in this case all n^2 breaks of

$$[n]^*(\mathrm{Kl}(\psi, \chi' s) \otimes \mathrm{Kl}(\psi, \chi'' s))$$

are equal to 1.As $(p, n) = 1$, this means that $\mathrm{Kl}(\psi, \chi\text{'s}) \otimes \mathrm{Kl}(\psi, \chi'\text{'s})$ has all breaks $1/n$, as asserted by the corollary.

If $(-1)^n = 1$ in $\mathbf{F}_q$, then $(\zeta_1 + \zeta_2)/n = 0$ for precisely the n pairs $(\zeta, -\zeta)$. So in this case we find the additional

10.4.6. Corollary. *If $(p, n) = 1$ and $(-1)^n = 1$ in $\mathbf{F}_q$, then for any non-trivial additive character ψ, and any multiplicative characters $\chi_1, \ldots, \chi_n, \chi'_1, \ldots, \chi'_n$,*

$$\mathrm{Kl}(\psi; \chi_1, \ldots, \chi_n) \otimes \mathrm{Kl}(\psi; \chi'_1, \ldots, \chi'_n)$$

has its breaks given by

break 0 with multiplicity n

break $1/n$ with multiplicity $n^2 - n$.

10.5. Question. Does this last corollary remain true for n's which are *not* prime to p?

CHAPTER 11

Global Monodromy of Kloosterman Sheaves

11.0. Formulation of the Theorem

11.0.1. Let us fix

an integer $n \geq 2$

a prime number p

a finite field $\mathbf{F}_q$ of characteristic p

a finite extension E of $\mathbf{Q}$ containing $q^{\frac{n-1}{2}}$

a non-trivial additive character $\psi : (\mathbf{F}_q, +) \to E^\times$

a prime number $l \neq p$

an l-adic place λ of E.

As in previous chapters, we denote by

$$\mathrm{Kl}_n(\psi) = \mathrm{Kl}(\psi; \underbrace{\mathbf{1}, \ldots, \mathbf{1}}_{n \text{ times}}; \quad \underbrace{1, \ldots, 1}_{n \text{ times}})$$

the lisse E_λ-sheaf on $\mathbf{G}_m \otimes \mathbf{F}_q$ whose existence is guaranteed by 4.1.1. Recall that $\mathrm{Kl}_n(\psi)$ has trace function given by (cf. 4.1.1) the "unadorned" Kloosterman sums

$$\mathrm{trace}\big(F_{a,k} \mid (\mathrm{Kl}_n(\psi))_{\bar{a}}\big) = (-1)^{n-1} \sum_{\substack{x_1 \ldots x_n = a \\ \text{all } x_i \in k}} (\psi \circ \mathrm{trace}_{k/\mathbf{F}_q}) \left(\sum_1^n x_i \right).$$

Because E and hence E_λ contains $q^{\frac{n-1}{2}}$, which necessarily lies in $(\mathcal{O}_\lambda)^\times$ because $l \neq p$, we may form the lisse rank-one E_λ-sheaf

$$E_\lambda\left(\frac{n-1}{2}\right)$$

first on $\mathrm{Spec}(\mathbf{F}_q)$ itself, then by inverse image on $\mathbf{G}_m \otimes \mathbf{F}_q$, where it is characterized up to isomorphism by

$$F_x \left| \left(E_\lambda \left(\frac{n-1}{2} \right) \right)_{\bar{x}} = \left(q^{\frac{1-n}{2}} \right)^{\deg(x)} \right.$$

for every closed point x of $\mathbf{G}_m \otimes \mathbf{F}_q$.

11.0.2. Let us denote simply by $\mathcal{F}$ the lisse E_λ-sheaf

$$\mathcal{F} = \mathrm{Kl}_n(\psi)\left(\frac{n-1}{2}\right) = \mathrm{Kl}_n(\psi) \bigotimes_{E_\lambda} E_\lambda\left(\frac{n-1}{2}\right).$$

Let us recall the basic facts we have established about $\mathrm{Kl}_n(\psi)$, but recast in terms of $\mathcal{F}$ (all these facts but the last appear already in [De-3]).

1. $\mathcal{F}$ is lisse on $\mathbf{G}_m \otimes \mathbf{F}_q$ of rank n, and pure of weight zero (cf. 4.1.1).
2. $\mathcal{F}$ is unipotent as I_0-representation, with a single Jordan block (cf. 7.4.1).
3. $\mathcal{F}$ is totally wild at ∞, with $\mathrm{Swan}_\infty(\mathcal{F}) = 1$ (cf. 4.1.1).
4. $\det \mathcal{F}$ is *trivial*, i.e. $\simeq E_\lambda$ as lisse sheaf on $\mathbf{G}_m \otimes \mathbf{F}_q$ (cf. 7.4.3).
5. If $(-1)^n = 1$ in $\mathbf{F}_q$, there exists a perfect pairing

   $$\langle , \rangle : \mathcal{F} \otimes \mathcal{F} \to E_\lambda$$

   of lisse sheaves on $\mathbf{G}_m \otimes \mathbf{F}_q$, which is *alternating* if n is even and which is *symmetric* if $p = 2$ and n is odd (cf. 4.2.1).
6. If $(-1)^n \neq 1$ in $\mathbf{F}_q$, then $\mathcal{F} \otimes \mathcal{F}$ is totally wild at ∞, with all breaks $1/n$ (cf. 10.4.4). In particular, even as P_∞-representations there is no non-zero P_∞-equivariant bilinear form $\mathcal{F} \otimes \mathcal{F} \to E_\lambda$ (compare 4.1.7, where we proved this for I_∞).

11.0.3. Now let us choose

> a geometric point $\bar{x} : \mathrm{Spec}(\Omega) \to \mathbf{G}_m \otimes \mathbf{F}_q$, with Ω a separably closed field, and
>
> "chemins" connecting $\bar{x}$ to the spectra $\bar{\eta}_0$ and $\bar{\eta}_\infty$ of separable closures of the fraction fields of the henselizations of $\mathbf{P}^1 \otimes \mathbf{F}_q$ at 0 and ∞ resp.

Using the chosen chemins, we obtain continuous homomorphisms of profinite groups

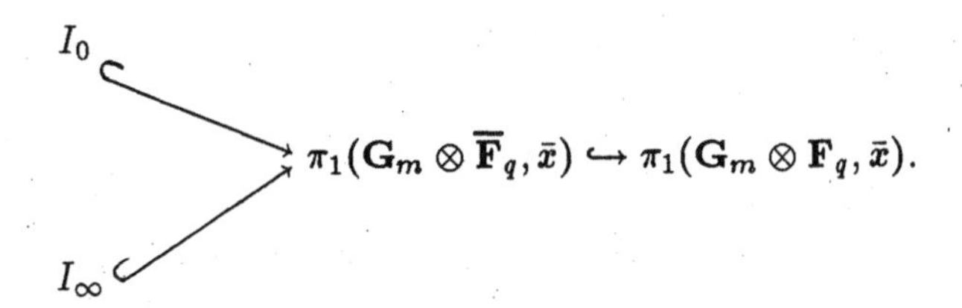

The monodromy representation of $\mathcal{F}$ at $\bar{x}$ is denoted

$$\rho : \pi_1(\mathbf{G}_m \otimes \mathbf{F}_q, \bar{x}) \to \mathrm{Aut}_{E_\lambda}(\mathcal{F}_{\bar{x}}) \simeq \mathrm{GL}(n, E_\lambda).$$

11.0.4. In view of fact 4 recalled in 11.0.2 above, we have

$$\mathrm{Image}(\rho) \subset \mathrm{SL}(n, E_\lambda).$$

In view of fact 5, we have the supplementary information

$$\mathrm{Image}(\rho) \subset \begin{cases} \mathrm{Sp}(n, E_\lambda) & \text{if } n \text{ even} \\ \\ \mathrm{SO}(n, E_\lambda) & \text{if } p = 2 \text{ and } n = \text{odd} \end{cases}$$

where the Sp and SO are with respect to the perfect pairing of 6.

Denoting by

$$\pi_1^{\mathrm{geom}} = \pi_1(\mathbf{G}_m \otimes \overline{\mathbf{F}}_q, \bar{x})$$

$$\pi_1^{\mathrm{arith}} = \pi_1(\mathbf{G}_m \otimes \mathbf{F}_q, \bar{x})$$

the "geometric" and "arithmetic" fundamental groups, we have a diagram of subgroups of $\mathrm{GL}(n, E_\lambda)$:

$$\begin{matrix} \rho(I_0) & \subset & \\ & & \rho(\pi_1^{\mathrm{geom}}) \subset \rho(\pi_1^{\mathrm{arith}}) \subset \mathrm{SL}(n, E_\lambda), \\ \rho(I_\infty) & \subset & \end{matrix}$$

with the additional information

$$\rho(\pi_1^{\mathrm{arith}}) \subset \begin{cases} \mathrm{Sp}(n, E_\lambda) & \text{if } n \text{ even} \\ \\ \mathrm{SO}(n, E_\lambda) & \text{if } p = 2 \text{ and } n \text{ odd.} \end{cases}$$

11.1. Main Theorem. *The Zariski closure G_{geom} of $\rho(\pi_1^{\mathrm{geom}})$ in $\mathrm{GL}(n)$, as algebraic group over E_λ, is given for $n \geq 2$ by*

$$G_{\mathrm{geom}} = \begin{cases} \mathrm{Sp}(n) & n \textit{ even} \\ \mathrm{SL}(n) & pn \textit{ odd} \\ \mathrm{SO}(n) & p = 2, n \textit{ odd}, n \neq 7 \\ G_2 & p = 2,\ n = 7. \end{cases}$$

11.2. Remarks.

(1) In the above list, the groups $\mathrm{Sp}(n)$ and $\mathrm{SO}(n)$ are with respect to the perfect pairings of 11.0.2.6 carried by the sheaf $\mathcal{F} = \mathrm{Kl}_n(\psi)(\frac{n-1}{2})$.

(2) The G_2 in the above list is viewed as lying inside $\mathrm{SO}(7)$ by its unique irreducible seven-dimensional representation. As subgroup of $\mathrm{SO}(7)$, G_2 is the fixer of some non-zero alternating trilinear form $\Lambda^3\mathcal{F} \to E_\lambda$. We will indirectly prove the *existence* of such a trilinear form, but we do *not* know an a priori description of it.

(3) Because G_{geom} is defined as the Zariski closure of a set of E_λ-valued points in an affine E_λ-scheme, it is immediate that for any overfield K of E_λ, we have

$$G_{\mathrm{geom}} \bigotimes_{E_\lambda} K = \text{ the Zariski closure of } \rho(\pi_1^{\mathrm{geom}}) \text{ in } \mathrm{GL}(n) \bigotimes_{E_\lambda} K.$$

Therefore it suffices to prove the Main Theorem with E_λ replaced by any convenient overfield (in occurrence $\overline{E}_\lambda$).

11.3. Corollary. *Under the monodromy representation ρ, we have*

$$\rho(\pi_1^{\mathrm{arith}}) \subset G_{\mathrm{geom}}(E_\lambda).$$

Proof. Except in the last case, this is just 11.0.4, given the explicit determination of G. In the last case, we notice that because π_1^{geom} is a normal subgroup of π_1^{arith}, $\rho(\pi_1^{\mathrm{arith}})$ normalizes G_2 in $\mathrm{SO}(7)$. But every automorphism of G_2 over $\overline{E}_\lambda$ is inner (the Dynkin diagram has no non-trivial automorphisms), and the inclusion $G_2 \hookrightarrow \mathrm{SO}(7)$ is an absolutely irreducible representation, so any element of $\mathrm{SO}(7, \overline{E}_\lambda)$ normalizing G_2 may be written in the form

$$(\text{an element of } G_2(\overline{E}_\lambda)) \times (\text{a scalar in } \mathrm{SO}(7, \overline{E}_\lambda)).$$

But as $\mathrm{SO}(7)$ contains no non-trivial scalars, this shows that over $\overline{E}_\lambda$, G_2 is its own normalizer in $\mathrm{SO}(7)$. Therefore

$$\rho(\pi_1^{\mathrm{arith}}) \subset G_2(\overline{E}_\lambda) \cap \mathrm{SO}(7, E_\lambda) = G_2(E_\lambda). \quad \blacksquare$$

As explained in Chapter 3, this last corollary leads to a rather concrete equidistribution theorem (concrete because we have determined the group G_{geom}).

11.4. Equidistribution Theorem. *For any embedding $\overline{E}_\lambda \hookrightarrow \mathbf{C}$, and any maximal compact subgroup K of the complex Lie group $G_{\mathrm{geom}}(\mathbf{C})$, the generalized "angles of Frobenius" for the sheaf $\mathcal{F} = \mathrm{Kl}_n(\psi)(\frac{n-1}{2})$ on $\mathbf{G}_m \otimes \mathbf{F}_q$,*

$$\theta(a) = \textit{the } K\textit{-conjugacy class of any element of } K \textit{ which is } G_{\mathrm{geom}}(\mathbf{C})\textit{-conjugate to } \rho(F_a)^{\mathrm{ss}},$$

as a runs over the closed points of $\mathbf{G}_m \otimes \mathbf{F}_q$, are equidistributed in the space $K^\natural$ of conjugacy classes in K with respect to the direct image $\mu^\natural$ of Haar measure, in any of the three standard senses (X, Y, Z, cf. 3.5).

EXPLICITATION: For the sequence of measure X_i, the explicit extimate 3.6.3 proven earlier gives, for the sheaf $\mathcal{F} = \mathrm{Kl}_n(\psi)(\frac{n-1}{2})$ on $\mathbf{G}_m \otimes \mathbf{F}_q$, the estimate

$$\left| \int_{K^\natural} \mathrm{trace}(\Lambda)\, dX_i \right| \leq \frac{\dim(\Lambda)\sqrt{q}^i}{n(q^i - 1)}$$

for any irreducible non-trivial continuous representation Λ of K. Further explicitation is given in Chapter 13.

11.5. Axiomatics of the proof; Classification Theorems

11.5.1. Let K be an algebraically closed field of characteristic zero, $n \geq 2$ an integer, and p a prime number. We suppose given

① an n-dimensional vector space V over K,

② a unipotent element $U_0 \in \mathrm{SL}(V)$ with a single Jordan block,

③ a Zariski-closed subgroup $\Gamma_\infty \subset \mathrm{SL}(V)$,

④ a Zariski-closed subgroup $H_\infty \subset \Gamma_\infty$,

and we suppose that the following conditions are satisfied:

a) Γ_∞ acts irreducibly on V.
b) Γ_∞ admits no faithful K-linear representation of dimension $< n$.
c) If pn is odd, we assume that

1) the space $(V^* \otimes V^*)^{H_\infty}$ of H_∞-invariant bilinear forms $V \otimes_K V \to K$ is zero.

2) any character $\chi : \Gamma_\infty \to \mathbf{G}_m \otimes K$ of Γ_∞ is trivial on H_∞.

11.5.2. The actual situation we have in mind is

$$\begin{cases} K = \overline{E}_\lambda,\ V = (\mathcal{F}_{\bar{x}}) \bigotimes_{E_\lambda} K, \\ U_0 = \rho(\gamma_0), \text{ for } \gamma_0 \text{ a topological generator of } I_0^{\text{tame}} \\ \Gamma_\infty = \text{ the finite subgroup } \rho(I_\infty) \subset \mathrm{SL}(V) \\ H_\infty = \text{ the finite subgroup } \rho(P_\infty) \subset \rho(I_\infty), \end{cases}$$

but it will be useful for other applications (e.g. to calculating differential galois groups) to work in the axiomatic situation.

11.5.2.1. Lemma. *In the axiomatic situation 11.5.1 above, let $G \subset \mathrm{SL}(V)$ be any Zariski-closed subgroup which contains U_0 and Γ_∞. Then the identity component G^0 of G is semisimple, and the action of G^0 on V is irreducible.*

Proof. The group G acts irreducibly on V, because its subgroup Γ_∞ does. Therefore G acts completely reducibly on V. Because G^0 is normal in G, G^0 acts completely reducibly on V.

The Zariski closure in $\mathrm{SL}(V)$ of the abstract subgroup generated by U_0 is the $\mathbf{G}_a$ given by $t \mapsto \exp(t\log(U_0))$, so it is connected. Because $U_0 \in G$, this Zariski closure lies in G, so lies in G^0. Therefore U_0 lies in G^0. But U_0 has only a single Jordan block acting in V, so V must be indecomposable as a $K[G^0]$-module, because it is already so under U_0. But V is completely reducible as a G^0-representation, so being G^0-indecomposable is equivalent to being G^0-irreducible. The semi-simplicity of G^0 is given by

11.5.2.2. Lemma. *Let K be an algebraically closed field of characteristic zero, V a finite-dimensional K-space, and $G^0 \subset \mathrm{SL}(V)$ a connected Zariski-closed subgroup which operates irreducibly on V. Then G^0 is semisimple.*

Proof. Because G^0 is given with a faithful completely reducible representation, it is certainly reductive. So it suffices to show that its center $Z(G^0)$ is finite. But $G^0 \subset \mathrm{SL}(V)$ acts irreducibly on V, so $Z(G^0)$ lies in both the scalars and in $\mathrm{SL}(V)$, so in $\mathrm{SL}(V) \cap$ scalars which is finite. ∎

11.5.2.3. Lemma. *Hypotheses and notations as in Lemma 11.5.2.1 above, let*

$$\mathcal{G} = Lie(G^0) \subset \mathrm{End}(V)$$

be the Lie algebra of G^0, and put

$$N = \log(U_0) \in \mathcal{G}.$$

Then the inclusion $\mathcal{G} \subset \mathrm{End}(V)$ is a faithful irreducible representation of $\mathcal{G}$ in which N acts nilpotently with a single Jordan block, and $\mathcal{G}$ is a simple Lie algebra.

Proof. By 11.5.2.1, G^0 is semisimple, and $G^0 \hookrightarrow \mathrm{SL}(V)$ is a faithful irreducible representation. By hypothesis, the element U_0 acts unipotently with a single Jordan block. Therefore $\mathcal{G}$ is a semisimple Lie algebra, the given representation $\mathcal{G} \hookrightarrow \mathrm{End}(V)$ is faithful and irreducible, and N acts nilpotently with a single Jordan block. If $\mathcal{G}$ is not simple, then

$$\mathcal{G} = \mathcal{G}_1 \oplus \mathcal{G}_2$$

where $\mathcal{G}_1$ and $\mathcal{G}_2$ are non-zero semisimple Lie algebras, and the given faithful irreducible representation of $\mathcal{G}$ is of the form

$$V = V_1 \underset{K}{\otimes} V_2$$

where V_1 and V_2 are faithful irreducible representations of $\mathcal{G}_1$ and $\mathcal{G}_2$ respectively. In particular we have $\dim(V_i) \geq 2$ for $i = 1, 2$, because these are non-trivial representations of semisimple Lie algebras. Writing $N = N_1 \oplus N_2$, we have N_1 and N_2 nilpotent in $\mathcal{G}_1$ and $\mathcal{G}_2$ respectively, and N acts on $V = V_1 \otimes V_2$ by

$$N|V = (N_1 \mid V_1) \otimes (id_{V_2}) + (id_{V_1}) \otimes (N_2 \mid V_2).$$

Because N_i is nilpotent on V_i, we have

$$(N_i)^{\dim V_i} = 0 \text{ on } V_i,$$

whence by the binomial theorem we find that $N = N_1 \otimes id + id \otimes N_2$ satisfies

$$(N)^{\dim V_1 + \dim V_2 - 1} = 0.$$

But N has a single Jordan block, of size $\dim(V) = \dim(V_1)\dim(V_2)$, so we know that

$$(N)^{\dim(V_1)\dim(V_2) - 1} \neq 0.$$

Therefore we must have

$$\dim(V_1) + \dim(V_2) > \dim(V_1)\dim(V_2).$$

But this is impossible, because both $\dim(V_1)$ and $\dim(V_2)$ are finite and ≥ 2. ∎

11.5.2.4. Lemma. *Hypotheses as in Lemma 11.5.2.1 suppose that there exists a K-isomorphism $\mathcal{G} \simeq sl(2)$ of Lie algebras. Then we are in one of the following two cases:*

(1) $n = 2$, $G^0 = G = \mathrm{SL}(V) = \mathrm{SL}(2)$.

(2) $n = 3, p = 2$, *and with respect to some non-degenerate quadratic form on V we have*

$$G^0 = \mathrm{SO}(3) \subset G \subset \mathrm{SO}(3) \times \boldsymbol{\mu}_3 \subset \mathrm{SL}(3) = \mathrm{SL}(V).$$

Proof. If $\mathcal{G} \simeq sl(2)$, then G^0 is isomorphic to either SL(2) or to SL(2)/± 1 as algebraic group, and the given representation $G \subset \mathrm{SL}(V)$, being irreducible on G^0, has no choice but to be $\mathrm{Symm}^{n-1}(\underline{\mathrm{std}})$ on G^0 where $\underline{\mathrm{std}}$ denotes the standard 2-dimensional representation of SL(2). Because $\mathrm{Symm}^{n-1}(\underline{\mathrm{std}})$ as SL(2)-representation has kernel $(\pm 1)^{n-1}$, and is faithful on G^0, we have

$$\mathrm{SL}(2)/(\pm 1)^{n-1} \xrightarrow[\mathrm{symm}^{n-1}(\underline{\mathrm{std}})]{\sim} G^0 \subset G \subset \text{ normalizer of } G^0 \text{ in } \mathrm{SL}(V).$$

Because every automorphism of G^0 is inner, and $G^0 \subset \mathrm{SL}(V)$ is irreducible, we have

$$\begin{aligned}\text{normalizer of } G^0 \text{ in } \mathrm{SL}(V) &= G^0 \cdot (\text{scalars in } \mathrm{SL}(V)) \\ &= (G^0 \times \boldsymbol{\mu}_n)/G^0 \cap \boldsymbol{\mu}_n \\ &\simeq \begin{cases} (\mathrm{SL}(2) \times \boldsymbol{\mu}_n)/\ (\text{the diagonal } \pm 1) \\ \text{if } n \text{ is even.} \\ \\ \mathrm{SO}(3) \times \boldsymbol{\mu}_n \text{ if } n \text{ is odd.} \end{cases}\end{aligned}$$

If n is even, we have

$$\mathrm{SL}(2) \times \boldsymbol{\mu}_n \ /\ (\pm 1) \hookrightarrow \mathrm{SL}(2) \times \mathbf{G}_m \ /\ (\pm 1) \xrightarrow{\sim} \mathrm{GL}(2),$$

while if n is odd we have

$$\mathrm{SO}(3) \times \boldsymbol{\mu}_n \hookrightarrow \mathrm{GL}(3).$$

Therefore we have inclusions of K-algebraic groups

$$\begin{cases} G \hookrightarrow \mathrm{GL}(2) \text{ if } n \text{ even} \\ G \hookrightarrow \mathrm{SO}(3) \times \boldsymbol{\mu}_n \hookrightarrow \mathrm{GL}(3) \text{ if } n \text{ odd.} \end{cases}$$

Consider the composite homomorphism

$$\Gamma_\infty \hookrightarrow G \hookrightarrow \begin{cases} \mathrm{GL}(2) & \text{if } n \text{ even} \\ \\ \mathrm{GL}(3) & \text{if } n \text{ odd.} \end{cases}$$

By hypothesis, Γ_∞ has no faithful representation of dimension $< n$. Therefore we have $n \leq 2$ if n is even, and $n \leq 3$ if n is odd. By hypothesis, $n \geq 2$. If $n = 2$, there is nothing to prove. If $n = 3$, then

$$G^0 \simeq \mathrm{SO}(3) \subset G \subset \mathrm{SO}(3) \times \boldsymbol{\mu}_3 \subset \mathrm{SL}(3) = \mathrm{SL}(V).$$

We must show that $p = 2$. If not, pn is odd, and any character of Γ_∞ is trivial on H_∞. Consider the character

$$\Gamma_\infty \hookrightarrow G \subset \mathrm{SO}(3) \times \boldsymbol{\mu}_3 \xrightarrow{pr_2} \boldsymbol{\mu}_3;$$

it is trivial on H_∞, so H_∞ lies in SO(3), and this contradicts the axiom c),1) of 11.5.1. ∎

11.5.2.5. Lemma. *Hypotheses and notations as in Lemma 11.5.2.1 suppose that n is odd and that $G^0 = \mathrm{SO}(n)$ for some non-degenerate quadratic form on V. Then $p = 2$.*

Proof. Just as in the above proof, we know that every automorphism of $\mathrm{SO}(n)$ is inner (the Dynkin diagram has no non-trivial automorphisms), and that $\mathrm{SO}(n)$ operates irreducibly on V, whence

$$G \subset \text{normalizer of } \mathrm{SO}(n) \text{ in } \mathrm{SL}(V) = \mathrm{SO}(n) \times \boldsymbol{\mu}_n,$$

the last equality because $\mathrm{SO}(n)$ contains no non-trivial scalars. Again the composite

$$\Gamma_\infty \subset G \subset \mathrm{SO}(n) \times \boldsymbol{\mu}_n \xrightarrow{pr_2} \boldsymbol{\mu}_n$$

is a character of Γ_∞, so it must be trivial on H_∞, whence $H_\infty \subset \mathrm{SO}(n)$, and this contradicts axiom c),1) unless $p = 2$. ∎

11.5.2.6. Lemma. *Hypotheses and notations as in Lemma 11.5.2.1, suppose that $n = 7$ and that the subgroup $G \subset \mathrm{SL}(V)$ has $G^0 = G_2$ where $G_2 \subset \mathrm{SO}(7) \subset \mathrm{SL}(7) = \mathrm{SL}(V)$ by "the" irreducible seven-dimensional representation of G_2. Then $p = 2$.*

Proof. Just as above, we have

$$G \subset \text{normalizer of } G_2 \text{ in } \mathrm{SL}(7) = G_2 \times \boldsymbol{\mu}_7,$$

and the argument proceeds as above. ∎

Let us temporarily admit the truth of the following Classification Theorem.

11.6. Classification Theorem. *Let K be an algebraically closed field of characteristic zero, $n \geq 2$ an integer, $\mathcal{G}$ a simple Lie algebra over K given with a faithful irreducible representation*

$$\mathcal{G} \xrightarrow{\rho} sl(n).$$

Suppose that there exists a nilpotent element $N \in \mathcal{G}$ such that $\rho(N)$ has a single Jordan block. Then the pair $(\mathcal{G}, \rho)$ is K-isomorphic to one of the following

(1) $\mathcal{G} = sl(2), \rho = \mathrm{Symm}^{n-1}$ *(standard* $2 - \mathrm{dim}$ *rep).*
(2) $\mathcal{G} = sl(n),\ n \geq 3,\ \rho =$ *standard n-dimensional representation (or its contragredient, but these are isomorphic).*
(3) $\mathcal{G} = sp(n)$ *if* $n \geq 4$ *<u>even</u>,* $\rho =$ *standard n-dimensional rep.*
(4) $\mathcal{G} = so(n)$ *if* $n \geq 5$ *odd,* $\rho =$ *standard n-dimensional rep.*
(5) $\mathcal{G} = \mathrm{Lie}(G_2)$ *if* $n = 7$*, with its unique 7-dimensional irreducible rep.*

11.6.1. Lemma. *Hypotheses and notations as in Lemma 11.5.2.1, we have the following list of possible values for the subgroup* G^0 *of* $\mathrm{SL}(V) = \mathrm{SL}(n)$*:*

$$\begin{cases} n \textit{ even} : G^0 = \mathrm{Sp}(n) \textit{ or } \mathrm{SL}(n) \\ pn \textit{ odd} : G^0 = \mathrm{SL}(n) \\ p = 2, n \textit{ odd} \neq 7 : G^0 = \mathrm{SO}(n) \textit{ or } \mathrm{SL}(n) \\ p = 2, n = 7 : G^0 = G_2 \textit{ or } \mathrm{SO}(7) \textit{ or } \mathrm{SL}(7). \end{cases}$$

Proof. This is immediate from the Classification Theorem and Lemmas 11.5.2.4-5-6. ∎

11.7. Axiomatic Classification Theorem. *In the axiomatic situation (11.5.1), suppose in addition that if pn is even, we are given a non-degenerate bilinear form* $\langle , \rangle : V \otimes_K V \to K$ *which is alternating if n is even and symmetric if n is odd. Let*

$$G \subset \begin{cases} \mathrm{Sp}(n) & n \textit{ even} \\ \mathrm{SL}(n) & pn \textit{ odd} \\ \mathrm{SO}(n) & p = 2, n \textit{ odd} \end{cases}$$

be a Zariski closed subgroup which contains U_0 *and* Γ_∞*. Then*

$$G = \begin{cases} \mathrm{Sp}(n) & n \textit{ even} \\ \mathrm{SL}(n) & pn \textit{ odd} \\ \mathrm{SO}(n) & p = 2, n \textit{ odd} \neq 7 \\ G_2 \textit{ or } \mathrm{SO}(7) & p = 2,\ n = 7. \end{cases}$$

Proof. Simply apply Lemma 11.6.1. In view of the a priori inclusions we assume, we find that

$$G^0 \subset G \subset \begin{cases} \mathrm{Sp}(n) & n \text{ even} \\ \mathrm{SL}(n) & pn \text{ odd} \\ \mathrm{SO}(n) & p = 2,\ n \text{ odd}, \end{cases}$$

and except for $(p,n) = (2,7)$, Lemma 11.6.1 forces equality everywhere above. In case $(p,n) = (2,7)$, either $G^0 = G_2$ or $G^0 = \mathrm{SO}(7)$. In the second case we have $G^0 \subset G \subset \mathrm{SO}(7)$ with $G^0 = \mathrm{SO}(7)$ forcing $G = \mathrm{SO}(7)$. In case $G^0 = G_2$, G lies in the normalizer of G_2 in SO(7), but this normalizer is, as already noted (cf. 11.3), G_2 itself. ∎

Applying this axiomatic classification theorem to G_{geom}, we find the Main Theorem except in the case $p = 2$, $n = 7$, in which case G_{geom} is either G_2 or SO(7). Let us temporarily admit the truth of

11.8. $\mathbf{G_2}$ Theorem. *For $p = 2$, $n = 7$, we have $G_{\text{geom}} = G_2$.*

Since even this exceptional case gives the smallest possible group for G_{geom} as allowed by the axiomatic classification theorem we find the following result, suggested to us by O. Gabber.

11.9. Density Theorem. *The subgroup of $\rho(\pi_1^{\text{geom}})$ generated by $U_0 = \rho(\gamma_0)$, γ_0 a topological generator of I_0^{tame}, and by the finite subgroup $\rho(I_\infty)$ is Zariski dense in G_{geom}.*

11.10. Proof of the Classification Theorem

11.10.1. In this section we will prove the Classification Theorem 11.6. Thus K is an algebraically closed field of characteristic zero, $\mathfrak{G}$ is a finite-dimensional simple Lie algebra over K, $n \geq 2$ is an integer, $N \in \mathfrak{G}$ is a nilpotent element, and $\rho : \mathfrak{G} \hookrightarrow sl(n)$ is a faithful irreducible representation in which $\rho(N)$ has a single Jordan block.

In order to classify such pairs $(\mathfrak{G}, \rho)$, we will make use of the theory of "principal $sl(2)$-triples," the Weyl character formula, and the classification of simple Lie algebras.

11.10.2. Let $\mathfrak{b}$ be a Borel subalgebra of $\mathfrak{G}$ such that N lies in $\mathfrak{b}$, and let $\mathfrak{h} \subset \mathfrak{b}$ be a Cartan subalgebra of $\mathfrak{b}$; by (Bouraki Lie VIII, §3, Cor. of Prop 9), $\mathfrak{h}$ is a Cartan subalgebra of $\mathfrak{G}$. Let R be the set of roots of $\mathfrak{G}$ with respect to $\mathfrak{h}$; an element $\alpha \in R$ is by definition a non-zero element of the K-linear dual $\mathfrak{h}^*$ of h which occurs in the restriction to $\mathfrak{h}$ of the adjoint representation of $\mathfrak{G}$ on itself. For $\alpha \in R$, the corresponding eigenspace $\mathfrak{G}^\alpha \subset \mathfrak{G}$ is one-dimensional, and as K-vector spaces we have

$$\mathfrak{G} = \mathfrak{h} \oplus \left(\bigoplus_{\alpha \in R} \mathfrak{G}^\alpha \right).$$

The set R_+ of positive roots with respect to $\mathfrak{b}$ is the set of those $\alpha \in R$ for which $\mathfrak{G}^\alpha$ lies in $\mathfrak{b}$, and we have

$$\mathfrak{b} = \mathfrak{h} \oplus \left(\bigoplus_{\alpha \in R_+} \mathfrak{G}^\alpha \right).$$

Let $B \subset R_+$ denote the set of simple roots, i.e., B consists of those elements $\alpha \in R_+$ which cannot be expressed as a sum of two non-zero elements of R_+. One knows that if $\alpha \in R$ is a root, then exactly one of α or $-\alpha$ lies in R_+, and one knows that every element of R_+ is a finite sum of elements of B. One also knows that the $\alpha \in B$ form a K-basis of $\mathfrak{h}^*$.

For any root α, the commutator $[\mathfrak{G}^\alpha, \mathfrak{G}^{-\alpha}]$ is a one-dimensional subspace of $\mathfrak{G}$ which lies in $\mathfrak{h}$, and on which α, viewed as an element of $\mathfrak{h}^*$, is non-zero. There is consequently a unique element

$$H_\alpha \in [\mathfrak{G}^\alpha, \mathfrak{G}^{-\alpha}] \subset \mathfrak{h}$$

which satisfies

$$\alpha(H_\alpha) = 2.$$

The set $\{H_\alpha\}_{\alpha \in R}$ form a root system R' in $\mathfrak{h}$, for which $\{H_\alpha\}_{\alpha \in B}$ is a base and for which $\{H_\alpha\}_{\alpha \in R_+}$ is the corresponding set of positive roots. It is possible to choose for each root $\alpha \in R$ a non-zero vector $X_\alpha \in \mathfrak{G}^\alpha$ such that

$$[X_\alpha, X_{-\alpha}] = -H_\alpha.$$

With this choice, one knows that the assignment

$$X_+ = \begin{pmatrix} 0 & 1 \\ 0 & 0 \end{pmatrix} \mapsto X_\alpha$$

$$X_- = \begin{pmatrix} 0 & 0 \\ -1 & 0 \end{pmatrix} \mapsto X_\alpha$$

$$H = \begin{pmatrix} 1 & 0 \\ 0 & -1 \end{pmatrix} \mapsto H_\alpha$$

defines a non-zero homomorphism of K-Lie algebras

$$sl(2) \to \mathfrak{G},$$

or, as one says, $(X_\alpha, H_\alpha, X_{-\alpha})$ is an $sl(2)$-triple in $\mathfrak{G}$.

Now let us denote by $\mathfrak{n} \subset \mathfrak{b}$ the set of nilpotent elements of $\mathcal{G}$ which lie in $\mathfrak{b}$. Then $\mathfrak{n}$ is equal to

$$\mathfrak{n} = \bigoplus_{\alpha \in R_+} \mathfrak{G}^\alpha .$$

As K-scheme, $\mathfrak{n}$ is thus the affine space $\mathbf{A}^{R_+}$.

By hypothesis, the set

$$\{N \in \mathfrak{n} \text{ such that } \rho(N) \text{ has a single Jordan block}\}$$

is non-empty. It is clearly Zariski-open in n, because, $\rho(N)$ being an $n \times n$ nilpotent matrix in any case, we have for $N \in \mathfrak{n}$ the equivalences

$$\begin{aligned} &\rho(N) \text{ has a single Jordan block} \\ &\iff \dim\ \mathrm{Ker}(\rho(N)) = 1 \\ &\iff \dim \mathrm{Im}(\rho(N)) = n - 1 \\ &\iff \text{some } (n-1) \times (n-1) \text{ minor of } \rho(N) \text{ is invertible,} \end{aligned}$$

and this last condition is Zariski open.

Recall that under the adjoint representation, for every $N \in \mathfrak{n}$ the endomorphism $\mathrm{ad}(N)$ has $\geq \dim(\mathfrak{h})$ Jordan blocks. Those with exactly $\dim(\mathfrak{h})$ Jordan blocks are called the principal nilpotent elements. One knows that

the set of principal nilpotent elements in $\mathfrak{n}$ is non-empty and Zariski open in $\mathfrak{n}$

(cf. Bourbaki Lie VIII, p. 167).

Because the intersection of two non-empty Zariski open subsets of $n \simeq \mathbf{A}^{R_+}$ is again non-empty, it follows that

there exists a principal nilpotent element $N_1 \in \mathfrak{n}$ such that $\rho(N_1)$ has a single Jordan block.

Now N_1, being nilpotent and non-zero, may be completed to an $sl(2)$-triple $(x, h, y) = (N_1, ?, ?)$ in $\mathfrak{G}$. Because N_1 is a principal nilpotent, any such $sl(2)$-triple is principal (Lie VIII, §11, 4, Prop. 7). By (Lie VIII, §11, 4, Prop. 8 and 9), any principal $sl(2)$-triple in $\mathfrak{G}$ is $\mathrm{Aut}_e(\mathfrak{G})$-conjugate to one of the form $(?, h^0, ?)$ where

$h^0 \in \mathfrak{h}$ is the unique element in $\mathfrak{h}$ for which $\alpha(h^0) = 2$ for all $\alpha \in B$.

Furthermore, any $sl(2)$-triple (x, h, y) with $h = h^0$ is principal.

By the standard representation theory of $sl(2)$, a finite dimensional representation (V, π) of $sl(2)$ is irreducible if and only if $\pi(X_+)$ has a single Jordan block. Therefore the restriction of ρ to any $sl(2)$-triple of the form

$(N_1, ?, ?)$ is irreducible. But such an $sl(2)$-triple is principal, and all principal $sl(2)$-triples are $\mathrm{Aut}_e(\mathfrak{G})$-conjugate (cf. above), so we conclude that the *restriction of ρ to any principal sl(2)-triple is irreducible.* In particular:

> the restriction of ρ to any $sl(2)$-triple (x, h, y) with $h = h^0 =$ the unique element $h \in \mathfrak{h}$ with $\alpha(h^0) = 2$ for all $\alpha \in B$, is irreducible.

11.10.3. Now let us recall briefly the representation theory of $\mathfrak{G}$. For any finite-dimensional representation λ of $\mathfrak{G}$, the weights of λ are those elements of $\mathfrak{h}^*$ which occur in the restriction of λ to $\mathfrak{h}$. The weights of $\mathfrak{G}$ are the elements of $\mathfrak{h}^*$ which are weights of some finite-dimensional λ. The weights are each $\mathbf{Q}$-linear combinations of the $\alpha \in B$ (which form a K-basis of $\mathfrak{h}^*$). We say that two weights w and w' satisfy $w \geq w'$ if

$$w - w' = \sum_{\alpha \in B} n_\alpha \alpha \text{ with rational coefficients } n_\alpha \geq 0.$$

A weight w of $\mathfrak{G}$ is said to be dominant if

$$w(H_\alpha) \text{ is a non-negative integer for all } \alpha \in B.$$

Because the vectors $(H_\alpha)_{\alpha \in B}$ form a K-basis of $\mathfrak{h}$, if we denote by $(w_\alpha)_{\alpha \in B}$ the dual basis of $\mathfrak{h}^*$, we see that the dominant weights are precisely the finite sums

$$\sum_{\alpha \in B} n_\alpha w_\alpha \text{ with integral coefficients } n_\alpha \geq 0.$$

The weights $(w_\alpha)_{\alpha \in B}$ are called the fundamental weights.

One knows that every finite-dimensional irreducible representation has a highest weight (in the sense of the above-defined partial order $\geq$ on weights), that its highest weight is a dominant weight, and that the map

$$\begin{array}{c} \text{isomorphism classes} \\ \text{of finite-dimensional} \\ \text{irreducible} \\ \text{representation of } \mathfrak{G} \end{array} \xrightarrow{\text{highest weight}} \begin{array}{c} \text{the set of dominant} \\ \text{weights} \end{array}$$

is bijective. Furthermore, given a dominant weight w, if we denote by V_w the unique finite-dimensional representation of $\mathfrak{G}$ whose highest weight is w, we have the Weyl dimensional formula:

> let $\Lambda_0 = \sum_{\alpha \in B} w_\alpha$ be the sum of the fundamental weights; then
>
> $$\dim(V_w) = \prod_{\alpha \in R_+} \left(1 + \frac{w(H_\alpha)}{\Lambda_0(H_\alpha)}\right).$$

Notice that because each H_α with $\alpha \in R_+$ is an $\mathbf{N}$-linear and non-zero linear combination of the H_α with $\alpha \in B$, the denominator $\Lambda_0(H_\alpha)$ for each $\alpha \in R_+$ is a strictly positive integer.

11.10.4. Lemma. *Let w be a dominant weight of $\mathfrak{G}$, V_w the corresponding irreducible representation. Then for any $sl(2)$-triple (x,h,y) in $\mathfrak{G}$ with $h = h^0 =$ the unique element $h^0 \in \mathfrak{h}$ with $\alpha(h^0) = 2$ for all $\alpha \in B$, the highest weight of the composite representation*

$$sl(2) \xrightarrow{(x,h^0,y)} \mathfrak{G} \longrightarrow \mathrm{End}(V_w)$$

is the integer $w(h^0)$.

Proof. The weights of the composite representation are the integers $w'(h^0)$, where w' ranges over the weights of V_w. Because w is the highest weight, we have for any weight w' of V_w

$$w = w' + \sum_{\alpha \in B} n_\alpha \alpha, \ n_\alpha \text{ rational } \geq 0.$$

Therefore as $\alpha(h^0) = 2$ for all $\alpha \in B$ we find

$$w(h^0) = w'(h^0) + 2\sum_{\alpha \in B} n_\alpha \geq w'(h^0). \quad \blacksquare$$

From the elementary theory of $sl(2)$-representation (or by the Weyl dimension formula in this trivial case), a finite dimensional irreducible representation (V, π) of $sl(2)$ satisfies

$$\dim(V) = 1 + \text{ the highest weight of } \pi \text{ on } sl(2).$$

So for any finite dimensional representation (V, π) of $sl(2)$, we have

$$\begin{cases} \dim(V) - 1 \geq \text{ highest weight of } \pi \text{ on } sl(2) \\ \text{with equality if and only if } (V, \pi) \text{ is } sl(2)\text{-irreducible.} \end{cases}$$

Combining this with the preceding discussion, we find

11.10.5. Criterion. Let w be a dominant weight of $\mathfrak{G}$. Then

$$\dim(V_w) - 1 \geq w(h^0),$$

with equality if and only if the restriction of V_w to every principal $sl(2)$-triple in $\mathfrak{G}$ is irreducible.

In the trivial case when $\mathfrak{G} = sl(2)$, there is nothing of 11.6 to prove.

11.10.6. Lemma. *If $\mathfrak{G} \neq sl(2)$, i.e., if* $\dim \mathfrak{h} \geq 2$, *and if w is a non-zero dominant weight of $\mathfrak{G}$ which satisfies*

$$\dim(V_w) - 1 = w(h^0),$$

then w is a fundamental weight.

Proof. If w is not fundamental, then we can write

$$w = w_1 + w_2$$

with w_1 and w_2 both non-zero dominant weights. As noted above, we always have the inequality

$$\dim(V_{w_i}) - 1 \geq w_i(h^0), \quad i = 1, 2.$$

so

$$\dim(V_{w_1}) - 1 + \dim(V_{w_2}) - 1 \geq w_1(h^0) + w_2(h^0) = w(h^0).$$

Thus it suffices to show that

$$\dim(V_{w_1+w_2}) - 1 > \dim(V_{w_1}) - 1 + \dim(V_{w_2}) - 1,$$

or equivalently

$$1 + \dim(V_{w_1+w_2}) > \dim(V_{w_1}) + \dim(V_{w_2}).$$

This we will do by looking at the general shape of the Weyl dimension formula.

11.10.6.1. Sublemma. *For $\mathfrak{G}$ simple of rank (=* $\dim \mathfrak{h} = \#(B)$*)* ≥ 2, *and w any non-zero dominant weight, there exist at least two distinct elements $\alpha \in R_+$ for which $w(H_\alpha) > 0$.*

Proof. Because $\mathfrak{G}$ is simple of rank ≥ 2, its Dynkin diagram is connected and contains ≥ 2 points. Write $w = \sum_{\alpha \in B} n_\alpha w_\alpha$ with integral coefficients $n_\alpha \geq 0$. Because $w \neq 0$, there is some $\alpha_0 \in B$ with $n_{\alpha_0} \neq 0$. Let $\alpha_1 \in B$ be an *adjacent* point to α_0 in the Dynkin diagram. Then both α_0 and $\alpha_0 + \alpha_1$ are positive roots, and the dual root to $\alpha_0 + \alpha_1$ is

$$H_{\alpha_0+\alpha_1} = A_0 H_{\alpha_0} + A_1 H_{\alpha_1}, \text{ with } A_0, A_1 > 0$$

(in terms of a W-invariant scalar product $(,)$ on the $\mathbf{Q}$-span of the roots, $H_\alpha = 2\alpha/(\alpha, \alpha)$, so the constants A_i are given by

$$A_i = (\alpha_i, \alpha_i)/(\alpha_0 + \alpha_1, \alpha_0 + \alpha_1)).$$

Therefore we find

$$w(H_{\alpha_0}) = n_{\alpha_0} > 0$$
$$w(H_{\alpha_0+\alpha_1}) = A_0 n_{\alpha_0} + A_1 n_{\alpha_1} \geq A_0 n_{\alpha_0} > 0,$$

as required. ∎

Now let us return to the proof of Lemma 11.10.6. For each $\alpha \in R_+$, let

$$x_\alpha = \frac{w_1(H_\alpha)}{\Lambda_0(H_\alpha)},$$

$$y_\alpha = \frac{w_2(H_\alpha)}{\Lambda_0(H_\alpha)}.$$

Then the quantities x_α, y_α are non-negative rational numbers, because for $\alpha \in R_+$ each H_α is a non-negative integral combination of the H_α with $\alpha \in B$. So we have, thanks to the sublemma above,

all $x_\alpha \geq 0$, and $x_\alpha > 0$ for at least two values of α

all $y_\alpha \geq 0$, and $y_\alpha > 0$ for at least two values of α.

By the Weyl dimension formula, we have

$$\dim(V_{w_1}) = \prod_{\alpha \in R_+} (1 + x_\alpha)$$
$$\dim(V_{w_2}) = \prod_{\alpha \in R_+} (1 + y_\alpha)$$
$$\dim(V_{w_1+w_2}) = \prod_{\alpha \in R_+} (1 + x_\alpha + y_\alpha).$$

So to prove the lemma, we are reduced to proving the following.

11.10.6.2. Sublemma. *Given a finite set R_+ of ≥ 2 elements, and real numbers $x_\alpha \geq 0$ and $y_\alpha \geq 0$ for each $\alpha \in R_+$, suppose that at least one x_α and at least two y_β's are > 0. Then*

$$1 + \prod_\alpha (1 + x_\alpha + y_\alpha) > \prod_\alpha (1 + x_\alpha) + \prod_\alpha (1 + y_\alpha).$$

Proof. Fix the y_α's, and consider the function of t given by

$$f(t) = 1 + \prod_\alpha (1 + tx_\alpha + y_\alpha) - \prod_\alpha (1 + tx_\alpha) - \prod_\alpha (1 + y_\alpha).$$

This function vanishes at $t = 0$, and its derivative is equal to

$$f'(t) = \sum_\alpha x_\alpha \left\{ \prod_{\beta \neq \alpha} (1 + tx_\beta + y_\beta) - \prod_{\beta \neq \alpha} (1 + tx_\beta) \right\}.$$

For $t \geq 0$, the coefficient of each x_α in this expression is visibly non-negative. There exists some value α_0 of α with $x_{\alpha_0} > 0$. Because at least two y_β's are > 0, there exists a value β_0 for which $\beta_0 \neq \alpha_0$ and $y_{\beta_0} > 0$, whence the coefficients of x_{α_0} in the above expression is strictly positive. Therefore $f'(t) > 0$ for all $t \geq 0$, while $f(0) = 0$. Thus we find

$$f(1) = f(1) - f(0) = \int_0^1 f'(t)\,dt > 0. \quad \blacksquare$$

11.10.7. Thus it remains to look in the tables to see which of the fundamental weights w of which of the simple Lie algebras

$$A_n, B_n, C_n, D_n, E_6, E_7, E_8, F_4, G_2$$

satisfy the equality

$$1 + \dim(V_w) = w(h^0).$$

In the tables (Bourbaki *Lie* VI, Planches I-IX, pp. 250-275), we find for each fundamental w of each $\mathfrak{G}$ its explicit expression

$$w = \sum_{\alpha \in B} m_\alpha \alpha \quad \text{with} \quad m_\alpha \in \mathbf{Q}_{\geq 0}$$

as a sum of simple roots. In terms of this expression, we have

$$w(h^0) = 2 \sum_{\alpha \in B} m_\alpha.$$

In (Bourbaki Lie VIII, Table 2, page 214), we find the explicit formula for

$$\dim(V_w),$$

for each fundamental weight w of each simple Lie algebra $\mathfrak{G}$.

With some perseverance, one finds that the equality

$$\dim(V_w) - 1 = w(h^0)$$

holds only for the following list of fundamental representations of simple Lie algebras of rank ≥ 2:

$$\begin{cases} A_l, l \geq 2 & ; \quad w_1 \text{ and } w_l \\ B_2 & ; \quad w_1 \text{ and } w_2 \\ B_l, l \geq 3 & ; \quad w_1 \\ C_2 & ; \quad w_1 \text{ and } w_2 \\ C_l, l \geq 3 & ; \quad w_1 \\ G_2 & ; \quad w_1. \end{cases}$$

We will only sketch the requisite calculations. For $A_2, B_2 \simeq C_2$, and the exceptional groups, one checks case by case and learns very little in the process except that the list is correct for these. It remains to treat $A_{l \geq 3}, B_{l \geq 3}, C_{l \geq 3}$, and $D_{l \geq 4} (D_3 \simeq A_3)$.

11.10.8. *Case* $A_l, l \geq 3$. We have

$$w_i = \frac{1}{l+1}\left[\sum_{1 \leq j \leq i} j(l+1-i)\alpha_j + \sum_{i+1 \leq j \leq l} i(l+1-j)\alpha_j\right]$$

$$w_i(h^0) = \frac{2}{l+1}\left[\sum_{1 \leq j \leq i} j(l+1-i) + \sum_{i+1 \leq j \leq l} i(l+1-j)\right]$$

$$= \frac{1}{l+1}[i(i+1)(l+1-i) + i(l-i)(l+1-i)]$$

$$= i(l+1-i)$$

$$\dim(V_{w_i}) = \binom{l+1}{i}.$$

We must see for which $1 \leq i \leq l$ we have

$$\binom{l+1}{i} = i(l+1-i) + 1.$$

This clearly holds for $i = 1$ and for $i = l$. The question is to show that

$$1 + i(l+1-i) < \binom{l+1}{i} \quad \text{if } 2 \leq i \leq l-1.$$

But the identity $(x-a)(x+a) = x^2 - a^2 \leq x^2$ shows that,

$$1 + i(l+1-i) \leq 1 + \left(\frac{l+1}{2}\right)^2$$

while for $2 \leq i \leq l-1$, the fact that binomial coefficients increase toward the middle shows

$$\binom{l+1}{i} \geq \binom{l+1}{2}.$$

So we are reduced to checking that for $l \geq 3$ we have

$$\binom{l+1}{2} > 1 + \left(\frac{l+1}{2}\right)^2,$$

i.e.

$$\frac{l(l+1)}{2} > 1 + \frac{(l+1)^2}{4},$$

i.e.

$$2l^2 + 2l > 4 + l^2 + 2l + 1,$$
$$l^2 > 5. \blacksquare$$

11.10.9. *Case $B_l, l \geq 3$.*

$$\text{for } l \leq i \leq l-1, \qquad w_i = \sum_{j=1}^{i} j\alpha_j + \sum_{j=i+1}^{l} i\alpha_j$$

and for $i = l$ we have

$$w_l = \frac{1}{2}(\alpha_1 + 2\alpha_2 + \cdots + l\alpha_l).$$

Therefore

$$w_i(h^0) = \begin{cases} i(2l+1-i) & \text{if} \quad 1 \leq i \leq l-1 \\ \frac{l(l+1)}{2} & \text{if} \quad i = l. \end{cases}$$

We have

$$\dim(V_{w_i}) = \begin{cases} \binom{2l+1}{i} & \text{for} \quad 1 \leq i \leq l-1 \\ 2^l & \text{for} \quad i = l. \end{cases}$$

For $i = l$, the inequality

$$2^l > 1 + \frac{l(l+1)}{2}$$

holds for $l = 3$ by inspection ($8 > 7$), and then for higher l by comparing the first differences of both sides. For $1 \leq i \leq l-1$, the equality

$$\binom{2l+1}{i} = 1 + i(2l+1-i)$$

holds only for $i = 1$ (as we saw in checking A_{2l}). $\blacksquare$

11.10.10. *Case $D_l, l \geq 4$.* Here we have

$$\text{for } 1 \leq i \leq l-2, w_i = \sum_{j=1}^{i} j\alpha_j + i \sum_{j=i+1}^{l-2} \alpha_j + \frac{i}{2}(\alpha_{l-1} + \alpha_l)$$

$$w_{l-1} = \frac{1}{2}\left(\sum_{j=1}^{l-2} j\alpha_j + \frac{l}{2}\alpha_{l-1} + \frac{l-2}{2}\alpha_l\right)$$

$$w_l = \frac{1}{2}\left(\sum_{j=1}^{l-2} j\alpha_j + \frac{l-2}{2}\alpha_{l-1} + \frac{l}{2}\alpha_l\right).$$

Thus we find

$$w_i(h^0) = \begin{cases} i(2l-1-i) & \text{for } 1 \leq i \leq l-2 \\ \frac{l(l-1)}{2} & \text{for } i = l-1 \text{ and } i = l. \end{cases}$$

We have

$$\dim(V_{w_i}) = \begin{cases} \binom{2l}{i} & \text{for } 1 \leq i \leq l-2 \\ 2^{l-1} & \text{for } i = l-1 \text{ and } i = l. \end{cases}$$

For $i = l-1$ and l, we easily check that

$$2^{l-1} > 1 + \frac{l(l-1)}{2},$$

first by inspection for $l = 4$, then for higher l by considering the first differences of both sides. For $1 \leq i \leq l-2$, we need

$$\binom{2l}{i} > 1 + i(2l-1-i).$$

But the universal inequality for A_{2l-1} and w_i gives

$$\binom{2l}{i} \geq 1 + i(2l-i),$$

which is more than adequate.

11.10.11. *Hard Case $C_l, l \geq 3$.* We have

$$w_i = \sum_{j=1}^{i-1} j\alpha_j + i\left(\alpha_i + \cdots + \alpha_{l-1} + \frac{1}{2}\alpha_l\right),$$

so

$$w_i(h^0) = i(2l - i).$$

We have

$$\dim(V_{w_i}) = \begin{cases} \binom{2l}{i} & \text{for } i = 1 \\ \\ \binom{2l}{i} - \binom{2l}{i-2} & \text{for } 2 \le i \le l. \end{cases}$$

Clearly we have the required equality for $i = 1$. For $2 \le i \le l$ with $l \ge 3$, we are to prove that

$$\binom{2l}{i} - \binom{2l}{i-2} > 1 + i(2l - i).$$

We will prove this in the form

$$\binom{2l}{i} - \binom{2l}{i-2} \ge 2 + i(2l - i).$$

Write $j = l - i$; we need

$$\binom{2l}{l-j} - \binom{2l}{l-j-2} \ge 2 + (l+j)(l-j) \quad \text{for } 0 \le j \le l-2.$$

For this, notice first that from equating coefficients in $(1+T)^x(1+T)^y = (1+T)^{(x+y)}$, we have the identity

$$\binom{x+y}{n} = \sum_{a+b=n} \binom{x}{a}\binom{y}{b}.$$

For $n = y$, we have

$$\binom{x+n}{n} = \sum_{a+b=n} \binom{x}{a}\binom{n}{b} = \sum_{a=0}^{n} \binom{x}{a}\binom{n}{a}.$$

Applying this with $x = l + j$, $n = l - j$, we have

$$\binom{2l}{l-j} = \sum_{a=0}^{l-j} \binom{l+j}{a}\binom{l-j}{a}.$$

Applying it with $x = l + j + 2$, $n = l - j - 2$, we have

$$\binom{2l}{l-j-2} = \sum_{a=0}^{l-j-2} \binom{l+j+2}{a}\binom{l-j-2}{a}.$$

Subtracting, we find

$$\binom{2l}{l-j} - \binom{2l}{l-j-2} = \binom{l+j}{l-j}\binom{l-j}{l-j} + \binom{l+j}{l-j-1}\binom{l-j}{l-j-1}$$

$$+ \sum_{a=0}^{l-j-2} \left[\binom{l+j}{a}\binom{l-j}{a} - \binom{l+j+2}{a}\binom{l-j-2}{a}\right].$$

11.10.11.1. Lemma. *For $1 \leq a \leq l-j-2$, we have*

$$\binom{l+j}{a}\binom{l-j}{a} - \binom{l+j+2}{a}\binom{l-j-2}{a} \geq 2.$$

Proof. Proceeding in two steps $j \mapsto j+1$, $j+1 \mapsto j+2$, it suffices to show that

$$\binom{l+j}{a}\binom{l-j}{a} > \binom{l+j+1}{a}\binom{l-j-1}{a} \quad \text{if } 1 \leq a \leq l-j-1.$$

Multiplying by $a!a!$, this amounts to

$$(l+j)(l+j-1)\cdots(l+j)-(a-1))\cdot(l-j)(l-j-1)\cdots(l-j-(a-1))$$
$$>(l+j+1)(l+j)\cdots(l+j+1-(a-1))\cdot(l-j-1)(l-j-2)\cdots(l-j-1-(a-1)).$$

Cancelling the common factors, this reduces to

$$(l+j-(a-1))(l-j) > (l+j+1)(l-j-1-(a-1)).$$

And this with patience reduces to

$$a(2j+1) > 0. \quad \blacksquare$$

Thus we have

$$\binom{2l}{l-j} - \binom{2l}{l-j-2} \geq \binom{l+j}{l-j} + \binom{l+j}{l-j-1}(l-j) + 2(l-j-2),$$

and so we are reduced to showing that

$$\binom{l+j}{l-j} + \binom{l+j}{l-j-1}(l-j) + 2(l-j-2) \geq 2 + (l+j)(l-j)$$

for $0 \leq j \leq l-2$, $l \geq 3$.

If $l-j \geq 3$, then because binomial coefficients increase toward the middle we have

$$\binom{l+j}{l-j-1} \geq l+j$$

so in this case the last two terms already suffice:

$$\binom{l+j}{l-j-1}(l-j)+2(l-j-2)\geq(l+j)(l-j)+2,$$

as required.

If $l-j=2$, we need

$$\binom{l+j}{2}+(l+j)(l-j)\geq 2+(l+j)(l-j),$$

i.e. we need

$$\binom{l+j}{2}\geq 2$$

which certainly holds, because $l+j\geq 3$.

It remains only to point out that the representations w_1 and w_2 of $B_2=\mathrm{Lie}(\mathrm{SO}(5))$ are the standard and spin representations respectively, but that under the isomorphism $B_2\simeq C_2=\mathrm{Lie}(\mathrm{Sp}(4))$, the representations w_1 and w_2 of B_2 become the representations w_2 and w_1 of C_2, with w_1 the standard four-dimensional representation of $\mathrm{Lie}(\mathrm{Sp}(4))$, and w_2 the 5-dimension representation $\Lambda^2(\text{standard})/(\text{the symplectic form})$. Therefore the list as given in the statement of the classification theorem, which lists B_2 and C_2 each with their "standard" representations w_1, is complete and non-redundant. ∎

11.10.12. Remark. In proving the Classification Theorem, we used a "general position" argument to see that if there exists $N\in\mathfrak{G}$ which is nilpotent such that $\rho(N)$ has a single Jordan block, then there exists a principal nilpotent $N_1\in\mathfrak{G}$ with the same property. In fact, we can see with hindsight that any such N is automatically itself a principal nilpotent. For by the theory of $sl(2)$-triples, any $sl(2)$-triple in $\mathfrak{G}$ is $\mathrm{Aut}_e(\mathfrak{G})$-conjugate to one of the form (x,h,y) where $h\in\mathfrak{h}$ is an element satisfying

$$\text{for every } \alpha\in B,\quad \alpha(h)\in\{0,1,2\},$$

and such an $sl(2)$-triple is principal if and only if $\alpha(h)=2$ for all $\alpha\in B$, i.e., if and only if $h=h^0$.

In general, an $sl(2)$-triple is principal if and only if its x-component is a principal nilpotent. So we must show that if $\mathfrak{G}$ is a simple Lie algebra, and if w is a fundamental weight of $\mathfrak{G}$ such that the restriction of the representation V_w to an $sl(2)$-triple (x,h,y) as above is irreducible, then $h=h^0$. To see this, we notice that the highest weight of V_w restricted to the $sl(2)$-triple (x,h,y) is $w(h)$, because $\alpha(h)\geq 0$ for all $\alpha\in B$ (compare the proof of 11.10.4). Therefore the irreducibility of V_w on (x, h, y) is equivalent to

$$\dim(V_w) = 1 + w(h).$$

Combining this with the inequality

$$\dim(V_w) \geq 1 + w(h^0)$$

we see that

$$w(h) \geq w(h^0).$$

But for $\mathfrak{G}$ simple, the expression of any fundamental weight as a **Q**-sum of simple roots

$$w = \sum_{\alpha \in B} m_\alpha \alpha, \quad m_\alpha \in Q$$

has *all* its coefficients *strictly positive*, i.e. $m\alpha > 0$ for all $\alpha \in B$. (This may be checked case by case by looking at Bourbaki *Lie* VI, Planches, pp. 250-275, or remembered in the form "the inverse of the Cartan matrix of a simple Lie algebra has all entries strictly positive.") Therefore

$$\begin{cases} w(h) = \sum m_\alpha \alpha(h) \\ \\ w(h^0) = 2 \sum m_\alpha \end{cases}$$

whence

$$0 \leq w(h) - w(h^0) = -\sum_\alpha (2 - \alpha(h)) m_\alpha.$$

Since each $m_\alpha > 0$ and each $2 - \alpha(h) \geq 0$, the inequality above is possible if and only if $2 = \alpha(h)$ for all $\alpha \in B$. ∎

11.11. Proof of the G_2 Theorem

11.11.1. We must prove that for $p = 2$ and $n = 7$, we have $G_{\text{geom}} = G_2$. We first reduce to the case when $\mathbf{F}_q = \mathbf{F}_2$. Indeed, let us denote by

$$\psi : (\mathbf{F}_2, +) \to \{\pm 1\} = \mathbf{Z}^\times \subset \mathbf{Q}^\times \subset E^\times$$

the unique non-trivial additive character of $\mathbf{F}_2$: explicitly, we have

$$\psi(x) = (-1)^x \quad \text{for} \quad x \in \mathbf{Z}/2\mathbf{Z} \simeq \mathbf{F}_2.$$

Given a finite overfield $\mathbf{F}_q$ of $\mathbf{F}_2$, any non-trivial additive character ψ_1 of $\mathbf{F}_q$ may be written uniquely in the form

$$\psi_1(y) = \psi(\text{trace}_{\mathbf{F}_q/\mathbf{F}_2}(ay)) \text{ for a unique } a \in \mathbf{F}_q^\times.$$

Therefore the E_λ-sheaf $\mathcal{L}_{\psi_1}$ on $\mathbf{G}_m \otimes \mathbf{F}_q$ is related to $\mathcal{L}_\psi$ (= the inverse image on $\mathbf{G}_m \otimes \mathbf{F}_q$ of the sheaf $\mathcal{L}_\psi$ on $\mathbf{G}_m \otimes \mathbf{F}_2$) by

$$\mathcal{L}_{\psi_1} \simeq (\mathrm{trans}_a)^*(\mathcal{L}_\psi),$$

where trans_a is the automorphism "translation by $a \in \mathbf{G}_m(\mathbf{F}_q)$" of $\mathbf{G}_m \otimes \mathbf{F}_q$. By property (11) of convolution (cf. 5.1), it follows that for every $n \geq 1$ we have

$$\mathrm{Kl}_n(\psi_1) \simeq (\mathrm{trans}_{a^n})^*(\mathrm{Kl}_n(\psi)),$$

an isomorphism of lisse E_λ-sheaves on $\mathbf{G}_m \otimes \mathbf{F}_q$. Taking $n = 7$ and twisting, this yields an isomorphism of lisse E_λ-sheaves on $\mathbf{G}_m \otimes \mathbf{F}_q$,

$$\mathrm{Kl}_7(\psi_1)(3) \simeq (\mathrm{trans}_b)^*(\mathrm{Kl}_7(\psi)(3)),\ b = a^7.$$

(This could also be checked by comparing trace functions, cf. Lemma 4.1.9.)

So if we pick a chemin in $\mathbf{G}_m \otimes \mathbf{F}_q$ from any given geometric point $\bar{x}$ to its translate $b\bar{x}$, and denote by $\mathcal{F}_1,\ \mathcal{F}$ the sheaves

$$\mathcal{F}_1 = \mathrm{Kl}_7(\psi_1)(3), \quad \mathcal{F} = \mathrm{Kl}_7(\psi)(3),$$

we have a commutative diagram of monodromy representations

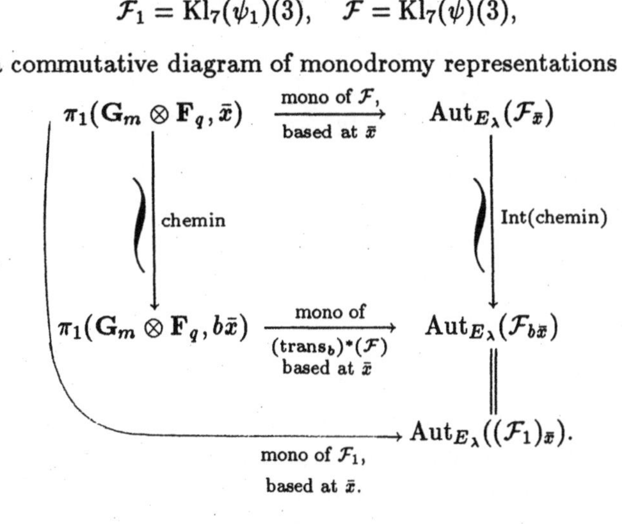

In particular, the monodromy representations ρ_1 of $\mathcal{F}_1$ and ρ of $\mathcal{F}$ of $\pi_1(\mathbf{G}_m \otimes \mathbf{F}_q, \bar{x})$ are E_λ-isomorphic, as are their restrictions to π_1^{geom}. Therefore up to E_λ-isomorphism, the E_λ-algebraic group G_{geom} we are trying to compute is the same for $\mathcal{F}_1$ and for $\mathcal{F}$.

11.11.2. Henceforth, we will deal with the lisse sheaf

$$\mathcal{F} = \mathrm{Kl}_7(\psi)(3) \quad \text{on} \quad \mathbf{G}_m \otimes \mathbf{F}_2.$$

11.11.3. We know that G_{geom} is either G_2 or SO(7). Let us first explain how to distinguish these groups by means of their representation theory. For the group SO(7), the representation

$$\Lambda^3(\text{standard seven-dimensional representation})$$

is irreducible (cf. Bourbaki Lie VIII, §13, 2, p. 195). Therefore $\Lambda^3(\text{std})$ has no non-zero invariants or coinvariants under SO(7). We will use this fact to show that if $G_{\text{geom}} = \text{SO}(7)$, then we arrive at a contradiction.

For consider the sheaf

$$\mathcal{G} = \Lambda^3(\mathcal{F}).$$

If $G_{\text{geom}} = \text{SO}(7)$ for $\mathcal{F}$, then $\mathcal{G}$ as representation of π_1^{geom} has no non-zero invariants or coinvariants (by Zariski density of $\rho(\pi_1^{\text{geom}})$ in SO(7)). Thus if $G_{\text{geom}} = \text{SO}(7)$, the following condition (H) is satisfied.

$$\text{(H)} \qquad H^0(\mathbf{G}_m \otimes \overline{\mathbf{F}}_2, \mathcal{G}) = 0 = H_c^2(\mathbf{G}_m \otimes \overline{\mathbf{F}}_2, \mathcal{G}).$$

We will show that (H) lends to a contradiction.

11.11.4. Lemma. *As representation of I_0, $\mathcal{G}$ is unipotent with 5 Jordan blocks, of dimensions $\{13, 9, 7, 5, 1\}$. The space $\mathcal{G}^{I_0}$ has dimension 5, and the eigenvalues of F_0 on $\mathcal{G}^{I_0}$ are $\{q^{-6}, q^{-4}, q^{-3}, q^{-2}, 1\}$, with $q = 2$.*

Proof. We have $\mathcal{G} = \Lambda^3(\mathcal{F}) = \Lambda^3(\text{Kl}_7(\psi)(3)) = (\Lambda^3(\text{Kl}_7(\psi)))(9)$.

We know (cf. 7.3.2) that $\text{Kl}_7(\psi)$ as representation of I_0 is unipotent with a single seven-dimensional Jordan block, that any element $F_0 \in D_0$ of degree one acts on $(\text{Kl}_7(\psi))^{I_0}$ as the identity, and that its eigenvalues on all of $\text{Kl}_7(\psi)$ as D_0-representation are

$$\{1, q, q^2, \ldots, q^6\}.$$

Therefore the eigenvalues of any such F_0 on $\Lambda^3(\text{Kl}_7(\psi))$ are all positive integral powers of q, because each is a product of three distinct of the above eigenvalues. Therefore the eigenvalues of F_0 on $\mathcal{G}$ as D_0-representation, obtained from those on $\Lambda^3(\text{Kl}_7(\psi))$ by dividing by q^9, are all integral powers of q. Because $\mathcal{G}$ is lisse on $\mathbf{G}_m \otimes \mathbf{F}_2$ and pure of weight zero with unipotent monodromy at zero, we have (cf. 7.0.7(2))

$$\dim \mathcal{G}^{I_0} = \text{number of Jordan blocks of } I_0 \text{ acting on } \mathcal{G}$$

and if

$$\mathcal{G} \simeq \bigoplus_{u=1}^{k} (\text{a Jordan block of dimension } 1 + i_u),$$

then after renumbering, we have

the eigenvalues $\alpha_1, \ldots, \alpha_k$ of F_0 on $\mathcal{G}^{I_0}$
have complex absolute values
$$|\alpha_u| = (\sqrt{q})^{-i_u} \quad \text{for} \quad u = 1, \ldots, k.$$

Combining this with the knowledge that all the α_u above are integral powers of q, we see that the i_u's are necessarily *even* integers, and that the eigenvalues of F_0 on $\mathcal{G}^{I_0}$ are the numbers

$$\{q^{-i_u/2}\} \quad \text{for} \quad u = 1, \ldots, k.$$

Thus in order to prove the lemma, it suffices to show that as I_0-representation, $\mathcal{G}$ is unipotent with Jordan blocks of sizes $\{13, 9, 7, 5, 1\}$. Because $\mathcal{G} = \Lambda^3(\mathcal{F})$ with $\mathcal{F}$ a unipotent 7-dimensional I_0-representation with a single Jordan block, this is the assertion

$$\Lambda^3\begin{pmatrix}\text{a unipotent Jordan} \\ \text{block of size 7}\end{pmatrix} = \begin{pmatrix}\bigoplus \text{ of unipotent Jordan blocks} \\ \text{of sizes } \{13, 9, 7, 5, 1\}\end{pmatrix}.$$

Equivalently, if we see the unipotent as the image of $\begin{pmatrix}1 & 1 \\ 0 & 1\end{pmatrix}$ in Symm^6 (the standard 2-dim rep of SL(2)), this is the assertion that as representations of SL(2), we have

$$\Lambda^3(\mathrm{Symm}^6) = \mathrm{Symm}^{12} + \mathrm{Symm}^8 + \mathrm{Symm}^6 + \mathrm{Symm}^4 + \mathrm{Symm}^0,$$

an identity certainly known in the last century.

Here is an alternate approach to the question. In terms of the numbers $\alpha_1, \ldots, \alpha_k$ and their weight-drops $i_1, \ldots, i_k$ we know that the totality of the 35 eigenvalues of F_0 on $\mathcal{G}$ are the numbers

$$\{\alpha_u, q\alpha_u, \ldots, q^{i_u}(\alpha_u)\}, \quad u = 1, \ldots, k.$$

Because i_u is even ≥ 0 and $\alpha_u = q^{-i_u/2}$, these are sequences of successive powers of q, each sequence symmetric around $q^0 = 1$. So the smallest (in the archimedean sense) of any of the 35 eigenvalues is certainly among the α_u, say α_1: cancelling $\alpha_1, q\alpha_1, q^2\alpha_1, \ldots, 1/\alpha_1$ from our list, the smallest remaining eigenvalue is another α_u, say $\alpha_2; \ldots.$

Let us carry this out explicitly. The seven eigenvalues of F_0 on $\mathcal{F} = \mathrm{Kl}_7(\psi)(3)$ are $\{q^{-3}, q^{-2}, q^{-1}, 1, q, q^2, q^3\}$. The set of all triple products of three distinct of these is the following set of powers of q, in "lexicographic order"

$$\begin{array}{l} -6,-5,-4,-3,-2 \\ -4,-3,-2,-1 \\ -2,-1,0 \\ \ \ 0,1 \\ \ \ 2 \\ -3,-2,-1,0 \\ -1,0,1 \\ \ \ 1,2 \\ \ \ 3 \\ \ \ 0,1,2 \\ \ \ 2,3 \\ \ \ 4 \\ \ \ 3,4 \\ \ \ 5 \\ \ \ 6 \end{array}$$

The smallest power is q^{-6}. Striking out its cycle $\{q^{-6}, q^{-5}, \ldots, q^5, q^6\}$, the smallest power remaining is q^{-4}. Striking out its cycle $\{q^{-4}, q^{-3}, \ldots q^3, q^4\}$, the smallest remaining power is q^3. Striking out its cycle $\{q^{-3}q^{-2}, \ldots, q^2, q^3\}$, the smallestremaining power is q^{-2}. Striking out its cycle$\{q^{-2}, q^{-1}, 1, q, q^2\}$, we are left with $q^0 = 1$. ∎

11.11.5. Corollary. *We have* $\dim H^0(\mathbf{G}_m \otimes \overline{\mathbf{F}}_2, \mathcal{G}) \leq 1$.

Proof. We know that G_{geom} is a semi-simple algebraic group (being either G_2 or SO(7)). Therefore by Zariski density of $\rho(\pi_1^{\text{geom}})$ in G_{geom}, $\mathcal{G}$ is completely reducible as representation of π_1^{geom}. In particular, the invariants are a direct factor, so on $\mathbf{G}_m \otimes \overline{\mathbf{F}}_2$ we have

$$\mathcal{G} \simeq \begin{pmatrix} \text{the constant sheaf} \\ H^0(\mathbf{G}_m \otimes \overline{\mathbf{F}}_2, \mathcal{G}) \end{pmatrix} \bigoplus \mathcal{H}$$

where $\mathcal{H}$ is a lisse sheaf with

$$H^0(\mathbf{G}_m \otimes \overline{\mathbf{F}}_2, \mathcal{H}) = 0.$$

Restricting the above direct-sum decomposition to I_0, we see that

$$\dim H^0(\mathbf{G}_m \otimes \overline{\mathbf{F}}_2, \mathcal{G}) \quad \leq \begin{pmatrix} \text{the number of 1-dimensional} \\ \text{Jordan blocks in the action of } I_0 \text{ on} \\ \mathcal{G}. \end{pmatrix}$$

By the above lemma, the number of such Jordan blocks is equal to one. ∎

Suppose now that (H) holds. The inclusion $j : \mathbf{G}_m \hookrightarrow \mathbf{P}^1$ gives a short exact sequence of sheaves on $\mathbf{P}^1 \otimes \mathbf{F}_2$

$$0 \to j_!\mathcal{G} \to j_*\mathcal{G} \to \mathcal{G}^{I_0} \oplus \mathcal{G}^{I_\infty} \to 0,$$

where $\mathcal{G}^{I_0}$ and $\mathcal{G}^{I_\infty}$ are viewed as punctual sheaves at zero and ∞ respectively. If (H) holds, the long exact cohomology sequence on $\mathbf{P}^1 \otimes \overline{\mathbf{F}}_2$ gives a short exact sequence

$$0 \to \mathcal{G}^{I_0} \oplus \mathcal{G}^{I_\infty} \to H^1_c(\mathbf{G}_m \otimes \overline{\mathbf{F}}_2, \mathcal{G}) \to H^1(\mathbf{P}^1 \otimes \overline{\mathbf{F}}_2, j_*\mathcal{G}) \to 0.$$

11.11.6. Lemma. *If (H) holds, then* $\dim H^1_c(\mathbf{G}_m \otimes \overline{\mathbf{F}}_2, \mathcal{G}) \leq 5$.

Proof. Because $\mathcal{G}$ is lisse, $H^0_c = 0$, and by (H) $H^2_c = 0$. Therefore the Euler–Poincaré formula gives

$$\dim H^1_c(\mathbf{G}_m \otimes \overline{\mathbf{F}}_2, \mathcal{G}) = \mathrm{Swan}_0(\mathcal{G}) + \mathrm{Swan}_\infty(\mathcal{G}).$$

Now $\mathcal{G} = \Lambda^3(\mathcal{F})$ with $\mathcal{F}$ tame at zero and with all breaks of $\mathcal{F}$ at ∞ equals to $1/7$. Therefore $\mathcal{G}$ is tame at zero, and all its breaks at ∞ are $\leq 1/7$, whence

$$\begin{cases} \mathrm{Swan}_0(\mathcal{G}) = 0, \\ \mathrm{Swan}_\infty(\mathcal{G}) \leq \mathrm{rk}(\mathcal{G}) \times (\text{ biggest break at } \infty) \leq 35 \times (1/7) \leq 5. \end{cases}$$ ∎

Combining these last two lemmas, we find that, under the hypothesis (H), we have, for dimension reasons,

$$\begin{cases} \mathcal{G}^{I_0} \xrightarrow{\sim} H^1_c(\mathbf{G}_m \otimes \overline{\mathbf{F}}_2, \mathcal{G}) \text{ has } \dim 5, \\ \text{eigenvalues of } F_0 \text{ are } \{q^{-6}, q^{-4}, q^{-3}, q^{-2}, 1\} \end{cases}$$

with $q = 2$. Replacing $\mathcal{G} = \Lambda^3(\mathcal{F})$ by

$$\mathcal{G}(-9) = \Lambda^3(\mathrm{Kl}_7(\psi)),$$

we see that, under (H), we have

$$H^i_c(\mathbf{G}_m \otimes \overline{\mathbf{F}}_2, \Lambda^3(\mathrm{Kl}_7(\psi))) = \begin{cases} 0 \text{ for } i \neq 1 \\ \\ \text{for } i = 1, \text{ a 5-dimensional space on} \\ \text{which } F \text{ has eigenvalues } \{q^3, q^5, q^6, q^7, q^9\}. \end{cases}$$

11.11.7. We next apply the Lefschetz trace formula to the lisse sheaf $\Lambda^3(\mathrm{Kl}_7(\psi))$ on $\mathbf{G}_m \otimes \mathbf{F}_2$, which very conveniently has just a single rational point, "1." Under hypothesis (H), we find

$$\mathrm{trace}(F_1 \mid \Lambda^3(\mathrm{Kl}_7(\psi))_1) = -(q^3 + q^5 + q^6 + q^7 + q^9).$$

Now the stalk at "1" $\in \mathbf{G}_m(\mathbf{F}_2)$ of $\mathrm{Kl}_7(\psi)$ has, for all $n \geq 1$,

$$\mathrm{trace}((F_1)^n \mid (\mathrm{Kl}_7(\psi))_1) = S_n$$

where

$$S_n = \sum_{\substack{x_1 \ldots x_7 = 1 \\ x_i \in \mathbf{F}_{2^n}}} (\psi \circ \mathrm{trace}_{\mathbf{F}_{2^n}/\mathbf{F}_2})\left(\sum x_i\right),$$

by the fundamental trace property of Kloosterman sheaves. Writing this in the form

$$\det(1 - TF_1 \mid (\mathrm{Kl}_7(\psi))_1) = \exp\left(-\sum_{n \geq 1} \frac{S_n T^n}{n}\right)$$

and expanding

$$\det(1 - TF_1 \mid (\mathrm{Kl}_7(\psi))_1) = \sum (-1)^i \,\mathrm{trace}(F_1 \mid \Lambda^i(\mathrm{Kl}_7(\psi))_1) T^i,$$

we see by equating coefficients that

$$\begin{aligned}
&- \mathrm{trace}(F_1 \mid \Lambda^3(\mathrm{Kl}_7(\psi))_1) \\
&= \text{coef of } T^3 \text{ in } \exp\left(-S_1 T - \frac{S_2 T^2}{2} - \frac{S_3 T^3}{3}\right) \\
&= \text{coef of } T^3 \text{ in} \\
&\quad \left(1 - S_1 T + \frac{(S_1)^2 T^2}{2} - \frac{(S_1)^3 T^3}{6}\right)\left(1 - \frac{S_2 T^2}{2}\right)\left(1 - \frac{S_3 T^3}{3}\right) \\
&= \frac{S_1 S_2}{2} - \frac{S_3}{3} - \frac{(S_1)^3}{6}.
\end{aligned}$$

Substituting our putative value for this trace, we find that if (H) holds, then

$$\frac{S_1 S_2}{2} - \frac{S_3}{3} - \frac{(S_1)^3}{6} = q^3 + q^5 + q^6 + q^7 + q^9,$$

with $q = 2$. The sums S_1, S_2, S_3 are all *integers* (ψ takes values ± 1 only), so in order to prove that the above identity is *false*, it suffices to show that it is false modulo some integer M which is prime to 6. For this purpose, we take $M = 7$.

11.11.8. The sum

$$S_n = \sum_{\substack{x_1 \ldots x_7 = 1 \\ \text{all } x_i \in \mathbf{F}_{2^n}}} (\psi \circ \mathrm{trace}_{\mathbf{F}_{2^n}/\mathbf{F}_2})\left(\sum x_i\right)$$

is invariant under permutation of the variables $x_1, \ldots, x_7$. In particular, it is invariant under $\mathbf{Z}/7\mathbf{Z}$ acting by cyclic permutation of the variables. The only fixed points under $\mathbf{Z}/7\mathbf{Z}$ are those $(x_1, \ldots, x_7)$ with $x_1 = x_2 = \cdots = x_7$, and any non-fixed point lies in a $\mathbf{Z}/7\mathbf{Z}$-orbit of precisely seven elements. Therefore for each $n \geq 1$ we have a congruence modulo 7

$$S_n \equiv \sum_{\substack{x^7 = 1 \\ x \in \mathbf{F}_{2^n}}} (\psi \circ \mathrm{trace}_{\mathbf{F}_{2^n}/\mathbf{F}_2})(7x) \qquad \mathrm{mod}\ 7$$

simply because the function being summed in S_n is constant and $\mathbf{Z}$-valued on each $\mathbf{Z}/7\mathbf{Z}$ orbit.

The sum S_1 is readily computed mod 7, since $\mathbf{F}_2$ contains only one 7'th root of unity, namely $x = 1$:

$$S_1 \equiv \sum_{x=1} \psi(7x) = \psi(7) = (-1)^7 \equiv -1 \quad \mathrm{mod}\ 7.$$

The sum S_2 is also easy, because $\mathbf{F}_4^\times = \boldsymbol{\mu}_3$ contains only one 7th root of unity, $x = 1$:

$$S_2 \equiv \sum_{x=1} (\psi \circ \mathrm{trace}_{\mathbf{F}_4/\mathbf{F}_2})(7x) = \psi(\mathrm{trace}_{\mathbf{F}_4/\mathbf{F}_2}(7)) = \psi(14) \equiv 1 \quad \mathrm{mod}\ 7.$$

As for $S_3, \mathbf{F}_8^\times = \boldsymbol{\mu}_7$, so by non-triviality of $\psi \circ \mathrm{trace}$,

$$S_3 \equiv \sum_{x \in \mathbf{F}_8^\times} (\psi \circ \mathrm{trace}_{\mathbf{F}_8/\mathbf{F}_2})(x) = -1 \ \mathrm{mod}\ 7.$$

Substituting the mod 7 values $-1, 1, -1$ for S_1, S_2, S_3, we find

$$-\frac{1}{2} + \frac{1}{3} + \frac{1}{6} \equiv 2^3 + 2^5 + 2^6 + 2^7 + 2^9 \ \mathrm{mod}\ 7,$$

i.e.,

$$0 \equiv 2^3 + 2^5 + 2^6 + 2^7 + 2^9 \ \mathrm{mod}\ 7.$$

But $2^3 = 8 \equiv 1(7)$, so (H) leads to

$$0 \equiv 1 + 2^2 + 1 + 2 + 1 \ \mathrm{mod}\ (7),$$

which is visibly false. Therefore (H) is false, whence $G_{\mathrm{geom}} \neq \mathrm{SO}(7)$, and so $G_{\mathrm{geom}} = G_2$ by default. ∎

11.11.9. We will now explain why the method we employed "had" to work. Because G_{geom} lies in SL(7), the sheaf $\mathcal{G} = \Lambda^3(\mathcal{F})$ is self-dual, so

$$\dim H^0(\mathbf{G}_m \otimes \overline{\mathbf{F}}_2, \mathcal{G}) = \dim H^2_c(\mathbf{G}_m \otimes \overline{\mathbf{F}}_2, \mathcal{G})$$

because the two spaces are, up to a Tate twist, duals of each other. In view of 11.11.5, this common dimension is ≤ 1. Suppose that it is equal to 1. Then $G_{\text{geom}} = G_2$, because SO((7) is excluded. In this case we have seen (cf. 11.3) that $\rho(\pi_1^{\text{arith}}) \subset G_{\text{geom}}(E_\lambda)$, so we have a direct-sum decomposition

$$\mathcal{G} \simeq E_\lambda \oplus \mathcal{H}$$

as lisse sheaves on $\mathbf{G}_m \otimes \mathbf{F}_2$ (reductivity of G_2), and by 11.11.5 we have

$$H^0(\mathbf{G}_m \otimes \overline{\mathbf{F}}_2, \mathcal{H}) = 0 = H^2_c(\mathbf{G}_m \otimes \overline{\mathbf{F}}_2, \mathcal{H}).$$

The cohomology sequence for $j_!\mathcal{H} \to j_*\mathcal{H}$ yields a short exact sequence

$$0 \to \mathcal{H}^{I_0} \oplus \mathcal{H}^{I_\infty} \to H^1_c(\mathbf{G}_m \otimes \overline{\mathbf{F}}_2, \mathcal{H}) \to H^1(\mathbf{P}^1 \otimes \overline{\mathbf{F}}_2, j_*\mathcal{H}) \to 0.$$

Because $\mathcal{G} \simeq E_\lambda \oplus \mathcal{H}$, the earlier Lemma 11.11.4 shows that

$$\dim(\mathcal{H}^{I_0}) = 4, \text{ and } F_0 \text{ acts with eigenvalues } \{q^{-6}, q^{-4}, q^{-3}, q^{-2}\}.$$

Because $\mathcal{H}$ has $H^0_c = H^2_c = 0$, we have, just as in 11.11.6,

$$\begin{aligned}
\dim H^1_c(\mathbf{G}_m \otimes \overline{\mathbf{F}}_2, \mathcal{H}) &= \mathrm{Swan}_0(\mathcal{H}) + \mathrm{Swan}_\infty(\mathcal{H}) \\
&= \mathrm{Swan}_\infty(\mathcal{H}) \\
&\leq [(\mathrm{rank}\,\mathcal{H})(\text{ biggest break at } \infty)] \\
&\leq [(34)(1/7)] = 4.
\end{aligned}$$

For dimension reasons, we find

$$\begin{cases} \mathcal{H}^{I_0} \xrightarrow{\sim} H^1_c(\mathbf{G}_m \otimes \overline{\mathbf{F}}_2, \mathcal{H}), \\ \text{eigenvalues of } F_0 \text{ are } \{q^{-6}, q^{-4}, q^{-3}, q^{-2}\}, q = 2. \end{cases}$$

Therefore the cohomology of $\mathcal{G} = \mathcal{H} \oplus E_\lambda$ is given by

$$H^i_c(\mathbf{G}_m \otimes \overline{\mathbf{F}}_2, \mathcal{G}) = \begin{cases} 0 \text{ for } i = 0 \\ \text{for } i = 1, \text{ 5-dimensional, } F \text{ acting with eigenvalues } \{q^{-6}, q^{-4}, q^{-3}, q^{-2}, 1\} \\ \text{for } i = 2, \text{ 1-dimensional, } F \text{ acts as } q. \end{cases}$$

The Lefschetz trace formula applied to $\mathcal{G}(-9) = \Lambda^3(\mathrm{Kl}_7(\psi))$ on $\mathbf{G}_m \otimes F_2$ gives

$$\mathrm{trace}(F_1 \mid \Lambda^3(\mathrm{Kl}_7(\psi))_1) = q^{10} - (q^3 + q^5 + q^6 + q^7 + q^9)$$

with $q = 2$ as always. Because

$$2^{10} \not\equiv 0 \bmod (7),$$

we may distinguish this case from the earlier one (where we assumed (H)) by computing mod 7, which is exactly what we did.

CHAPTER 12

Integral Monodromy of Kloosterman Sheaves (d'après O. Gabber)

12.0. Formulation of the theorem

Let us fix

an integer $n \geq 2$

a prime number p

a finite field $\mathbf{F}_q$ of characteristic p

a finite extension E of $\mathbf{Q}$, containing $q^{\frac{n-1}{2}}$.

a non-trivial additive character $\psi : (\mathbf{F}_q, +) \to E^{\times}$.

For each finite place λ of E whose residue characteristic is $l \neq p$, we choose an $\mathcal{O}_\lambda$-form $\mathcal{F}_\lambda$ of the lisse E_λ-sheaf $\mathrm{Kl}_n(\psi)(\frac{n-1}{2})$ on $\mathbf{G}_m \otimes \mathbf{F}_q$. For example, we may take for $\mathcal{F}_\lambda$ the twist by $\mathcal{O}_\lambda(\frac{n-1}{2})$ of the n-fold self-convolution of "$\mathcal{L}_\psi$ as lisse $\mathcal{O}_\lambda$-sheaf of rank one"; in view of 4.1.2, any other choice is $\mathcal{O}_\lambda$-isomorphic to this one by an isomorphism which is itself unique up to an $\mathcal{O}_\lambda^{\times}$ scalar. If pn is *even*, we also choose a bilinear form

$$\mathcal{F}_\lambda \underset{\mathcal{O}_\lambda}{\otimes} \mathcal{F}_\lambda \to \mathcal{O}_\lambda$$

which is non-zero $\mod \lambda$. By 4.1.11 and 4.2.1, we know that, for pn even, such forms exist, that any two are $\mathcal{O}_\lambda^{\times}$-proportional, and that any such form is an $\mathcal{O}_\lambda$-autoduality of $\mathcal{F}_\lambda$ which is alternating if n is even and symmetric if n is odd.

We also fix, exactly as in the previous chapter (cf. 11.0.3), a geometric point $\bar{x} : \mathrm{Spec}(\Omega) \to \mathbf{G}_m \otimes \mathbf{F}_q$, and "chemins" connecting $\bar{x}$ to the spectra $\bar{\eta}_0$ and $\bar{\eta}_\infty$ of separable closures of the fraction fields of the henselizations of $\mathbf{P}^1 \otimes \mathbf{F}_q$ at 0 and ∞ resp. The chemins give rise to a diagram

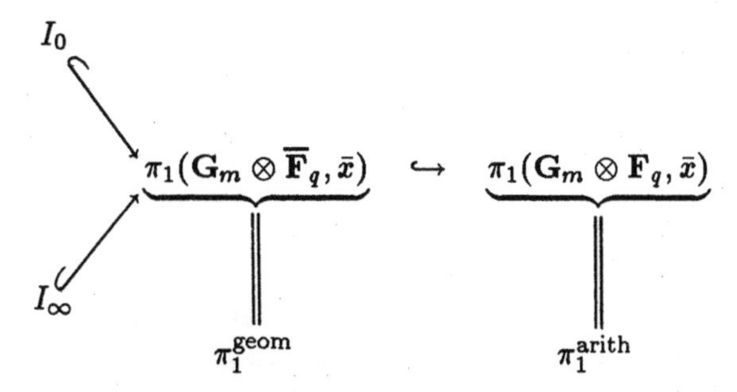

The monodromy representation of $\mathcal{F}_\lambda$ at $\bar{x}$ is denoted

$$\rho_\lambda : \pi_1^{\text{arith}} \to \text{Aut}_{\mathcal{O}_\lambda}((\mathcal{F}_\lambda)_{\bar{x}}) \simeq \text{GL}(n, \mathcal{O}_\lambda).$$

The existence of the pairing when pn is even, and 11.0.2 (4) in general, give the *à* priori inclusions

$$\rho_\lambda(\pi_1^{\text{geom}}) \subset \rho_\lambda(\pi_1^{\text{arith}}) \subset \begin{cases} \text{Sp}(n, \mathcal{O}_\lambda) & n \text{ even} \\ \\ \text{SL}(n, \mathcal{O}_\lambda) & pn \text{ odd} \\ \\ \text{SO}(n, \mathcal{O}_\lambda) & p = 2,\ n \text{ odd.} \end{cases}$$

12.1. Theorem (O. Gabber). *Suppose that n is even or $p \neq 2$. There exists an integer D depending upon (n, p, q, E, ψ) such that for any finite place λ of E of residue characteristic $l > D$, $l \neq p$, at which $\mathcal{O}_\lambda = \mathbf{Z}_l$, we have*

$$\rho_\lambda(\pi_1^{\text{geom}}) = \begin{cases} \text{Sp}(n, \mathcal{O}_\lambda) & \textit{if } n \textit{ even} \\ \\ \text{SL}(n, \mathcal{O}_\lambda) & \textit{if } n \textit{ odd.} \end{cases}$$

In fact, there is a more precise statement which clarifies the role of the "$\mathcal{O}_\lambda = \mathbf{Z}_l$" condition in the above theorem. As before, let us fix a topological generator γ_0 of I_0^{tame}. Then $\rho_\lambda(\gamma_0)$ is a unipotent $n \times n$ matrix. So for λ any finite place of residue characteristic $l \geq n$, the elements

$$\{\exp(t \log(\rho_\lambda(\gamma_0)))\}_{t \in \mathcal{O}_\lambda}$$

form a closed subgroup of $\text{Sp}(n, \mathcal{O}_\lambda)$ for n even, and of $\text{SL}(n, \mathcal{O}_\lambda)$ for n odd. If $\mathcal{O}_\lambda = \mathbf{Z}_l$, this group is just $\rho_\lambda(I_0)$.

12.2. Theorem (O. Gabber). *Suppose n is even or $p \neq 2$. There exists an integer D depending on (n, p, q, E, ψ) such that for any finite place λ of E of residue characteristic $l > D$, $l \neq p$, the group*

$$\begin{cases} \mathrm{Sp}(n, \mathcal{O}_\lambda) & \text{if } n \text{ even} \\ \mathrm{SL}(n, \mathcal{O}_\lambda) & \text{if } n \text{ odd} \end{cases}$$

is generated by the finite subgroup $\rho_\lambda(I_\infty)$ and by the one-parameter subgroup

$$\{\exp(t \log(\rho_\lambda(\gamma_0)))\}_{t \in \mathcal{O}_\lambda}.$$

In particular, if $\mathcal{O}_\lambda = \mathbf{Z}_l$, and $l > D$, then we have

$$\left.\begin{array}{l} \textit{subgroup generated by} \\ \rho(I_0) \textit{ and } \rho(I_\infty) \end{array}\right\} = \rho(\pi_1^{\mathrm{geom}}) = \rho(\pi_1^{\mathrm{arith}}) = \begin{cases} \mathrm{Sp}(n, \mathcal{O}_\lambda) & n \text{ even} \\ \\ \mathrm{SL}(n, \mathcal{O}_\lambda) & n \text{ odd.} \end{cases}$$

12.3. Reduction to a universal situation

12.3.1. Lemma. *For fixed (n, p, q, E, ψ), the finite group $\rho_\lambda(I_\infty)$ is canonically independent of the choice of the finite place λ of E of residue characteristic $l \neq p$.*

Proof. By Deligne, the character of ρ_λ on I_∞ is independent of λ (cf. 10.0). Because $\rho_\lambda(I_\infty)$ is finite (cf. 1.11), the kernel of ρ_λ on I_∞ consists precisely of those elements $\alpha \in I_\infty$ where $\mathrm{trace}(\rho_\lambda(\alpha)) = n$. Therefore $\mathrm{Ker}(\rho_\lambda \mid I_\infty)$ is independent of λ, and $\rho_\lambda(I_\infty) \xleftarrow{\sim} I_\infty / \mathrm{Ker}(\rho_\lambda \mid I_\infty)$. ∎

Let us denote by

$$\Gamma_\infty = \text{ the finite group } I_\infty/\mathrm{Ker}(\rho_\lambda \mid I_\infty)$$
$$H_\infty = P_\infty/\mathrm{Ker}(\rho_\lambda \mid P_\infty) = \text{ the } p\text{-Sylow subgroup of } \Gamma_\infty$$

and by

$$\mathrm{char}_\rho : P_\infty \to \mathcal{O}_E$$

the common trace function of the ρ_λ's. This data depends only on (n, p, q, E, ψ).

12.3.2. Lemma. *Let K be an algebraically closed over-field of E, V an n-dimensional vector space over K, and*

$$\psi : \Gamma_\infty \to \mathrm{GL}(V)$$

a group homomorphism such that for all $\gamma_\infty \in \Gamma_\infty$ we have

$$\mathrm{trace}(\psi(\gamma_\infty)) = \mathrm{char}_\rho(\gamma_\infty).$$

Then

(1) ψ *is injective*
(2) Γ_∞ *acts irreducibly on* V
(3) Γ_∞ *admits no faithful* K*-linear representation of dimension* $< n$.
(4) *if pn is odd, there exists no non-zero* H_∞*-invariant bilinear form* $V \underset{K}{\otimes} V \to K$
(5) *if pn is odd, any character* $\chi : \Gamma_\infty \to K^\times$ *is trivial on* H_∞.

Proof. Properties (3) and (5) are intrinsic properties of Γ_∞ as finite group, and have already been established (cf. 1.19). Properties (1), (2), (4) depend only on the character of ψ and on its restriction to H_∞, so they hold if and only if they hold for $\psi =$ some ρ_λ, in which case they have already been established (cf. 1.19, 10.4.4, 11.02). ∎

12.3.3. Lemma. *For* γ_0 *a topological generator of* I_0^{tame}, *and any finite place* λ *of* E *of residue characteristic* $l \neq p$, *the unipotent element*

$$\rho_\lambda(\gamma_0) \in \mathrm{Aut}_{\mathcal{O}_\lambda}((\mathcal{F}_\lambda)_{\bar{x}}) \simeq \mathrm{GL}(n, \mathcal{O}_\lambda)$$

has a single Jordan block "over $\mathcal{O}_\lambda$*" in the sense that*

$$\begin{cases} (\rho_\lambda(\gamma_0) - 1)^n = 0, \text{ but} \\ \rho_\lambda(\gamma_0) - 1 \text{ has an } n-1 \times n-1 \text{ minor which is invertible in } \mathcal{O}_\lambda. \end{cases}$$

Proof. This amounts to the statement that $\rho_\lambda(\gamma_0) \otimes \mathbf{F}_\lambda$ has a single Jordan block. It certainly has at least one, and the fact that it has at most one is, as explained in 7.5.1.3, a consequence of the fact that $\mathcal{F}_\lambda \otimes \mathbf{F}_\lambda$ is lisse on $\mathbf{G}_m$, tame at zero, and totally wild at ∞ with $\mathrm{Swan}_\infty = 1$. ∎

12.3.4. For given (n, p, q, E, ψ), we will define a moduli problem $\mathcal{M}$ (i.e. a covariant functor) on the category of all $\mathcal{O}[1/p]$-algebras, where $\mathcal{O} = \mathcal{O}_E$ denotes the ring of algebraic integers in E.

12.3.4.1. If *pn* is *odd*, then we define, for any $\mathcal{O}[1/p]$-algebra R, $\mathcal{M}(R) =$ the set of all pairs (ψ, U) where

$\psi : \Gamma_\infty \to \mathrm{SL}(n, R)$ is a group homomorphism
with $\mathrm{trace}(\psi(\gamma)) = \mathrm{char}_\rho(\gamma)$ for all $\gamma \in \Gamma_\infty$.
$U \in \mathrm{SL}(n, R)$ is an element satisfying
a)$(U - 1)^n = 0$
b) Zariski locally on $\mathrm{Spec}(R)$, some $(n-1) \times (n-1)$ minor of $U - 1$ is invertible.

12.3.4.2. If n is *even*, then for any $\mathcal{O}[1/p]$-algebra R, we define $\mathcal{M}(R) =$ the set of triples $(\langle\ ,\ \rangle, \psi, U)$ where

$\langle\ ,\ \rangle : R^n \times R^n \to R$ is a strongly alternating
(meaning $\langle v, v\rangle = 0$ for $v \in R^n$) R-bilinear
R-autoduality of R^n with itself.

$\psi : \Gamma_\infty \to \mathrm{Aut}_R(R^n, \langle\ ,\ \rangle) = \mathrm{Sp}(n, R)$ is a group homomorphism
with $\mathrm{trace}(\psi(\gamma)) = \mathrm{char}_\rho(\gamma)$ for all $\gamma \in \Gamma_\infty$.

$U \in \mathrm{Sp}(n, R)$ is an element satisfying

a) $(U - 1)^n = 0$

b) Zariski locally on $\mathrm{Spec}(R)$, some $(n-1) \times (n-1)$ minor of $U - 1$ is invertible.

12.3.4.3. Finally, if $p = 2$ and n is odd, we define $\mathcal{M}$ just as in 12.3.4.2 above but with $\langle,\rangle$ a symmetric autoduality, and with $\mathrm{Sp}(n, R)$ replaced by $\mathrm{SO}(n, R)$.

12.3.5. Lemma. *The moduli problem $\mathcal{M}$ is representable by an $\mathcal{O}[1/p]$-scheme of finite type (still denoted $\mathcal{M}$).*

Proof. Indeed if we specify in addition which $(n-1) \times (n-1)$ minor of $U-1$ is to be invertible, the corresponding open sub-problem is represented by the spec of a finitely generated $\mathcal{O}[1/p]$-algebra. For example, when pn is odd, we adjoin $n^2\#(\Gamma_\infty)$ indeterminates for the matrix coefficients of the elements $\psi(\gamma)$, $\gamma \in \Gamma_\infty$, and n^2 more indeterminates for those of U. The conditions that the $\psi(\gamma)$ be in $\mathrm{SL}(V)$, that $\psi(\gamma_1)\psi(\gamma_2) = \psi(\gamma_1\gamma_2)$ for all $\gamma_1, \gamma_2 \Gamma_\infty$, that $\mathrm{trace}\,\psi(\gamma) = \mathrm{char}_\rho(\gamma)$, and that $(U-1)^n = 0$, all are expressed by polynomial identities in the matrix coefficients. The condition that a given minor of $U - 1$ be invertible is obtained by adjoining its inverse. The case n even is similar, but here one needs also to add $\frac{1}{2}(n^2 - n)$ indeterminates for the skew-symmetric matrix of the form $\langle,\rangle$, to assure non-degeneracy by inverting its determinant, and to impose the additional polynomial identities that express that U and all the $\psi(\gamma)$, $\gamma \in \Gamma_\infty$ lie in the symplectic group for this form. The case $p = 2$, n odd is similar, but here one adds $\frac{1}{2}(n^2 + n)$ indeterminates for the symmetric matrix of the form $\langle,\rangle, \ldots$ ∎

12.3.6. Over the $\mathcal{O}[1/p]$-scheme $\mathcal{M}$ of finite type, we have the following "universal" structure, which represents the moduli problem:

a free $\mathcal{O}_{\mathcal{M}}$-module $\mathbf{V}$ of rank n, namely $(\mathcal{O}_{\mathcal{M}})^n$,

if pn is even, an $\mathcal{O}_{\mathcal{M}}$-linear autoduality $\mathbf{V} \underset{\mathcal{O}_{\mathcal{M}}}{\otimes} \mathbf{V} \to \mathcal{O}_{\mathcal{M}}$ which is strictly alternating if n is even, symmetric if n is odd,

$$\text{the corresponding } \mathcal{M}\text{-group scheme } G = \begin{cases} \mathrm{SL}(\mathbf{V}) & pn \text{ odd} \\ \mathrm{Sp}(\mathbf{V}, \langle , \rangle), n \text{ even} \\ \mathrm{SO}(\mathbf{V}, \langle , \rangle), n \text{ odd}, p = 2, \end{cases}$$

a unipotent element $U_{\text{univ}} \in G(\mathcal{M})$, with a single Jordan block ($(U-1)^n = 0$, and Zar-loc. on $\mathcal{M}$, some $(n-1) \times (n-1)$ minor of $U - 1$ is invertible),

a homomorphism $\psi_{\text{univ}} : \Gamma_\infty \to G(\mathcal{M})$ with character equal to char_ρ.

The point of considering this universal situation is that for each finite place λ of E of residue characteristic $l \neq p$, once we pick an $\mathcal{O}_\lambda$-basis of $(\mathcal{F}_\lambda)_{\bar{x}}$ as free $\mathcal{O}_\lambda$-module of rank n, the data

$$\begin{cases} \langle , \rangle, = \text{chosen } \langle , \rangle : \mathcal{F}_\lambda \underset{\mathcal{O}_\lambda}{\otimes} \mathcal{F}_\lambda \to \mathcal{O}_\lambda \text{ if } pn \text{ even} \\ \psi = \rho_\lambda \mid \Gamma_\infty \\ U = \rho_\lambda(\gamma_0) \end{cases}$$

defines an $\mathcal{O}_\lambda$-valued point of $\mathcal{M}$. This observation "reduces" us to studying the situation over $\mathcal{M}$ itself.

12.3.7. Lemma. *Suppose $(p, n) \neq (2, 7)$. Then for every geometric point of $\mathcal{M}$ in characteristic zero, i.e. for every z : $\mathrm{Spec}(K) \to \mathcal{M}$ with K an algebraically closed extension field of E, the subgroup of $G_z(K)$ generated by the unipotent element $U_{\text{univ}}(z)$ and by the subgroup $\psi_{\text{univ}}(\Gamma_\infty)(z) \subset G_z(K)$ is Zariski dense in G_z.*

Proof. This is "just" the Axiomatic Classification Theorem 11.7, in view of Lemmas 12.3.2 and 12.3.3. ∎

12.3.8. Lemma. *Let $\Gamma \subset G(\mathcal{M})$ be the subgroup generated by the element U_{univ} and by the finite subgroup $\psi_{\text{univ}}(\Gamma_\infty)$. Then there exist*

a finite set of elements $\gamma_1, \gamma_2, \ldots, \gamma_r$ of Γ, and an integer $D_1 \geq 1$ such that over $\mathcal{M}[1/(n-1)!D_1]$, the logarithms $\log(\gamma_i U_{\text{univ}} \gamma_i^{-1}) = \gamma_i \log(U_{\text{univ}}) \gamma_i^{-1}$ span $\underline{\mathrm{Lie}}(G)$ as an $\mathcal{O}_{\mathcal{M}}$-module.

Proof. Once we invert $(n-1)!$, $\log(U_{\text{univ}})$ makes sense as a nilpotent global section over $\mathcal{M}$ of $\underline{\text{Lie}}(G)$, as do its conjugates by any $\gamma \in \Gamma$. Because $\mathcal{M}$ is noetherian and $\underline{\text{Lie}}(G)$ is a free $\mathcal{O}_{\mathcal{M}}$-module of finite rank, it is enough to prove that over a suitable $\mathcal{M}[1/(n-1)!D_1]$, $\underline{\text{Lie}}(G)$ is spanned over $\mathcal{O}_{\mathcal{M}}$ by *all* the $\gamma \log(U_{\text{univ}})\gamma^{-1}$ for all $\gamma \in \Gamma$. Consider the support Z of the cokernel of the map

$$\bigoplus_{\gamma \in \Gamma} \mathcal{O}_{\mathcal{M}} \xrightarrow{\oplus \gamma \log(U_{\text{univ}})\gamma^{-1}} \underline{\text{Lie}}(G).$$

It is a scheme of finite type over $\mathbf{Z}$, being a closed subscheme of $\mathcal{M}$. So either Z is annihilated by some integer $D_1 \geq 1$, or it has some geometric point of characteristic zero. In the first case, the lemma "works" with this D_1. In the second case, let z be a characteristic-zero valued geometric point of the support, say $z : \text{Spec}(K) \to Z \subset \mathcal{M}$. The $\Gamma(z)$ is Zariski dense in G_z by Lemma 5, and, as G_z is *simple*, $\underline{\text{Lie}}(G_z)$ is an irreducible representation of G_z, so also an irreducible representation of $\Gamma(z)$ by Zariski density. The element $\log(U_{\text{univ}})(z) \in \underline{\text{Lie}}(G_z)$ is certainly non-zero, so by irreducibility $\underline{\text{Lie}}(G_z)$ is spanned over K by all the transforms $\gamma(z)(\log(U_{\text{univ}})(z))\gamma(z)^{-1}$ of $\log(U_{\text{univ}})(z)$ by $\gamma(z)$'s in $\Gamma(z)$. Therefore z is *not* in the support Z, contradiction. ∎

12.3.9. Lemma. *Notations as above, Zariski locally on* $\mathcal{M}[1/(n-1)!D_1]$ *there exist a finite set* $\gamma_1, \ldots, \gamma_{\dim G}$ *of* $\dim(G)$ *elements of* Γ *such that the logarithms* $\{\gamma_i \log(U_{\text{univ}})\gamma_i^{-1}\}_{i=1,\ldots,\dim(G)}$ *form an* $\mathcal{O}_{\mathcal{M}}$*-basis of* $\underline{\text{Lie}}(G)$.

Proof. Obvious from Lemma 12.3.8. ∎

12.3.10. Lemma. *There exists an integer* $D_0 \geq 1$ *such that* $\mathcal{M}[1/D_0]$ *is flat over* $\mathbf{Z}$.

Proof. By Lemma 12.3.5, $\mathcal{M}$ is of finite type over $\mathbf{Z}$, cf. ([A-K], V, 5.2). ∎

12.3.11. Theorem. *Suppose that either n is even, or that pn is odd, so that the group G over $\mathcal{M}$ is either* $\text{Sp}(n)$ *or* $\text{SL}(n)$. *There exists an integer* $D_2 = D_2(n, p, q, E, \psi)$ *with* $D_2 \geq 1$ *such that for any complete noetherian local* $\mathcal{O}_E[1/p(n-1)!D_0D_1D_2]$*-algebra R with finite residue field, and any point* $z \in \mathcal{M}(R)$, *the group* $G_z(R)$ *is generated by the subgroup* $\exp(R\log(U_{\text{univ}}(z)))$ *and by finitely many of its conjugates by elements of the subgroup* $\Gamma(z) \subset G_z(R)$.

This theorem, applied to $R = \mathcal{O}_\lambda$, $z \in \mathcal{M}(\mathcal{O}_\lambda)$ the point defined by the Kloosterman sheaf $\mathcal{F}_\lambda$, yields Theorem 12.2. It is itself a special case of the following theorem, applied to X running over a finite covering of

$\mathcal{M}[1/(n-1)!D_0D_1]$ by open sets over which Lemma 12.3.9 holds, to $G =$ $\mathrm{Sp}(n)$ or $\mathrm{SL}(n)$ over X, and to the unipotent elements $U_i = \gamma_i U_{\mathrm{univ}} \gamma_i^{-1}$ whose logarithms form an $\mathcal{O}_X$-basis of $\underline{\mathrm{Lie}}(G)$.

12.4. Generation by unipotent elements

12.4.1. Theorem (O. Gabber). *Let $n \geq 2$ be an integer, X a flat $\mathbf{Z}[1/(n-1)!]$-scheme of finite type, and*

$$G \subset \mathrm{GL}(n)_X$$

a Zariski closed subgroup scheme which is smooth over X with geometrically connected fibres of constant dimension $d = \dim(G)$. Suppose that

1. *we are given unipotent elements $U_1, \ldots, U_{\dim G} \in G(X)$, each satisfying $(U_i - 1)^n = 0$ in $M(n)(X)$, whose logarithms form an $\mathcal{O}_X$-basis of $\underline{\mathrm{Lie}}(G)$.*
2. *for every characteristic-zero valued geometric point $z : \mathrm{Spec}(K) \to X$ with K an algebraically closed field of characteristic zero, the geometric fibre G_z is simply connected (in the sense that it has no non-trivial finite etale coverings).*

Then there exists an integer $D_2 \geq 1$ such that for any complete noetherian local ring R with finite residue field in which $(n-1)!D_2$ is invertible, and for any R-valued point $z : \mathrm{Spec}(R) \to X$ of X,

$$\begin{array}{ccc} G_z & \longrightarrow & G \\ \downarrow & & \downarrow \\ \mathrm{Spec}(R) & \xrightarrow{z} & X \end{array}$$

the group $G_z(R)$ is generated by the one-parameter subgroups

$$\exp(R \log(U_i(z))), \qquad i = 1, \ldots, \dim G.$$

Proof. We first explain why the endomorphisms $\log(U_i)$ lie in $\underline{\mathrm{Lie}}(G)$, and why $t \mapsto \exp(t \log(U_i))$ defines an X-homomorphism from $\mathbf{G}_a$ to G. Both statements are true over X with G replaced by $\mathrm{GL}(n)_X$, and by Lie theory they are both true as stated over $X \underset{\mathbf{Z}}{\otimes} \mathbf{Q}$. Because G is smooth over X and X is flat over $\mathbf{Z}$, G is flat over $\mathbf{Z}$, so G is necessarily equal to the schematic closure of $G \underset{\mathbf{Z}}{\otimes} \mathbf{Q}$ in $\mathrm{GL}(n)_X$. Similarly, $\underline{\mathrm{Lie}}(G)$ is the schematic closure of $\underline{\mathrm{Lie}}(G) \otimes \mathbf{Q}$ in $\underline{\mathrm{Lie}}(\mathrm{GL}(n))$. This second fact shows that $\log(U_i)$, à priori a global section of $\underline{\mathrm{Lie}}(\mathrm{GL}(n))$ which after $\otimes \mathbf{Q}$ falls into $\underline{\mathrm{Lie}}(G)$, already lies in $\underline{\mathrm{Lie}}(G)$. The first fact shows that the morphism of X-groups

$\mathbf{G}_a \to \mathrm{GL}(n)$ defined by $t \to \exp(t\log(U_i))$ must land in G, because it does so after $\otimes\mathbf{Q}$.

Consider the morphism of X-schemes

$$\mathbf{A}^{\dim(G)} \longrightarrow G$$

$$(T_1,\ldots,T_{\dim(G)}) \longmapsto \prod_{i=1}^{\dim G} \exp(T_i \log(U_i)),$$

the product taken in the order $1 \times 2 \times 3 \times \cdots \times \dim G$. Because the $\{\log(U_i)\}_{i=1,\ldots,\dim(G)}$ form an $\mathcal{O}_X$-basis of $\underline{\mathrm{Lie}}(G)$, this map is etale along the zero-section of $\mathbf{A}^{\dim(G)}$. Therefore if we denote by

$$\mathcal{U} \subset \mathbf{A}^{\dim(G)}$$

the open set on which this map is etale, then $\mathcal{U}$ is a smooth X-scheme of finite type, all of whose geometric fibres are geometrically connected of dimension $\dim(G)$, and we have an etale morphism of X-schemes

$$f : \mathcal{U} \to G.$$

Because G is smooth over X, the "multiplication in G" morphism $G \times_X G \to G$ is smooth (by means of the "shearing" automorphism $(x,y) \mapsto (x,xy)$ of $G \times_X G$, it's isomorphic to $\mathrm{pr}_2 : G \times_X G \to G$). Therefore the composite map

$$\underbrace{\mathcal{U} \times_X \mathcal{U} \xrightarrow{f\times f} G \times_X G \xrightarrow{\text{mult}} G}_{f(2)}$$

is smooth, being an etale map followed by a smooth map. Similarly, the composite map

$$\underbrace{\mathcal{U} \times_X \mathcal{U} \times_X \mathcal{U} \times_X \mathcal{U} \xrightarrow{f\times f\times f\times f} G \times_X G \times_X G \times_X G \xrightarrow{\text{mult}} G}_{f(4)}$$

is smooth.

We will prove that, under the hypotheses of the theorem

(*) there exists an integer $D_2 \geq 1$ such that for any complete noetherian local $\mathbf{Z}[1/(n-1)!D_2]$-algebra R with finite residue field, and any R-valued point $z : \mathrm{Spec}(R) \to X$, the map $f(4)_z$ obtained by base change

$$f(4)_z : \mathcal{U}_z \times_R \mathcal{U}_z \times_R \mathcal{U}_z \times_R \mathcal{U}_z \longrightarrow G_z$$

is *surjective* on R-valued points.

Because $f(4)_z$ is a smooth morphism between smooth R-schemes, and R is complete noetherian local, $f(4)_z$ will be surjective on R-valued points if and only if it is surjective on k-valued points where k is the residue field of R. Therefore (*) is equivalent to

(**) there exists an integer $D_2 \geq 1$ such that for any finite field k whose characteristic is prime to $(n-1)!D_2$ and any k-valued point $z : \mathrm{Spec}(k) \to X$, the map $f(4)_z$ is surjective on k-valued points.

The map $f(4)$ is obtained from $f(2)$ by doubling" it: we have a factorization of $f(4)$ as

$$(\mathcal{U} \times_X \mathcal{U}) \times_X (\mathcal{U} \times_X \mathcal{U}) \xrightarrow{f(2)\times f(2)} G \times_X G \xrightarrow{\text{mult}} G.$$

Therefore for a given finite field k and a given k-valued point $z : \mathrm{Spec}(k) \to X$, if we denote by $\mathrm{Im}(2; k, z) \subset G_z(k)$ the image of

$$f(2)_z : \mathcal{U}_z \times_k \mathcal{U}_k \to G_z$$

on k-valued points, then the image of $f(4)_z$ on k-valued points is equal to the image of

$$\mathrm{Im}(2, z, k) \times \mathrm{Im}(2, z, k) \xrightarrow{\text{mult}} G_z(k).$$

But given any subset S (e.g., $\mathrm{Im}(2, z, k)$) of any finite group H (e.g., $G_z(k)$), the multiplication map $S \times S \to H$ is surjective if $\#(S) > (1/2)\,\#(H)$ (because if $h \in H$ cannot be written $s_1 s_2$ with s_1, s_2 in S, then the two subsets S and hS^{-1} of H are disjoint, so the cardinality of their union is $2\#(S) > \#(H)$, contradiction). Therefore (**) is implied by

(***) there exists an integer $D_2 \geq 1$ such that for any finite field k, and any k-valued point $z : \mathrm{Spec}(k) \to X[1/D_2]$, we have

$$\#\,\mathrm{Im}(2; k, z) > (1/2)\,\#(G_z(k)).$$

For any X-scheme $Y \to X$, the X-morphism

$$f(2) : \mathcal{U} \underset{X}{\times} \mathcal{U} \to G$$

gives by inverse image a Y-morphism

$$f(2)_Y : \mathcal{U}_Y \underset{Y}{\times} \mathcal{U}_Y \to G_Y.$$

Given an X-scheme $Y \to X$, and an integer $d \geq 1$, we denote by

$$(***)(Y, d)$$

the statement

for any finite field k and any k-valued point $z : \mathrm{Spec}(k) \to Y[1/d]$, we have

$$\#\,\mathrm{Im}(2; k, z) > (1/2)\,\#(G_z(k)).$$

Clearly we have

$$(***)(Y, d) \Longleftrightarrow (***)(Y^{\mathrm{red}}, d).$$

Moreover, if

$$Y^{\mathrm{red}} = \coprod_{i=1,\ldots,r} Z_i$$

is a partition of Y^{red} into a disjoint union of finitely many reduced locally closed subschemes, then

$$(***)(Z_i, d) \quad \text{for } i = 1, \ldots, r \Longleftrightarrow (***)(Y, d),$$

and if $d_1, \ldots, d_r$ are r possibly distinct integers each ≥ 1, then

$$(***)(Z_i, d_i) \quad \text{for } i = 1, \ldots, r \Longrightarrow (***)(Y, d_1 \ldots d_r).$$

Because X is of finite type over $\mathbf{Z}$, X^{red} certainly admits a finite constructible partition

$$X^{\mathrm{red}} = \coprod Z_i$$

where each Z_i is affine, irreducible, of finite type over $\mathbf{Z}$, and either smooth over $\mathbf{Z}$ or killed by some prime number p and smooth over $\mathbf{F}_p$. If Z_i is an $\mathbf{F}_p$-scheme, then $(***)(Z_i, p)$ holds trivially. So we are reduced to proving $(***)$ under the additional hypothesis

X is an irreducible smooth affine $\mathbf{Z}$-scheme of finite type.

By induction on $\dim X$, it suffices to prove, for given X as above, that

> there exists an open neighborhood $\mathcal{U} \subset X$ of the generic point of X such that $(***)(\mathcal{U}, d)$ holds for some integer $d \geq 1$.

Consider the diagram

$$\begin{array}{c} \mathcal{U} \times_X \mathcal{U} \\ \downarrow f(2) \\ G \\ \downarrow \\ X. \end{array}$$

12.4.2. Lemma. *The morphism $f(2)$ is smooth of relative dimension $\dim(G)$, and is surjective on geometric points.*

Proof. The factorization of $f(2)$ as

$$\mathcal{U} \times_X \mathcal{U} \xrightarrow{f \times f} G \times_X G \xrightarrow{\text{mult}} G$$

exhibits it as the composition of the etale map $\mathcal{U} \times \mathcal{U} \to G \times G$ with the smooth map $G \times G \to G$ of relative dimension $\dim(G)$. For any geometric point $z : \operatorname{Spec}(K) \to X$ with K an algebraically closed field,

$$f_z : \mathcal{U}_z \to G_z$$

is an etale morphism, whose image contains the identity section by construction. Therefore $f_z(\mathcal{U}_z) \subset G_z$ is a non-void open set in G_z. Now G_z is by hypothesis a connected smooth K-scheme, so it is irreducible. Therefore for any K-valued point $g \in G_z(K)$, the two non-void open sets of G_z

$$f_z(\mathcal{U}_z) \quad \text{and} \quad g(f_z(\mathcal{U}_z))^{-1}$$

must have a non-void intersection. As K is algebraically closed, this intersection contains a K-point, i.e., there exist u_1, u_2 in $\mathcal{U}_z(K)$ such that

$$f_z(u_1) = g(f_z(u_2))^{-1},$$

whence

$$g = f_z(u_1)\, f_z(u_2),$$

as required. ∎

Let us admit temporarily the truth of the following key lemma.

12.4.3. Key Lemma. *The set of points in G over which the fibre of*

$$f(2):\mathcal{U}\times_X\mathcal{U}\to G$$

is geometrically connected contains the generic point of G, i.e., the generic point of the generic fibre G_η of $G\to X$.

The assertion "makes sense," because G is normal and connected (being smooth with geometrically connected fibres over X which is itself normal and connected, being irreducible and smooth over $\mathbf{Z}$). Therefore G has a generic point, and as $G\to X$ is dominating, the generic point of G lies over the generic point of X.

By (EGA IV, 9.7.7), the set of points in G where the fibre of $f(2)$ is geometrically connected is constructible, so by the Key Lemma it contains an open set $V\subset G$ which is non-empty. The image of V in X under the structural map $G\to X$ is constructible, and it contains the generic point of X, so it contains some non-empty open set $\mathcal{U}\subset X$. By our earlier reductions, it suffices to prove that $(***)(\mathcal{U},d)$ holds for some integer $d\geq 1$. Further shrinking $\mathcal{U}$, we may assume that there exists a prime number l which is invertible on $\mathcal{U}$. Renaming $\mathcal{U}$ "X", we are thus reduced to proving that $(***)$ holds under the following additional hypotheses:

a) X is a scheme of finite typė over $\mathbf{Z}$,
b) a prime number l is invertible on X,
c) there is an open set $V\subset G$ such that,

(1) The intersection of V with every fibre of $G\to X$ is non-empty.
(2) Over V, the fibres of $f(2)$ are geometrically connected.

We will do this by the "Lang-Weil method." Let us denote by

$$Z=G-V$$

the closed complement of V in G, say with its reduced structure. By hypothesis, Z meets every fibre of $G\to X$ in a *proper* closed subset. Because the fibres of $G\to X$ are smooth and geometrically connected of dimension $\dim(G)$, the fibres of $Z\to X$ all have dimension strictly less than $\dim(G)$, while the fibres of $V=G-Z\to X$ are all smooth and geometrically connected of dimension $\dim(G)$.

Consider these morphisms

$$\begin{array}{ccc} Z & \xrightarrow{\dim(G)-2} & V = G - Z \\ \downarrow \alpha & & \downarrow \beta \\ X & \xrightarrow{\dim(G)-2} & X, \end{array}$$

and the cohomology sheaves on X

$$R^i\alpha_!\mathbf{Q}_l, \qquad R^i\beta_!\mathbf{Q}_l.$$

These are each constructible sheaves on X, mixed of weight $\leq i$ by Weil II, whose formation commutes with passage to fibres. We have, for dimension reasons,

$$\begin{aligned} R^i\alpha_!\mathbf{Q}_l &= 0 \qquad \text{for } i > 2\dim(G) - 2, \\ R^i\beta_!\mathbf{Q}_l &= 0 \qquad \text{for } i > 2\dim(G). \end{aligned}$$

Because β is smooth and surjective with geometrically connected fibres of dimension $\dim(G)$, the trace morphism provides an isomorphism of $\mathbf{Q}_l$-sheaves on X

$$R^{2\dim(G)}\beta_!\mathbf{Q}_l \xrightarrow{\sim} \mathbf{Q}_l(-\dim(G)).$$

By constructibility, there exist integers $A > 0,\ B > 0$, such that for any geometric point $\bar{x}$ of X,

$$\sum_{i=0}^{2\dim(G)-2} \operatorname{rank}(R^i\alpha_!\mathbf{Q}_l)_{\bar{x}} \leq A$$

$$\sum_{i=0}^{2\dim(G)-1} \operatorname{rank}(R^i\beta_!\mathbf{Q}_l)_{\bar{x}} \leq B.$$

Therefore if k is any finite field, and if

$$z : \operatorname{Spec}(k) \to X$$

is any k-valued point of X, the cardinalities of the partition

$$G_z(k) = Z_z(k) \coprod V_z(k)$$

satisfy

$$\begin{aligned} \#V_z(k) &\geq (\#(k))^{\dim(G)} - B(\#(k))^{\dim(G)-1/2} \\ \#Z_z(k) &\leq A(\#(k))^{\dim(G)-1}. \end{aligned}$$

So there exists a constant C, depending only on A and B (explicitly, $\sqrt{C} = (B/2) + \sqrt{A + (1/4)B^2})$) such that

$$\text{if } \#(k) > C, \qquad \text{then } \#(V_z(k) > \#Z_z(k),$$
$$\text{i.e., } \#V_z(k) > (1/2)\,\#G_z(k).$$

Now consider the morphism γ defined by the cartesian diagram

$$\begin{array}{ccc} \mathcal{U} \underset{X}{\times} \mathcal{U} & \hookleftarrow & f(2)^{-1}(V) \\ \downarrow f(2) & & \downarrow \gamma \\ G & \hookleftarrow & V \end{array}$$

By construction of V, the morphism γ is smooth and surjective with geometrically connected fibres of dimension $\dim(G)$. So we have

$R^i\gamma_!\mathbf{Q}_l$ is constructible on V, mixed of weight $\leq i$, zero for $i > 2\dim(G)$,

and

$$R^{2\dim(G)}\gamma_!\mathbf{Q}_l \xrightarrow{\sim} \mathbf{Q}_l(-\dim(G)).$$

So if we denote by $D \geq 1$ a bound for the sum of the ranks of the $R^i\gamma_!\mathbf{Q}_l$ for $i \leq 2\dim(G) - 1$, we have

for any finite field k, and any k-valued point $v : \mathrm{Spec}(k) \to V$, the k-scheme $\gamma^{-1}(v)$ has

$$\#(\gamma^{-1}(v))(k) \geq (\#(k))^{\dim(G)} - D(\#(k))^{\dim(G)-1/2}.$$

In particular, we have

if $\#(k) > D^2$, then for any k-valued point $v \in V(k)$, the fibre $\gamma^{-1}(v)$ has a k-rational point.

Thus if $\#(k) > \max(C, D^2)$, then for any $z \in X(k)$ we have

$$\begin{cases} \#V_z(k) > (1/2)\,\#G_z(k), \\ \mathrm{Im}((\mathcal{U}_z \underset{k}{\times} \mathcal{U}_z)(k) \xrightarrow{f(2)_z} G_z(k)) \text{ contains } V_z(k); \end{cases}$$

so in particular if $\#(k) > \max(C, D^2)$, then we have

$$\#\,\mathrm{Im}(2; z, k) > (1/2)\,\#G_z(k).$$

Therefore we have proven that $(***)$ holds on X, with D_2 the product of all prime numbers $l \leq \max(C, D^2)$ (because if $\mathrm{char}(k)$ is prime to D_2, then $\mathrm{char}(k) > \max(C, D^2)$, and $\#(k) \geq \mathrm{char}(k)$). This concludes the proof of the theorem, modulo the Key Lemma. Because the generic point η of X is of characteristic zero, the geometric generic fibre $G_{\bar{\eta}}$ is by hypothesis simply connected. Therefore the Key Lemma 12.4.2 results from the following variant in which $G_{\bar{\eta}}$ becomes "G" and in which the two copies of $\mathcal{U}_{\bar{\eta}}$ becomes U and V respectively. ∎

12.4.4. Key Lemma (O. Gabber). *Let K be an algebraically closed field, G a smooth connected K-group scheme of finite type, and*

$$f : U \to G, \qquad g : V \to G$$

two etale morphisms. Suppose that

(1) *U and V are each non-empty smooth connected K-schemes of finite type,*
(2) *G is simply connected (in the sense that it has no non-trivial finite etale coverings).*

Then the morphism $\pi : U \underset{K}{\times} V \to G$, defined as the composite

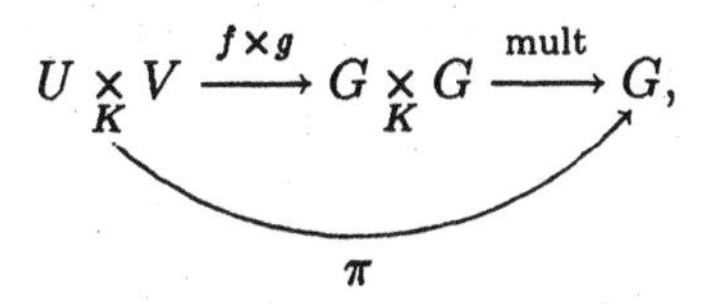

$$U \underset{K}{\times} V \xrightarrow{f\times g} G \underset{K}{\times} G \xrightarrow{\text{mult}} G,$$

is smooth of relative dimension $\dim(G)$, *surjective on geometric points, and its generic fibre is geometrically connected.*

Proof. Just as in the proof of 12.4.1, we see that the morphism π is smooth of relative dimension $\dim(G)$, and surjective on geometric points. Let us denote by ξ the generic point of G, i.e., $\xi = \operatorname{Spec}(K(G))$. To show that $\pi^{-1}(\xi)$ is geometrically connected, it suffices to show that $\pi^{-1}(\xi) \underset{K(G)}{\otimes} L$ is connected for any finite extension L of $K(G)$. As connectedness is invariant under purely inseparable field extension, it suffices to prove that $\pi^{-1}(\xi) \otimes L$ is connected for any finite separable extension L of $K(G)$. So if we fix

$$\begin{aligned} L_\infty &= \text{ a separable closure of } K(G) \\ \operatorname{Gal} &= \operatorname{Gal}(L_\infty/K(G)) \\ \bar{\xi} &= \operatorname{Spec}(L_\infty) \to \xi, \end{aligned}$$

it suffices to prove that $\pi^{-1}(\bar{\xi})$ is connected.

Let us denote by

$$S = \text{ the set of connected components of } \pi^{-1}(\bar{\xi}).$$

Then S is a finite set on which $\mathrm{Gal}(L_\infty/K(G))$ acts continuously. Because $\pi^{-1}(\xi)$ is connected (because irreducible, as a localization of the irreducible scheme $U \underset{K}{\times} V$), Gal acts *transitively* on S.

Let us now fix a point $v \in V(k)$, and consider the commutative diagram

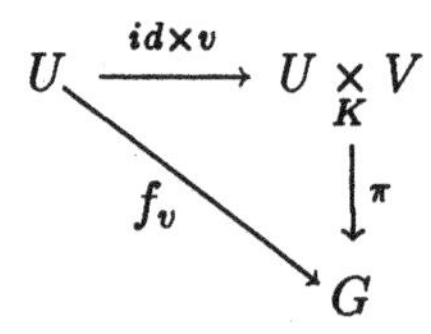

where $f_v : U \to G$ is the *etale* morphism $u \to f(u)\,g(v)$. Let us denote by

$$\begin{aligned} T_v = &\text{ the set of connected components of } f_v^{-1}(\bar{\xi}) \\ = &\text{ the set of points of } f_v^{-1}(\bar{\xi}). \end{aligned}$$

Then just as above T_v is a finite set upon which $\mathrm{Gal}(L_\infty/K(G))$ acts continuously and transitively and the above commutative diagram, read over $\bar{\xi}$, gives a $\mathrm{Gal}(L_\infty/K(G))$-equivariant morphism $f_v^{-1}(\bar{\xi}) \to \pi^{-1}(\bar{\xi})$, which induces on the respective sets of connected components a $\mathrm{Gal}(L_\infty/K(G))$-equivariant map

$$T_v \to S.$$

Because Gal acts transitively on S, and T_v is non-empty, this map is automatically surjective:

$$T_v \twoheadrightarrow S.$$

Now the original morphism $f : U \to G$ is etale of finite type, so there exists an open neighborhood $U_1 \subset G$ of the generic point ξ of G over which the induced morphism $f^{-1}(U_1) \to U_1$ is finite etale. For this non-empty open set $U_1 \subset G$, we have, by translation,

$$f_v : U \to G \text{ is finite etale over } U_1 g(v) \subset G.$$

Therefore T_v as representation of $\mathrm{Gal}(L_\infty/K(G))$ is the generic fibre of a finite etale sheaf of sets on $U_1 g(v)$ (namely of the sheaf represented by the finite etale map $f_v^{-1}(U_1 g(v)) \to U_1 g(v)$). Because S is a *quotient* of T_v as finite Gal-set, it follows that

> for every $v \in V(K)$, there exists a finite etale sheaf of sets on $U_1 g(v)$ whose generic fibre is S.

Because G is normal and connected, the sheaves in question are unique on the open sets $U_1 g(v)$. For variable $v \in V(K)$, the open sets $U_1 g(v)$ cover G

(because U_1 and $g(V)$ are two non-empty open sets in G), so we may patch together these sheaves to conclude

> there exists a finite etale sheaf of sets on G whose generic fibre is S.

But as G is simply connected, any such sheaf on G is constant, i.e., $\mathrm{Gal}(L_\infty/K(G))$ acts trivially on S. As it also acts transitively on S, we must have $\#(S) = 1$. ∎

12.5. Analysis of The Special Case $p = 2$, n odd $\neq 7$

12.5.1. In this section we fix data (n, p, q, E, ψ) as in the first section, with

$$p = 2, \quad n \text{ odd }, \; n \neq 3, \quad E = \mathbf{Q}.$$

Let us temporarily fix a prime number $l \neq 2$. Because $E = \mathbf{Q}$, we write $\mathcal{F}_l$ instead of $\mathcal{F}_\lambda$. Because $\mathcal{F}_l$ carries a unique-up-to-$\mathbf{Z}_l^\times$ symmetric bilinear autoduality

$$\langle , \rangle : \mathcal{F}_l \underset{\mathbf{Z}_l}{\otimes} \mathcal{F}_l \to \mathbf{Z}_l$$

for $p = 2$, n odd, the monodromy representation ρ_l of $\mathcal{F}_l$ at the chosen geometric point $\bar{x}$ of $\mathbf{G}_m \otimes \mathbf{F}_q$ is a homomorphism

$$\rho_l : \pi_1^{\mathrm{arith}} \to \mathrm{SO}(n, \mathbf{Z}_l),$$

the SO with respect to the symmetric bilinear autoduality $\langle , \rangle$. Let us denote by

$$\mathrm{Spin}(n)$$

the corresponding spin group, viewed as a smooth $\mathbf{Z}_l$-groupscheme. Then $\mathrm{Spin}(n)$ is a finite etale $\{\pm 1\}$-covering of $\mathrm{SO}(n)$, and on $\mathbf{Z}_l$-valued points we have an exact sequence of groups

$$1 \to \{\pm 1\} \to \mathrm{Spin}(n, \mathbf{Z}_l) \to \mathrm{SO}(n, \mathbf{Z}_l).$$

In fact the index in $\mathrm{SO}(n, \mathbf{Z}_l)$ of the image of $\mathrm{Spin}(n, \mathbf{Z}_l)$ is two (by Hensel's Lemma, this index is equal to its $\mathbf{F}_l$-analogue, and the groups $\mathrm{Spin}(n, \mathbf{F}_l)$ and $\mathrm{SO}(n, \mathbf{F}_l)$ have the same orders). Thus we have an exact sequence of groups

$$1 \to \{\pm 1\} \to \mathrm{Spin}(n, \mathbf{Z}_l) \to \mathrm{SO}(n, \mathbf{Z}_l) \xrightarrow{\chi_l} \{\pm 1\} \to 1.$$

12.5.2. Lemma. *The character* $\chi_l : \mathrm{SO}(n, \mathbf{Z}_l) \to \{\pm 1\}$ *is trivial on* $\rho_l(\pi_1^{\mathrm{geom}})$.

Proof. The composite character $\chi_l \circ \rho_l : \pi_1^{\text{geom}} \to \pm 1$ is tame at zero, because ρ_l is unipotent on I_0. By 1.19, χ_l is trivial on $\rho_l(P_\infty)$. Therefore $\chi_l \circ \rho_l$ is tame at ∞. But the "tame at zero and ∞" quotient of π_1^{geom} is $\prod_{l \neq p} \mathbf{Z}_l(1)$, a group of order prime to $p = 2$, so it has no non-trivial characters to $\{\pm 1\}$. ∎

12.5.3. Theorem (O. Gabber). *If $p = 2$, $n \neq 3$ odd and $n \neq 7$, then for all sufficiently large l we have*

$$\left.\begin{array}{l}\textit{the subgroup in } \mathrm{SO}(n, \mathbf{Z}_l) \\ \textit{generated by } \rho_l(I_0) \textit{ and} \\ \textit{by } \rho_l(I_\infty)\end{array}\right\} = \rho_l(\pi_1^{\text{geom}}) = \textit{ the image of } \mathrm{Spin}(n, \mathbf{Z}_l).$$

Proof. By the previous lemma, $\rho_l(\pi_1^{\text{geom}})$ lies in this image. Now consider the moduli space $\mathcal{M}$ attached to the data (n, p, q, E, ψ); it is a $\mathbf{Z}[1/2]$-scheme of finite type, over which we have a free $\mathcal{O}_{\mathcal{M}}$-module $\mathbf{V}$ of rank n, together with a symmetric bilinear autoduality

$$\langle , \rangle : \mathbf{V} \underset{\mathcal{O}_{\mathcal{M}}}{\otimes} \mathbf{V} \to \mathcal{O}_{\mathcal{M}},$$

the corresponding $\mathcal{M}$-groupscheme $\mathrm{SO}(n)$, a unipotent element

$$U_{\text{univ}} \in \mathrm{SO}(n)(\mathcal{M})$$

and a homomorphism

$$\psi_{\text{univ}} : \Gamma_\infty \to \mathrm{SO}(n)(\mathcal{M}),$$

satisfying various conditions as in 12.3.2.

Over $\mathcal{M}$, we can form the spin group $\mathrm{Spin}(n)$ attached to the symmetric bilinear autoduality on $\mathbf{V}$. Because $\mathrm{Spin}(n)$ is a finite etale $\{\pm 1\}$-covering of $\mathrm{SO}(n)$, they have canonically isomorphic Lie algebras. So once we invert $(n-1)!$ on $\mathcal{M}$, the nilpotent elements of Lemma 12.3.8

$$\log(\gamma_i U_{\text{univ}} \gamma_i^{-1}) = \gamma_i \log(U_{\text{univ}}) \gamma_i^{-1} \in \underline{\mathrm{Lie}}(\mathrm{SO}(n))$$

make sense in $\mathrm{Lie}(\mathrm{Spin}(n))$. Because $\mathrm{Spin}(n)$ has a faithful linear representation of dimension $2^{\frac{n-1}{2}}$, its "spin represention," once we further invert $(2^{\frac{n-1}{2}} - 1)!$, we have well defined $\mathcal{M}$-homomorphisms

$$\begin{aligned} \mathbf{G}_a &\longrightarrow \mathrm{Spin}(n) \\ T &\longmapsto \exp(T \log(\gamma_i U_{\text{univ}} \gamma_i^{-1})), \end{aligned}$$

which land in the unipotent elements of $\mathrm{Spin}(n)$.

By Lemmas 12.3.8-9, once we Zariski localize on $\mathcal{M}$ and invert a large integer, the nilpotent elements

$$\log(\gamma_i U_{\mathrm{univ}} \gamma_i^{-1})$$

will form a basis of $\underline{\mathrm{Lie}}(\mathrm{Spin}(n))$ for suitable $\gamma_i \in \Gamma$.

Because $\mathrm{Spin}(n)$ is simply connected over algebraically closed fields of characteristic zero, we may apply Theorem 12.4.0, to find in particular that for $l \gg 0$, and any $\mathbf{Z}_l$-valued point $z \in \mathcal{M}(\mathbf{Z}_l)$, the $\mathbf{Z}_l$-points of the corresponding Spin group are generated by finitely many of the one-parameter subgroups

$$\exp(\mathbf{Z}_l \log(\gamma_i U_{\mathrm{univ}} \gamma_i^{-1})(z)) \subset \mathrm{Spin}(n, \mathbf{Z}_l).$$

Apply this to the points $z \in \mathcal{M}(\mathbf{Z}_l)$ provided by the monodromy representations ρ_l of the $\mathcal{F}_l$. The images in $\mathrm{SO}(n, \mathbf{Z}_l)$ of the one-parameter subgroups in question all lie in $\rho_l(\pi_1^{\mathrm{geom}})$, and they are all conjugates of $\rho_l(I_0)$ by elements of the subgroup generated by $\rho_l(I_0)$ and by $\rho_l(I_\infty)$. In particular, for $l \gg 0$ we have

$$\begin{matrix}\text{image of } \mathrm{Spin}(n, \mathbf{Z}_l) \\ \text{in } \mathrm{SO}(n, \mathbf{Z}_l)\end{matrix} \subset \begin{matrix}\text{subgroup of } \mathrm{SO}(n, \mathbf{Z}_l) \text{ generated} \\ \text{by } \rho_l(I_0) \text{ and by } \rho_l(I_\infty).\end{matrix}$$

As $\rho_1^{\mathrm{geom}})$ always lies in the image of $\mathrm{Spin}(n, \mathbf{Z}_l)$ in $\mathrm{SO}(n, \mathbf{Z}_l)$, by Lemma 12.5.1, the above inclusion forces equality throughout. ∎

12.6. Analysis of The Special Case $p = 2$, $n = 7$

12.6.1. In this section we fix data (n, p, q, E, ψ) with

$$p = 2, \quad n = 7, \quad E = \mathbf{Q}.$$

For each prime number $l \neq 2$, we have the monodromy representation of $\mathcal{F}_l$,

$$\rho_l : \pi_1^{\mathrm{arith}} \to \mathrm{SO}(7, \mathbf{Z}_l),$$

the SO with respect to the unique-up-to-$\mathbf{Z}_l^\times$ symmetric bilinear autoduality $\langle , \rangle$ (cf. 4.1.11, 4.2.1).

We have proven that the Zariski closure of $\rho_l(\pi_1^{\mathrm{geom}})$ in $\mathrm{SO}(7)$, as algebraic group over $\mathbf{Q}_l$ is G_2 viewed as lying in $\mathrm{SO}(7)$ by its unique irreducible representation std_7 of dimension seven, and we have proven that the subgroup of $\rho_l(\pi_1^{\mathrm{geom}})$ generated by $\rho_l(I_0)$ and by $\rho_l(I_\infty)$ is already Zariski dense in this G_2.

From the representation theory of G_2, we know that $\Lambda^3(\mathrm{std}_7)$ has a one-dimensional space of invariants under G_2 (cf. 11.11.9).

12.6.2. Lemma. *Over an algebraically closed field of characteristic zero, the Zariski closed subgroup $G_2 \subset \mathrm{SO}(7)$ is exactly the fixer in* SO(7) *of any non-zero invariant of G_2 acting on $\Lambda^3(\mathrm{std}_7)$.*

Proof. If we denote this fixer by K, we have $G_2 \subset K \subset \mathrm{SO}(7)$. Because already G_2 contains a unipotent U_0 with a single Jordan block in std_7, we may apply the Axiomatic Classification Theorem to K, with unipotent U and with $\Gamma_\infty = G_2$. Then $K = G_2$ or $K = \mathrm{SO}(7)$, and the second case is impossible because $\Lambda^3(\mathrm{std}_7)$ is SO(7)-irredeucible. ∎

Because the subgroups $\rho_l(I_0)$ and $\rho_l(I_\infty)$ together generate a Zariski-dense subgroup of G_2, it follows that

> in $\Lambda^3((\mathcal{F}_l)_{\bar{\eta}})$, the $\mathbf{Z}_l$-module of simultaneous invariants under $\rho_l(I_0)$ and under $\rho_l(I_\infty)$ is a free $\mathbf{Z}_l$-module of rank 1; if we denote by
>
> $$T_l \in \Lambda^3((\mathcal{F}_l)_{\bar{\eta}})$$
>
> a $\mathbf{Z}_l$-basis of this space of invariants, then G_2 is the fixer of T_l in SO(7), as $\mathbf{Q}_l$-algebraic group.

12.6.3. Theorem (O. Gabber). *If $p = 2$ and $n = 7$, then for all sufficiently large l we have*

$$\left.\begin{array}{l}\textit{the subgroup generated by } \rho_l(I_0)\\ \textit{and by } \rho_l(I_\infty)\end{array}\right\} = \rho_l(\pi_1^{\mathrm{geom}}) = \left\{\begin{array}{l}\textit{the fixer of } T_l\\ \textit{in } \mathrm{SO}(7, \mathbf{Z}_l).\end{array}\right.$$

Proof. We will build a modified version $\mathcal{M}'$ of the moduli problem $\mathcal{M}$. For any $\mathbf{Z}[1/2]$-algebra R, we define $\mathcal{M}'(R) =$ the set of quadruples $(\langle,\rangle, \psi, U, T)$ where

$\langle\ ,\ \rangle, : R^n \times R^n \to R$ is a symmetric R-linear autoduality of R^n with itself.

$\psi : \Gamma_\infty \to \mathrm{SO}(n, R)$ is a group homomorphism, with $\mathrm{trace}(\psi(\gamma)) = \mathrm{char}_\rho(\gamma)$ for all $\gamma \in \Gamma_\infty$.

$U \in \mathrm{SO}(n, R)$ is an element satisfying

a) $(U-1)^n = 0$,

b) Zariski locally on R, some $(n-1) \times (u-1)$ minor of $U - 1$ is invertible.

$T \in \Lambda^3(R^n)$is a nowhere-vanishing section which is invariant under $\psi(\Gamma_\infty)$ and under U.

Just as in 12.3.5, we see that $\mathcal{M}'$ is represented by a $\mathbf{Z}[1/2]$-scheme of finite type, still noted $\mathcal{M}'$, over which we have the "universal" data

$$(\langle , \rangle_{\text{univ}}, \psi_{\text{univ}}, U_{\text{univ}}, T_{\text{univ}}).$$

We denote by

$$G \subset \mathrm{SO}(\mathbf{V}, \langle , \rangle_{\text{univ}})$$

the closed $\mathcal{M}'$-subgroup-scheme of $\mathrm{SO}(n)$ defined by

$$\begin{array}{l} G = \text{ the fixer of } T_{\text{univ}} \text{ in } \mathrm{SO}(\mathbf{V}, \langle , \rangle_{\text{univ}}). \\ \downarrow \\ \mathcal{M}'. \end{array}$$

For any geometric point $z : \mathrm{Spec}(K) \to \mathcal{M}'$ with K an algebraically closed field of characteristic zero, the K-algebraic group G_z is isomorphic to G_2, by the Axiomatic Classification Theorem (which makes it either G_2 or $\mathrm{SO}(7)$) and the fact that it has a non-zero invariant in $\Lambda^3(\mathrm{std}_7)$ (which rules out $\mathrm{SO}(7)$). In particular, the characteristic-zero geometric fibres of $G \to \mathcal{M}'$ are all smooth, connected and simply connected of constant dimension $\dim(G_2)$.

12.6.4. Lemma. *There exists an integer $D \geq 1$ such that $(\mathcal{M}'[1/D])^{\text{red}}$ is flat over $\mathbf{Z}$, and such that over $(\mathcal{M}'[1/D])^{\text{red}}$ the inverse image of $G \to \mathcal{M}'$ is smooth with geometrically connected fibres of constant dimension* $\dim(G_2)$.

Proof. Because $\mathcal{M}'$ is a scheme of finite type over $\mathbf{Z}$, and $G \to \mathcal{M}'$ is a morphism of finite type whose characteristic-zero geometric fibres are all smooth and connected of constant dimension $\dim(G)$, it follows by standard constructibility arguments that there exists an integer $D \geq 1$ such that $(\mathcal{M}'[1/D])^{\text{red}}$ is flat over $\mathbf{Z}$, and a finite constructible partition

$$(\mathcal{M}'[1/D])^{\text{red}} = \coprod_{i=1,\ldots,r} X_i$$

where each X_i is a smooth connected $\mathbf{Z}$-scheme of finite type, over which the induced group-scheme $G_{X_i} \to X_i$ is smooth with geometrically connected fibres of constant dimension $\dim(G_2)$.

We show that the lemma "works" with this choice of D. For if we denote by X the scheme $(\mathcal{M}'[1/D])^{\mathrm{red}}$, then the inverse image of $G \to \mathcal{M}'$ over X,

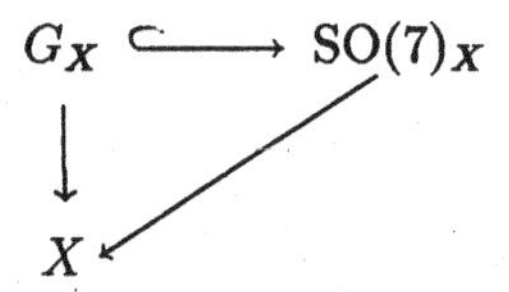

has all of its geometric fibres smooth and connected of constant dimension $\dim(G_2)$. We need only show that $G_X \to X$ is flat. By the valuative criterion for flatness (EGA IV, 11.8.1), it suffices to prove that after any base-change $Y \to X$ with Y the spectrum of a discrete valuation ring, the induced Y-group-scheme

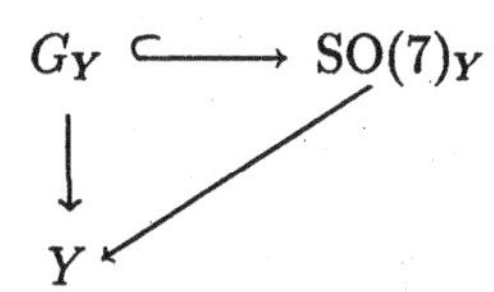

is flat over Y. Let us denote by η and s respectively the generic and the closed point of Y, and by

$$\begin{array}{l} \tilde{G}_Y = \text{ the schematic closure of } G_\eta \text{ in } \mathrm{SO}(7)_Y. \\ \big\downarrow \\ Y \end{array}$$

Concretely, a global function on the affine scheme $\mathrm{SO}(7)_Y$ vanishes on $\tilde{G}_Y$ if and only if it vanishes on G_η. Therefore the ideal defining G_Y in $\mathrm{SO}(7)_Y$ vanishes on $\tilde{G}_Y$, whence $\tilde{G}_Y$ is a closed Y-subgroup-scheme of G_Y. Therefore the special fibre $(\tilde{G}_Y)_s$ is a closed subgroup-scheme of $G_Y)_s$. Because $\tilde{G}_Y \to Y$ is flat, with generic fibre of dimension $\dim(G_2)$, the special fibre $(\tilde{G}_Y)_s$ is purely of dimension $\dim(G_2)$, (cf. Alt–Kl, Chapter VI, 2.10) and is non-empty, because it is a group-scheme. But $(G_Y)_s$ is smooth and geometrically connected of dimension $\dim(G_2)$, so any closed subscheme of the same dimension must be all of it:

$$(\tilde{G}_Y)_s = (G_Y)_s.$$

This implies that in fact

$$\tilde{G}_Y = G_Y.$$

(For their affine rings sit in a short exact sequence of $\mathcal{O}_Y$-modules

$$0 \to \mathrm{Ker} \to \mathrm{Aff}(G_Y) \to \mathrm{Aff}(\tilde{G}_Y) \to 0$$

with the last term flat over $\mathcal{O}_Y$, so reducing mod π, π a uniformizing parameter for $\mathcal{O}_Y$, the equality of special fibres gives

$$\mathrm{Ker}/\pi\,\mathrm{Ker} = 0, \qquad \text{so } \mathrm{Ker} = \pi\,\mathrm{Ker} = \pi^2\,\mathrm{Ker} = \cdots.$$

But the general fibres are also equal, so

$$\mathrm{Ker}[1/\pi] = 0.$$

Because Ker is an ideal in $\mathrm{Aff}(G_Y)$ it is finitely generated as $\mathrm{Aff}(G_Y)$-module, whence for some large integer N, π^N kills each generator, whence

$$\pi^N\,\mathrm{Ker} = 0.$$

There $\mathrm{Ker} = \pi\,\mathrm{Ker} = \cdots = \pi^N\,\mathrm{Ker} = 0$, whence $\tilde{G}_Y = G_Y$ as claimed.) Therefore G_Y is flat over Y. ∎

Thanks to the lemma, we may apply Theorem 12.4.0 to the inverse image of $G \to \mathcal{M}'$ on $(\mathcal{M}'[1/D])^{\mathrm{red}}$, and (once we invert a large integer and Zariski localize), a set of nilpotent elements

$$\log(\gamma_i U_{\mathrm{univ}} \gamma_i^{-1})$$

which form a basis of $\underline{\mathrm{Lie}}(G)$, for suitable $\gamma_i \in \Gamma$. Because the ring $\mathbf{Z}_l$ is reduced, all $\mathbf{Z}_l$-valued points of $\mathcal{M}'[1/D]$ factor through $(\mathcal{M}'[1/D])^{\mathrm{red}}$. So we may apply the conclusion of Theorem 12.4.0 to the $\mathbf{Z}_l$-valued points defined by the monodromy ρ_l of $\mathcal{F}_l$ and the chosen invariant T_l. ∎

12.6.5. Remark. As a by-product of the proof, we find that for $l \gg 0$, the $\mathbf{Z}_l$-group-scheme defined as the fixer of T_l in $\mathrm{SO}(7)_{\mathbf{Z}_l}$ is in fact smooth over $\mathbf{Z}_l$ with geometrically connected fibres (because for l prime to the unspecified integer $D \geq 1$, this group is the inverse image of $\mathbf{Z}_l$ of the universal group G over $(\mathcal{M}'[1/D])^{\mathrm{red}}$, by the $\mathbf{Z}_l$-valued point of $\mathcal{M}'$ defined by ρ_l and by T_l). Is there a more direct way of seeing this?

CHAPTER 13

Equidistribution of "Angles" of Kloosterman Sums

13.0. Uniform description of the space of conjugacy classes and its Haar measure

Fix an integer $n \geq 2$, and an n-dimensional complex vector space V. If n is *even*, fix an alternating **C**-linear autoduality $\langle , \rangle$ of V. Denote by G the complex Lie group

$$G = \begin{cases} \mathrm{SL}(V) \simeq \mathrm{SL}(n, \mathbf{C}) & \text{if } n \text{ odd} \\ \\ \mathrm{Sp}(V, \langle\, , \rangle) \simeq \mathrm{Sp}(n, \mathbf{C}) & \text{if } n \text{ even.} \end{cases}$$

Let

$$K \subset G$$

be a maximal compact subgroup of G. We denote by

$$P(n) \subset \mathbf{C}[T]$$

the set of all those monic polynomials $f(T)$ of degree n which satisfy the following conditions:

if n odd: there exist $\alpha_1, \ldots, \alpha_n$ in **C** with $\prod_i \alpha_i = 1$,

$$\| \alpha_i \| = 1 \quad \text{for } i = 1, \ldots, n, \quad \text{and } f(T) = \prod_i (T - \alpha_i).$$

if n even, say $n = 2g$; there exist $\alpha_1, \ldots, \alpha_g$ in **C** with

$$\| \alpha_i \| = 1 \quad \text{for } i = 1, \ldots, g, \quad \text{and } f(T) = \prod_{i=1}^{g} ((T - \alpha_i)(T - \alpha_i^{-1})).$$

When n is even, by writing $\alpha_i + \alpha_i^{-1} = 2\cos\theta_i$ with $\theta_i \in [0, \pi]$, we may rephrase this condition as:

if $n = 2g$ is even; there exist real numbers $\theta_1, \ldots, \theta_g$ in $[0, \pi]$

$$\text{such that } f(T) = \prod_{i=1}^{g} (T^2 - 2\cos(\theta_i)T + 1).$$

13.1. Lemma. *Notations as above, the map "characteristic polynomial"*

$$K \longrightarrow \mathbf{C}[T]$$
$$k \longmapsto \det(T - k|V)$$

maps K onto $P(n)$, and two elements of K are conjugate in K if and only if they have the same image in $P(n)$.

Proof. To see that K maps to and onto $P(n)$, we argue as follows. Because K is a compact subgroup of $\mathrm{GL}(V)$, any element $k \in K$ is semi-simple with all eigenvalues of absolute value one. Because all maximal compact subgroups of G are conjugate, any semi-simple element of G with all eigenvalues of absolute value one is G-conjugate to an element of K. Because the characteristic polynomial is invariant by G-conjugation (even by $\mathrm{GL}(V)$-conjugation), we are "reduced" to showing that any semi-simple element of G with eigenvalues of absolute value one has its characteristic polynomial in $P(n)$, and that every element of $P(n)$ is such a characteristic polynomial. For n odd, we have $G = \mathrm{SL}(n)$, and any semi-simple element is G-conjugate to a diagonal matrix of determinant one, so the above assertion is obvious in this case. For n even, say $n = 2m$, we have $G = \mathrm{Sp}(V, \langle,\rangle)$. If $g \in G$ is semi-simple, there exists a basis $v_1, \ldots, v_n$ of V and $\alpha_1, \ldots, \alpha_n \in \mathbf{C}^\times$ such that $gv_i = \alpha_i v_i$ for all i. Because $g \in \mathrm{Sp}$, we have

$$\langle v_i, v_j\rangle = \langle gv_i, gv_j\rangle = \alpha_i\alpha_j\langle v_i, v_j\rangle,$$

and therefore

$$\langle v_i, v_j\rangle \neq 0 \Longrightarrow \alpha_i\alpha_j = 1.$$

Because $\langle,\rangle$ is alternating and non-degenerate, we have $\langle v_1, v_1\rangle = 0$ but $\langle v_1, v_i\rangle \neq 0$ for some i. Renumbering and scaling the v_i, we may assume $\langle v_1, v_{m+1}\rangle = 1$, and consequently $\alpha_1\alpha_{m+1} = 1$. Then the subspace $W = \mathbf{C}v_1 + \mathbf{C}v_{m+1}$ of V is g-stable, and the restriction to it of $\langle,\rangle$ is non-degenerate. Therefore $V = W \oplus \perp(W)$ is a g-stable orthogonal direct sum decomposition. Repeating the above considerations with a diagonalization of g on $\perp(W)$, we conclude that V admits a basis $v_1, \ldots, v_n$ in which g is diagonal, $gv_i = \alpha_i$, and for which

$$\langle v_i, v_j\rangle \neq 0 \Longleftrightarrow |i - j| = n/2 = m.$$
$$\langle v_i, v_{m+i}\rangle = 1 \quad \text{for } i = 1, \ldots, n/2 = m.$$

In such a "standard symplectic" basis of V, the matrix of g is

$$\mathrm{Diag}(\alpha_1, \ldots, \alpha_{n/2}, \alpha_1^{-1}, \ldots, \alpha_{n/2}^{-1}).$$

This shows that if our semi-simple g has all eigenvalues of absolute value one, then its characteristic polynomial lies in $P(n)$. Applying the preceding discussion to $g = id$, we see that V admits a "standard symplectic" base $v_1, \ldots, v_n$ as above. With respect to such a base, for any $\alpha_1, \ldots, \alpha_{n/2}$ in $\mathbf{C}^\times$, the diagonal matrix $\mathrm{Diag}(\alpha_1, \ldots, \alpha_{n/2}, \alpha_1^{-1}, \ldots, \alpha_{n/2}^{-1})$ lies in $\mathrm{Sp}(V, \langle,\rangle)$. Therefore every element of $P(n)$ is the characteristic polynomial of a semi-simple element of G with all eigenvalues of absolute value one, as required.

Given two elements of K with the same characteristic polynomial, we must show they are K-conjugate. By representation theory (cf. 3.2), it suffices to show they are G-conjugate. So we are reduced to showing that two semi-simple elements g, g' of G with the same eigenvalues $\alpha_1, \ldots, \alpha_n$ are G-conjugate. To see this, choose bases $v_1, \ldots, v_n$ and $v'_1, \ldots, v'_n$ of V which diagonalize g and g' respectively, $gv_i = \alpha_i v_i, gv'_i = \alpha_i v'_i$ for $i = 1, \ldots, n$, and which are both "standard symplectic" bases of V for n even (resp. are bases giving equal volume elements $v_1 \wedge \cdots \wedge v_n = v'_1 \wedge \ldots v'_n$ in $\Lambda^n V$, for n odd). Then the automorphism γ of V which maps v_i to v'_i for $i = 1, \ldots, n$ lies in G, and it conjugates g into g' (i.e., $g' = \gamma g \gamma^{-1}$). ∎

13.2. Remark. Here is another way to see that for both $\mathrm{SL}(n, \mathbf{C})$ and $\mathrm{Sp}(n, \mathbf{C})$, the coefficients of the characteristic polynomial, i.e., the traces of the exterior powers $\Lambda^i(\mathrm{std})$ of the standard representation separate K-conjugacy classes. By Peter–Weyl, it suffices to show that the character of any finite-dimensional representation is a $\mathbf{C}$-polynomial in the $\mathrm{trace}(\Lambda^i(\mathrm{std}))$'s. In fact, for SL and Sp any such character is a $\mathbf{Z}$-polynomial in the $\mathrm{trace}(\Lambda^i(\mathrm{std}))$'s, i.e., the $\Lambda^i(\mathrm{std})$ generate the representation ring as $\mathbf{Z}$-algebra. To see this, it suffices to show that every fundamental representation is a $\mathbf{Z}$-polynomial in the $\Lambda^i(\mathrm{std})$, because the representation ring is a polynomial ring over $\mathbf{Z}$ in the fundamental representations. For $\mathrm{SL}(n)$, the $n-1$ fundamental representations are given by $\omega_i = \Lambda^i(\mathrm{std})$ for $i = 1, \ldots, n-1$. For $\mathrm{Sp}(n)$, the $n/2$ fundamental representations are given virtually by $\omega_1 = \mathrm{std}$, $\omega_2 = \Lambda^2(\mathrm{std}) - \mathbf{1}$, and $\omega_i = \Lambda^i(\mathrm{std}) - \Lambda^{i-2}(\mathrm{std})$ for $3 \leq i \leq n/2$.

13.3. Thanks to this lemma, we may view $P(n)$ as "the space $K^\natural$ of conjugacy classes in K," independently of the *particular* maximal compact K in the *particular* $G = \mathrm{SL}(V)$ or $\mathrm{Sp}(V, \langle,\rangle)$ which we chose. Our next task is to describe intrinsically the measure on $P(n)$ which is "the direct image on $K^\natural$ of normalized (total mass one) Haar measure on K." In order to do this, it is convenient to view $P(n)$ as the quotient of a suitable compact torus $T(n)$ by a suitable finite group $W(n)$. Here is the explicit description:

n odd $T(n)$ is the subgroup of $(\mathbf{R}/2\pi\mathbf{Z})^n$ of elements $(\theta_1, \ldots, \theta_n)$ with $\Sigma\theta_i \equiv 0 \bmod 2\pi\mathbf{Z}$.
$W(n)$ is the symmetric group S_n, acting by permutation of the θ_i's

$T(n)/W(n) \xrightarrow{\sim} P(n)$ is the map

$$(\theta_1, \ldots, \theta_n) \longmapsto \prod_{j=1}^{n}(T - e^{i\theta_j})$$

$n = 2g$ even: $T(n)$ is $(\mathbf{R}/2\pi\mathbf{Z})^g$, coordinates $\theta_1, \ldots, \theta_g$.
$W(n)$ is the semi-direct product $(\{\pm 1\})^g \ltimes S_g$, where S_g acts by permuting the θ_i's, and an element $(\varepsilon_1, \ldots, \varepsilon_g)$ in $(\{\pm 1\})^g$ maps $(\theta_1, \ldots, \theta_g)$ to $(\varepsilon_1\theta_1 \ldots, \varepsilon_g\theta_g)$.

$T(n)/W(n) \xrightarrow{\sim} P(n)$ is the map

$$(\theta_1, \ldots, \theta_g) \longmapsto \prod_{i=1}^{g}(T^2 - 2\cos(\theta_i)T + 1).$$

We denote by $\mu(n)$ the following measure on $T(n)$:

if n is odd: use $\theta_1, \ldots, \theta_{n-1}$ to identify $T(n)$ to $[0, 2\pi)^{n-1}$; then

$$\mu(n) = \frac{4^{n(n-1)/2}}{n!}\left(\frac{1}{2\pi}\right)^{n-1}\left(\prod_{1 \leq i < j \leq n} \sin^2\left(\frac{\theta_i - \theta_j}{2}\right)\right) d\theta_1 \ldots d\theta_{n-1}.$$

if $n = 2g$ is even: use $\theta_1, \ldots, \theta_g$ to identify $T(n)$ to $[0, 2\pi)^g$; then

$$\mu(n) = \frac{4^{g^2}}{2^g(g!)}\left(\frac{1}{2\pi}\right)^g\left(\prod_{i=1}^{g}\sin^2(\theta_i)\right) \times$$

$$\left(\prod_{1 \leq i < j \leq g} \sin^2\left(\frac{\theta_i - \theta_j}{2}\right)\sin^2\left(\frac{\theta_i + \theta_j}{2}\right)\right) d\theta_1 \ldots d\theta_g.$$

In terms of the measure $\mu(n)$, we define a measure $\mu(n)^\natural$ on $P(n)$ by decreeing that for any continuous $\mathbf{C}$-valued function f on $P(n)$ ("continuous" when we view $P(n)$ as a compact subset of the $\mathbf{C}^n$ of monic polynomials of degree n, or equivalently when we view $P(n)$ as $T(n)/W(n)$ with the quotient topology), with inverse image $\tilde{f}$ on $T(n)$, we have

$$\int_{P(n)} f\,d\mu(n)^\natural \stackrel{\text{dfn}}{=} \int_{T(n)} \tilde{f}\,d\mu(n).$$

13.4. Theorem (Weyl Integration Formula, cf. Bourbaki LIE IX, §6, 2, Cor. 2). *The measure $\mu(n)^\natural$ on $P(n)$ corresponds, via the isomorphism $K^\natural \xrightarrow{\sim} P(n)$ defined by the map "characteristic polynomial," to the direct image $\mu^\natural$ on $K^\natural$ of normalized (total mass one) Haar measure on K.*

EXAMPLE: For $n = 2$, $P(2)$ is the set of real polynomials $T^2 - 2\cos(\theta)T + 1$, with $\theta \in [0, \pi]$. Viewing $P(2)$ as the quotient of $T(2) = \mathbf{R}/2\pi\mathbf{Z}$ by $\theta \mapsto -\theta$, $\mu(2)^\natural$ is Sato–Tate measure $(2/\pi)\sin^2\theta\,d\theta$ on $[0, \pi]$.

13.5. Formulation of the theorem

13.5.1. As in the previous section, we fix an integer $n \geq 2$. For any triple $(\mathbf{F}_q, \psi, a)$ consisting of

$$\begin{cases} \text{a finite field } \mathbf{F}_q \\ \text{a non-trivial } \mathbf{C}\text{-valued additive character } \psi : (\mathbf{F}_q, +) \to \mathbf{C}^\times \\ \text{an element } a \in \mathbf{F}_q^\times, \end{cases}$$

we wish to define a point

$$\theta(\mathbf{F}_q, \psi, a) \in P(n),$$

which we think of as the "generalized angle" of the corresponding n-variable Kloosterman sum.

Its intrinsic description as a monic polynomial

$$\theta(\mathbf{F}_q, \psi, a) \sim \prod_{i=1}^{n} (T - \alpha_i)$$

is as follows: for $1 \leq k \leq n$, the Newton symmetric functions of the roots $\alpha_1, \ldots, \alpha_n$ are to be

$$(13.5.1.1) \quad \sum_{i=1}^{n} (\alpha_i)^k = (-1)^{n-1} \left(q^{\frac{1-n}{2}}\right)^k \sum_{\substack{x_1 \ldots x_n = a \\ \text{all } x_i \in \mathbf{F}_{q^k}}} \psi(\operatorname{trace}_{\mathbf{F}_{q^k}/\mathbf{F}_q}(\textstyle\sum x_i)).$$

To see that the polynomial thus defined lies in $P(n)$, we must appeal to the l-adic theory, as follows. The polynomial in question has coefficients in the subfield $\mathbf{Q}$ (values of ψ) $= \mathbf{Q}(\zeta_p)$ of $\mathbf{C}$, p denoting the characteristic of $\mathbf{F}_q$, because its Newton functions do. For any $l \neq p$, and any l-adic place λ of the field $\mathbf{Q}(\zeta_p, q^{\frac{n-1}{2}}) = E$, we may form the lisse rank n E_λ-sheaf $\mathcal{F}_\lambda = \mathrm{Kl}_\psi(n)(\frac{n-1}{2})$ on $\mathbf{G}_m \otimes \mathbf{F}_q$ studied at such length in the previous two chapters. Viewed as an E_λ-polynomial, θ is given by

$$\theta(\mathbf{F}_q, \psi, a)(T) = \det(T - F_a \mid (\mathcal{F}_\lambda)_{\bar{a}}) \tag{13.5.1.2}$$

where F_a denotes the geometric Frobenius at the rational point $a \in \mathbf{F}_q^\times = \mathbf{G}_m(\mathbf{F}_q)$ of $\mathbf{G}_m \otimes \mathbf{F}_q$ (both sides are monic of degree n, and their Newton symmetric functions agree, by the local trace property 4.1.1.(2) of Kloosterman sheaves). Because $\mathcal{F}_\lambda$ is pure of weight zero, with π_1^{arith} acting through SL (and through Sp for n even), it follows that $\theta(\mathbf{F}_q, \psi, a)$ does in fact lie in $P(n)$. Indeed, if we pick a complex embedding $\overline{E}_\lambda \hookrightarrow \mathbf{C}$ which agrees on E with the given inclusion $E \subset \mathbf{C}$, then if either n is even or if pn is odd, the element $\theta(\mathbf{F}_q, \psi, a) \in P(n)$ when viewed as a conjugacy class in $K^\natural$, is none other than that of $\rho_\lambda(F_a)^{\mathrm{ss}}$, i.e. the class denoted "$\theta(a)$" in Theorem 11.4.

An equivalent "à priori formula" for θ is the following (a rewriting of (13.5.1.2) above):

$$T^n \theta(\mathbf{F}_q, \psi, a)(1/T) = \exp\left((-1)^n \sum_{k=1}^{\infty} \frac{(\sqrt{q}^{1-n}T)^k}{k} S_k\right)$$

(13.5.1.3)

$$\text{where } S_k = \sum_{\substack{x_1 \ldots x_n = a \\ \text{all } x_i \in \mathbf{F}_{q^k}}} \psi(\mathrm{trace}_{\mathbf{F}_{q^k}/\mathbf{F}_q}(\textstyle\sum x_i)).$$

13.5.2. Having defined the points $\theta(\mathbf{F}_q, \psi, a)$ in $P(n)$, we now define a measure $\mu(\mathbf{F}_q, \psi)$ on $P(n)$, by

$$\mu(\mathbf{F}_q, \psi) \overset{\mathrm{dfn}}{=} \frac{1}{q-1} \sum_{a \in \mathbf{F}_q^\times} \begin{pmatrix} \text{Dirac delta measure at} \\ \theta(\mathbf{F}_q, \psi, a) \end{pmatrix}.$$

In terms of $\mathcal{F}_\lambda$ on $\mathbf{G}_m \otimes \mathbf{F}_q$, this is the measure "$X_1$" on $K^\natural$, provided that either n is even or that pn is odd.

In fact, the measure $\mu(\mathbf{F}_q, \psi)$ is *independent* of the auxiliary choice of non-trivial ψ. For any other is of the form

$$\psi_b(x) = \psi(bx), \quad \text{for some } b \in \mathbf{F}_q^\times.$$

But we have trivially the identity

$$\theta(\mathbf{F}_q, \psi_b, a) = \theta(\mathbf{F}_q, \psi, b^n a),$$

as is immediate from looking at the corresponding Kloosterman sums (cf. 11.11.1, where this also occurs). Therefore in the sum defining $\mu(\mathbf{F}_q, \psi)$, the effect of replacing ψ by ψ_b is just to permute the terms. We denote simply by

$$\mu(\mathbf{F}_q) = \mu(\mathbf{F}_q, \psi), \quad \text{for any non-trivial } \psi$$

the resulting measure on $P(n)$.

13.5.3. Theorem. *Fix $n \geq 2$. In any sequence of finite fields, possibly of varying characteristic, whose cardinalities tend to ∞, and which for n odd are all of odd characteristic, the measures $\mu(\mathbf{F}_q)$ on the space $P(n)$ tend weak $*$ to the "direct image of Haar measure" $\mu(n)^\natural$ on $P(n)$. More precisely, if Λ is an irreducible non-trivial continuous representation of K, and if $\mathbf{F}_q$ is any finite field (of odd characteristic, if n is odd), then we have the estimate*

$$\left| \int_{P(n)} \operatorname{trace}(\Lambda) d\mu(\mathbf{F}_q) \right| \leq \frac{\dim(\Lambda)}{n} \frac{\sqrt{q}}{(q-1)},$$

where trace(Λ), *à priori a central function on K, is viewed as a continuous function on $P(n)$ via the "characteristic polynomial" isomorphism $K^\natural \xrightarrow{\sim} P(n)$. (In fact,* trace$(\Lambda)$ *will be a $\mathbf{Z}$-polynomial in the functions "ith coefficient" on $P(n)$ viewed as a space of monic polynomials, cf. 13.2.)*

Proof. In view of 11.1, 11.3, and 11.4, this is just 3.6.3, restated from a slightly different perspective. Indeed for any $l \neq \operatorname{char}(\mathbf{F}_q)$, and any field embedding $\overline{\mathbf{Q}}_l \hookrightarrow \mathbf{C}$, $\mu(\mathbf{F}_q)$ is the measure X_1 attached to $\mathcal{F}_\lambda$ on $\mathbf{G}_m \otimes \mathbf{F}_q$. ∎

EXAMPLE 13.6: Take $n = 2$. Then $P(2) = [0, \pi]$, $\mu(2)^\natural = (2/\pi)\sin^2\theta\, d\theta$.
For each prime number p, use the character

$$\begin{aligned} \psi_p : (\mathbf{F}_p, +) &\longrightarrow \mathbf{C}^\times \\ x &\longmapsto \exp(2\pi i x/p). \end{aligned}$$

For each $a \in \mathbf{F}_p^\times$, we write

$$\operatorname{Kl}(p, a) = \sum_{\substack{xy=a \\ x,y\in\mathbf{F}_p}} \psi_p(x+y)$$

$$\frac{-1}{\sqrt{p}} \operatorname{Kl}(p, a) = 2\cos(\theta(p, a)), \qquad \theta(p, a) \in [0, \pi]$$

$$\mu(\mathbf{F}_p) = \text{ the measure } \frac{1}{p-1} \sum_{a\in\mathbf{F}_p^\times} \begin{pmatrix} \text{Dirac } \delta\text{-measure} \\ \text{at } \theta(p, a) \end{pmatrix}$$

on the interval $[0, \pi]$. The irreducible non-trivial representations of $K = \mathrm{SU}(2)$ are the $\mathrm{Symm}^n(\mathrm{std})$, $n = 1, 2, \ldots,$ whose traces are

$$\mathrm{trace}(\mathrm{Symm}^n(\mathrm{std})) = \frac{\sin((n+1)\theta)}{\sin\theta}, \quad n = 1, 2, 3, \cdots .$$

These functions, together with the constant function $\mathbf{1}$ $(n = 0$ above) form an orthonormal base of $L^2[0, \pi]$ for the Sato–Tate measure $(2/\pi)\sin^2\theta\, d\theta$. The estimate in this case is

$$\left| \int_0^\pi \frac{\sin((n+1)\theta)}{\sin\theta} d\mu(\mathbf{F}_p) \right| \leq \frac{n+1}{2} \frac{\sqrt{p}}{p-1} \quad \text{for } n \geq 1.$$

In words: "as $p \uparrow \infty$, the $p - 1$ Kloosterman angles $\{\theta(p, a)\}_{a \in \mathbf{F}_p^\times}$ become equidistributed in $[0, \pi]$ for Sato–Tate measure."

References

[A–K] Altman, A., and Kleiman, S., *Introduction to Grothendieck Duality Theory*, Springer Lecture Notes in Mathematics 146, 1970.

[B–B–D] Beilinson, A. A., Bernstein, I. N., and Deligne, P., Faisceaux pervers, in "*Analyse et Topologie sur les espaces singuliers*" I, Conférence de Luminy, juillet 1981, Astérisque 100, 1982.

[Be] Belyi, G. V., On galois extensions of a maximal cyclotomic field, Math. U.S.S.R. Izvestija vol. 14, 1980, no. 2, 247-256 (English translation page numbers).

[Bour–1] Bourbaki, N., *Groupes et Algebres de Lie*, Chapitres 4, 5 et 6, Masson, Paris, 1981.

[Bour–2] ———, *Groupes et Algebres de Lie*, Chapitres 7 et 8, Diffusion CCLS, Paris, 1975.

[Br] Brylinski, J. -L., Transformations canoniques, dualité projective, théorie de Lefschetz, transformation de Fourier et sommes trigonométriques, Astérisque 140–141 (1986), 3–134.

[C–R] Curtis, C. W., and Reiner, I., *Representation theory of Finite Groups and Associative Algebras*, Interscience Publ., New York and London, 1962.

[Dav] Davenport, H., On the distribution of quadratic residues (mod p), J. London Math. Soc. 6, 1931, 49–54, reprinted in *The Collected Works of Harold Davenport* (ed. Birch, Halberstam, Rogers), Academic Press, London, New York and San Francisco, 1977, vol. IV, 1451–1456.

[De–1] Deligne, P., Les constantes des équations functionelles des functions L, in *Proceedings 1972 Antwerp Summer School of Modular Forms*, Springer Lecture Notes in Mathematics 349, 1973.

[De–2] ———, Rapport sur la Formule des traces, in *Cohomologie Etale (SGA 4 1/2)*, Springer Lecture Notes in Mathematics 569, 1977, 76–109.

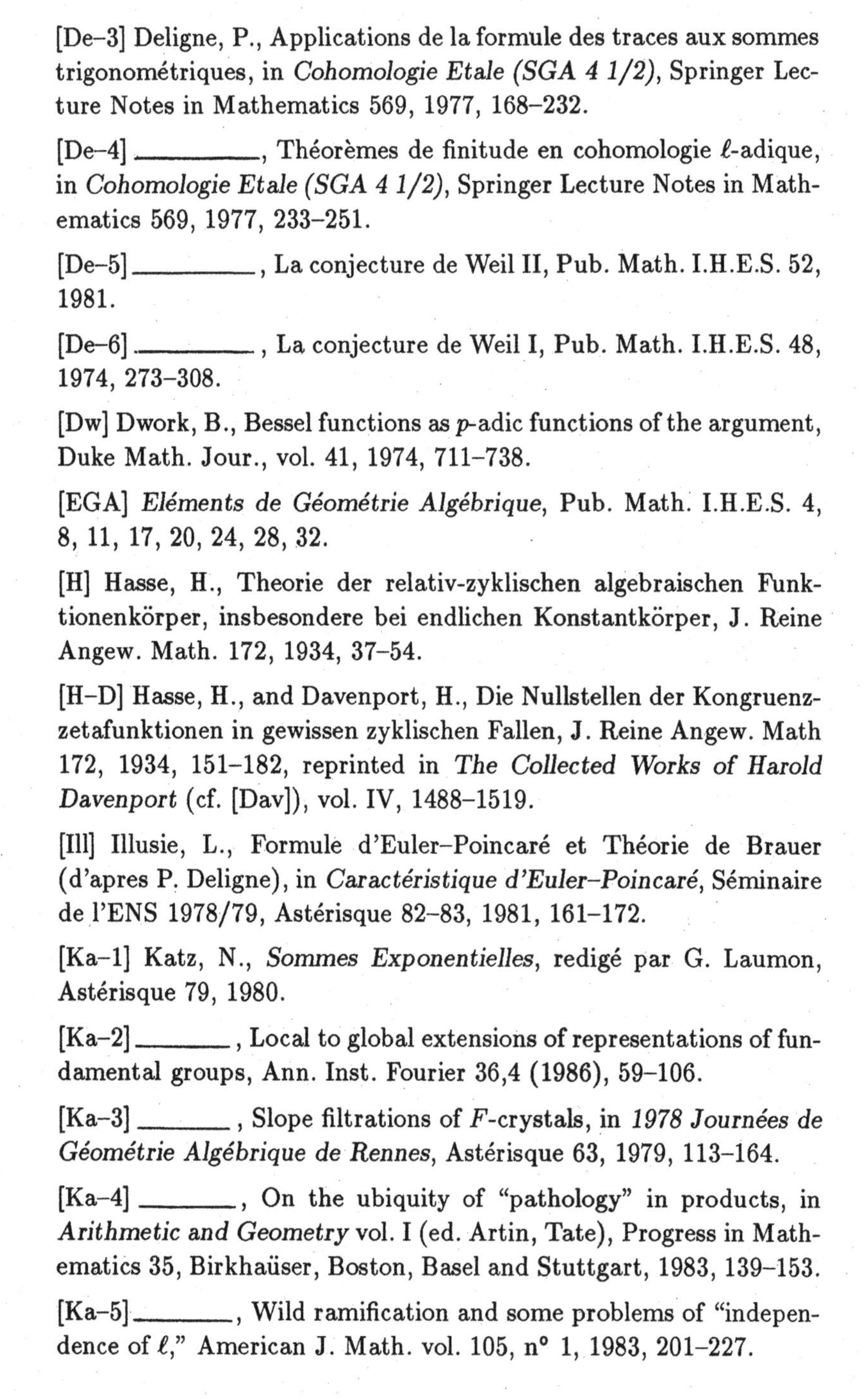

[De–3] Deligne, P., Applications de la formule des traces aux sommes trigonométriques, in *Cohomologie Etale (SGA 4 1/2)*, Springer Lecture Notes in Mathematics 569, 1977, 168–232.

[De–4] ________, Théorèmes de finitude en cohomologie ℓ-adique, in *Cohomologie Etale (SGA 4 1/2)*, Springer Lecture Notes in Mathematics 569, 1977, 233–251.

[De–5] ________, La conjecture de Weil II, Pub. Math. I.H.E.S. 52, 1981.

[De–6] ________, La conjecture de Weil I, Pub. Math. I.H.E.S. 48, 1974, 273–308.

[Dw] Dwork, B., Bessel functions as p-adic functions of the argument, Duke Math. Jour., vol. 41, 1974, 711–738.

[EGA] *Eléments de Géométrie Algébrique*, Pub. Math. I.H.E.S. 4, 8, 11, 17, 20, 24, 28, 32.

[H] Hasse, H., Theorie der relativ-zyklischen algebraischen Funktionenkörper, insbesondere bei endlichen Konstantkörper, J. Reine Angew. Math. 172, 1934, 37–54.

[H–D] Hasse, H., and Davenport, H., Die Nullstellen der Kongruenzzetafunktionen in gewissen zyklischen Fallen, J. Reine Angew. Math 172, 1934, 151–182, reprinted in *The Collected Works of Harold Davenport* (cf. [Dav]), vol. IV, 1488–1519.

[Ill] Illusie, L., Formule d'Euler–Poincaré et Théorie de Brauer (d'apres P. Deligne), in *Caractéristique d'Euler–Poincaré*, Séminaire de l'ENS 1978/79, Astérisque 82–83, 1981, 161–172.

[Ka–1] Katz, N., *Sommes Exponentielles*, redigé par G. Laumon, Astérisque 79, 1980.

[Ka–2] ________, Local to global extensions of representations of fundamental groups, Ann. Inst. Fourier 36,4 (1986), 59–106.

[Ka–3] ________, Slope filtrations of F-crystals, in *1978 Journées de Géométrie Algébrique de Rennes*, Astérisque 63, 1979, 113–164.

[Ka–4] ________, On the ubiquity of "pathology" in products, in *Arithmetic and Geometry* vol. I (ed. Artin, Tate), Progress in Mathematics 35, Birkhaüser, Boston, Basel and Stuttgart, 1983, 139–153.

[Ka–5] ________, Wild ramification and some problems of "independence of ℓ," American J. Math. vol. 105, n° 1, 1983, 201–227.

[Ka–6] Katz, N., Monodromy of families of curves; applications of some results of Davenport–Lewis, Séminaire D.P.P. 1979/80, Progress in Mathematics 12, Birkhaüser, 1983, 171–195.

[Ka–7] ______, On the calculation of some differential Galois groups, Inv. Math 87 (1987), 13–61.

[K–L] Katz, N., and Laumon, G.,Transformation de Fourier et majoration de sommes exponentielles, Pub. Math. I.H.E.S. 62 (1985), 361–418.

[K–M] Katz, N., and Mazur, B., *Arithmetic Moduli of Elliptic Curves*, Annals of Math. Studies, 108, Princeton University Press, Princeton, 1985.

[Lau–1] Laumon, G., Semicontinuité de conducteur de Swan (d'aprés Deligne), in *Caractéristique d'Euler–Poincaré*, Séminaire de l'ENS 1978/79, Astérisque 82–83, 1981, 173–219.

[Lau–2] ______, Transformation de Fourier, constantes d'équations fonctionelles et conjecture de Weil, Pub. Math. I.H.E.S. 65 (1987), 131–210.

[Maz] Mazur, B., Frobenius and the Hodge filtration, Bull. A.M.S. 78, 1972, 653–667.

[No] Nori, M. V., On subgroups of $SL(n, \mathbf{Z})$ and $SL(n, \mathbf{F}_p)$, preprint.

[Ray] Raynaud, M., Caractéristique d'Euler–Poincaré d'un Fais ceaux et cohomologie des variétés abéliennes, Séminaire Bourbaki 1964/65, n° 286, W. A. Benjamin, New York, 1966.

[Se–1] Serre, J.-P., *Corps Locaux*, deuxième édition, Hermann, Paris, 1968.

[Se–2] ______, *Représentations Linéaires des Groupes Finis*, troisième édition corrigée, Hermann, Paris, 1978.

[Se–Ta] Serre, J.-P., and Tate, J., Good reduction of abelian varieties, Annals of Math. 88, 1978, 492–517.

[SGA] *Séminaire deGéométrie Algébrique* I, IV, VII parts I and II, Springer Lecture Notes in Mathematics 224 (I), 269–270–305 (IV), 288 (VII, part I), 340 (VII, part II).

[Sp–1] Sperber, S., Congruence properties of the hyperkloosterman sum, Compositio Math., vol. 40, 1980, 3–33.

[Sp–2] ______, Newton polygons for general hyperkloosterman sums, in *Cohomologie P-Adique*, Astérisque 119–120, 1984, (volume dedicated to B. Dwork), pp. 267–330.

[Ver] Verdier, J.-L., Specialisation de Faisceaux et monodromie modérée, in "*Analyse et Topologie sur les espaces singuliers*," vol. II et III, Astérisque 101–102, 1982, 332–364.

[We] Weil, A., On some exponential sums, Proc. Nat. Acad. Sci. U.S.A. 34, 1948, 204–207.

www.ingramcontent.com/pod-product-compliance
Ingram Content Group UK Ltd.
Pitfield, Milton Keynes, MK11 3LW, UK
UKHW011709131224
452420UK00019B/301